ADVANCES IN CHEMICAL PHYSICS

VOLUME 103

EDITORIAL BOARD

Advances in
CHEMICAL PHYSICS

Edited by

I. PRIGOGINE

Center for Studies in
Statistical Mechanics and
Complex Systems
The University of Texas
Austin, Texas
and
International Solvay
Institutes, Université
Libre de Bruxelles
Brussels, Belgium

and

STUART A. RICE

Department of Chemistry
and
The James Franck Institute
The University of Chicago
Chicago, Illinois

VOLUME 103

AN INTERSCIENCE® PUBLICATION
JOHN WILEY & SONS, INC.
NEW YORK • CHICHESTER • WEINHEIM • BRISBANE • SINGAPORE • TORONTO

Library of Congress Catalog Number: 58-9935

ISBN 0-471-24752-9

Printed in the United States of America.

10 9 8 7 6 5 4 3 2 1

CONTRIBUTORS TO VOLUME 103

PAUL BLAISE, Centre d'Etudes Fondamentales, Université de Perpignan, Perpignan Cedex, France

WILLIAM COFFEY, School of Engineering, Dept. of Electronic & Electrical Engineering, Trinity College, Dublin, Ireland

D. S. F. CROTHERS, Theoretical and Computational Physics Reseach Division, Department of Applied Mathematics and Theoretical Physics, The Queen's University of Belfast, Belfast, Northern Ireland

K. W. FOREMAN, Department of Pharmaceutical Chemistry, University of California at San Francisco, San Francisco, CA

KARL F. FREED, James Franck Institute and Department of Chemistry, The University of Chicago, Chicago, IL

OLIVIER HENRI-ROUSSEAU, Centre d'Etudes Fondamentales, Université de Perpignan, Perpignan Cedex, France

R. M. HERMAN, Department of Physics, Pennsylvania State University, University Park, PA

B. S. NESBITT, Theoretical and Computational Physics Research Division, Department of Applied Mathematics and Theoretical Physics, The Queen's University of Belfast, Belfast, Northern Ireland

J. F. OGILVIE, Centre for Experimental and Constructive Mathematics, Simon Fraser University, Burnaby, BC V5A 156, Canada

S. F. C. O'ROURKE, Theoretical and Computational Physics Research Division, Department of Applied Mathematics and Theoretical Physics, The Queen's University of Belfast, Belfast, Northern Ireland

INTRODUCTION

Few of us can any longer keep up with the flood of scientific literature, even in specialized subfields. Any attempt to do more and be broadly educated with respect to a large domain of science has the appearance of tilting at windmills. Yet the synthesis of ideas drawn from different subjects into new, powerful, general concepts is as valuable as ever, and the desire to remain educated persists in all scientists. This series, *Advances in Chemical Physics*, is devoted to helping the reader obtain general information about a wide variety of topics in chemical physics, a field that we interpret very broadly. Our intent is to have experts present comprehensive analyses of subjects of interest and to encourage the expression of individual points of view. We hope that this approach to the presentation of an overview of a subject will both stimulate new reseach and serve as a personalized learning text for beginners in a field.

I. Prigogine
Stuart A. Rice

CONTENTS

ADVANCES IN CHEMICAL PHYSICS

VOLUME 103

THE INFRARED SPECTRAL DENSITY OF WEAK HYDROGEN BONDS WITHIN THE LINEAR RESPONSE THEORY

OLIVIER HENRI-ROUSSEAU AND PAUL BLAISE

*Centre d'Etudes Fondamentales, Université de Perpignan,
52 avenue de Villeneuve, 66860 Perpignan cedex, France*

CONTENTS

Advances in Chemical Physics, Volume 103, Edited by I. Prigogine and Stuart A. Rice.
ISBN 0-471-24752-9 © 1998 John Wiley & Sons, Inc.

I. INTRODUCTION

A. The Infrared Spectral Properties of Hydrogen Bonds

The wide range of changes in the hydrogen bond properties invites for a conventional classification into weak, medium, strong and very strong hydrogen bonds. More precisely, the usual classification from the experimental view point is the following:

1. The Usual Classification of Hydrogen Bonds

Weak Hydrogen Bonds. They are nonsymmetrical and involve long bond lengths $X–H\cdots Y$ around 2.9 Å, and hydrogen bonding enthalpies around 5 or 6 kcal mole^{-1}.

Medium-Strength Hydrogen Bonds. They are nonsymmetrical, moderately strong hydrogen bonds. They involve bond lengths that are shorter than the weak hydrogen bonds and lie about 2.75 Å. The corresponding enthalpies of bonding lie between 6 and 12 kcal mol^{-1}.

Strong Hydrogen Bonds. They are nonsymmetrical and involve short lengths near 2.5 Å and bonding enthalpies lying between 12 or 15 kcal mole^{-1}.

Superstrong Hydrogen Bonds. They are symmetrical or nearly symmetrical and correspond to very short lengths reaching 2.3 Å and important hydrogen bonding enthalpies about 20 kcal mole^{-1} or more.

However, there are no obvious discontinuities in the evolution of properties with increasing interaction energy, and the limits between the classes are set deliberately. For the purpose of theoretical treatments, a different discriminator is more useful; it is based on the one-dimensional proton potential function. Thus hydrogen bonds can be classified into those with a single, asymmetric potential minimum (roughly corresponding to weak and medium strong hydrogen bonds) and those with low (or without) barrier in the potential function (in other words, hydrogen bonds with appreciable tunneling effects and those in which tunneling may be neglected). That leads to distinguish between asymmetric hydrogen bonds where the $X–H$ bond length is very different from that of the $H\cdots Y$ that are of the kind $X–H\cdots Y$ and symmetric hydrogen bonds that are of the form $X\cdots H\cdots X$. The formation of symmetrical species is connected as a rule with the ionization and therefore is only likely in the condensed phase. Only strong hydrogen bonds form symmetrical complexes. Besides, hydrogen bond strength is closely related to $X–H\cdots Y$ bond length and to hydrogen bonding enthalpy, the stronger the hydrogen bond, the shorter the $X–H\cdots Y$ length and the greater the enthalpy.

2. The Infrared Spectra of Hydrogen-Bonded Species

Infrared spectra contain complete information on the electronic and, consequently, nuclear dynamic effects of hydrogen bonding. The same is true of Raman spectra; however, owing to the weakness of bands and experimental hurdles in obtaining good spectra, far less experimental material is available, and therefore most theories are focused on infrared spectra. Moreover, the essential ideas of the latter can be applied to Raman spectra with little adaptation.

From a spectroscopic viewpoint, the main vibrational modes of a hydrogen-bonded system are the $v_s(X–H)$ high-frequency stretching mode, the in-plane $\delta(X–H\cdots Y)$ bending mode, the $\gamma(X–H\cdots Y)$ out-of-plane bending mode, and the $v_s(X–H\cdots Y)$ stretching low-frequency mode.

The most evident effects of hydrogen bonding on the spectral density of the $v_s(X–H)$ stretching mode are the following, by comparison with a free hydrogen stretching band:

The frequencies are strongly shifted towards lower values;
The band shapes are extensively broadened with often peculiar sub-bands;
The intensities are strongly enhanced;
The frequencies and intensities involve an isotopic effect;
The band shapes often remain the same structure in the gas and condensed phases.

It is known that the other infrared transitions do not show any important change when a hydrogen bond is formed.

Depending on the properties of the hydrogen bond, the red shifts of the $v_s(X–H)$ stretching mode may be as large as some thousands of cm^{-1} and the intensity increases up to thousand-fold. The changes of the various properties are interrelated, and it is the interaction energy, often termed *strength*, that is usually taken as the reference parameter for correlations of the most conspicuous properties involving the bandwidth and, more generally, the band shape. This bandwidth is known to increase from around 15 cm^{-1} to several thousand. More precisely, for very weak hydrogen bonds, the half-width is near $100–300$ cm^{-1}, for weak hydrogen bonds, around $300–1000$ cm^{-1} for medum-strength hydrogen bonds, about $1000–1500$ cm^{-1} for strong asymmetric hydrogen bonds, and about 2000 cm^{-1} for symmetrical or nearly symmetrical superstrong hydrogen bonds. Moreover there is a reverse relationship between the frequency shift and the length $R(X\cdots Y)$ as well as between this length and the equilibrium distance $R(X–H)$ and between the latter and this same shift (Pimentel and Mac Clellan, 1960; Novak, 1974). Besides, the substitution of deuterium for hydrogen leads to a considerable lowering of the frequency. Moreover, the isotopic frequency ratio $v_s(X–H)/v_s(X–D)$ may be considerably lower than that of the free $X–H$ group (Novak, 1974).

On the other hand, the large increase of the high-frequency bandwidth is but one challenge for the nuclear dynamics theories of hydrogen bonding.

The other are the band asymmetry, the appearance of subsidiary absorption maxima and minima, such as windows, and the peculiar isotope effects on frequency shifts, that are the subject of this chapter.

There are several mechanisms that influence the evolution of the high-frequency band shape. They may be separated into intrinsic and medium related or extrinsic ones. Band shapes in the low-pressure gas phase are obviously owing exclusively to mechanisms determined by the nuclear dynamics of the isolated hydrogen-bonded complex; this also is true to a good approximation of dilute noble gas matrices. The only complementary broadening effect under these conditions is due to intermolecular collisions. In dilute solution, the medium (inert solvent, for instance) takes the important role of the thermal bath. In the more concentrated solution and pure liquid state, several other mechanisms arise such as differences in the metrics of association. These are absent in crystalline solids, but instead the correlation field splitting and phonon coupling effects have to be considered.

A theory that is susceptible to take into account the dramatic changes in the infrared spectrum may be considered to give some understanding of the dynamics of the hydrogen bond. One may classify theories dealing with the shape of the infrared spectrum as first and second generation, the first one being qualitative and the second one quantitative.

B.   Theories Dealing with the Spectral Density of Hydrogen Bond

1.   Old Qualitative Theories Dealing with the Spectral Density

Now, let us look at the old qualitative theories dealing with the $v_s(X-H)$ band shape. They are of four types: the fluctuation theory, the predissociation theory, the double minimum theory, and the anharmonicity theory.

The Fluctuation Theory of Lansberg and Baryshanskaya (1946). According to this theory, there is a considerable fluctuation in the $X-H \cdots Y$ bond distance in the liquid phase, because the hydrogen bond is produced by an intermolecular force that is weak compared to the genuine classical valence force. This fluctuation produces fluctuations of the $X-H$ force constant and thus, of the transition frequency of the band $v_s(X-H)$.

The Predissociation Theory of Stepanov (1945, 1946). Its basic postulate is that the vibrational excitation does not remain localized in the $X-H$ bond. This excitation is transferred to the $H \cdots Y$ bond and can break it when the quantum energy excitation of the oscillator is greater than the dissociation energy of the hydrogen bond. Thus the lifetime of the excited state is short-ened, and therefore, according to the uncertainty principle, the first excited state of the fast mode has a non-negligible width.

The Double-Minimum Theory. It assumes that the potential energy surface governing the proton motion in the X–H$\cdots Y$ complex, involves two minima that are not necessarily equivalent. That leads to a splitting of the vibrational levels so that the number of allowed spectral transitions is cor-respondingly increasing. If these transitions are blurred by molecular inter-actions, a broadening of the $v_s(X$–H) band results.

The Anharmonicity Theory of Bratos and Hadzi (1957). It assumes that the potential surface governing the X–H$\cdots Y$ vibration is strongly anharmonic. In this theory, it is obvious that the term *modes* cannot be understood in the sense of normal modes since, of course, the interactions are beyond the quadratic behavior. It may be observed that there are some correspon-dences between this theory and the other ones since, for instance, the double-minimum theory necessarily involves some anharmonicity.

2. *The Quantitative Theories within the Linear Response Theory*

The quantitative theories are subsequent to 1965. The physical basis idea that is common to all the quantitative theories is the assumption of an anharmonic coupling between the high-frequency mode and the hydrogen bond bridge. Related to this main mechanism must be quoted Fermi reso-nances, Davydov splitting, and the tunnel effect. This anharmonic theory will be discussed later. Among the quantitative theories there are a lot that involve the linear response theory and to which the present exposition is devoted. Thus it is suitable to give some information on the linear response theory dealing with the subject of spectroscopy.

Consider a group of molecules, the Hamiltonian of which is H, that are exposed to a monochromatic electromagnetic field $\mathbf{E}(t, \omega)$:

$$\mathbf{E}(t, \omega) = E_0 \cos(\omega t) \tag{1}$$

If the pulsation ω of this field is near that of a characteristic transition of the molecule, it is well known that the electromagnetic field will induce a transition between the corresponding energy levels of the molecule. If the interaction between the radiation and matter is approximated by the elec-tric dipole interaction, which is valid when the wavelength is large com-pared to the molecular dimensions, the interaction Hamiltonian describing the coupling $V(t, \omega)$ between the field and the dipole moment of the molecule is given by:

$$V(t, \omega) = \mathbf{E}(t, \omega) \cdot \boldsymbol{\mu} \qquad (2)$$

where $\boldsymbol{\mu}$ is the dipole moment operator of the molecule. If the perturbation of the molecule by the electromagnetic field is small, the time evolution operator $\{U(t)\}$ of the system may be expanded up to first order. That leads to the following equation, which is linear in the electromagnetic field:

$$\{U(t)\} = \{U^0(t)\} + [1/(i\hbar)]\{U^0(t)\}$$

$$\times \int_0^t \{U^0(t')\}^+ \{\mathbf{E}(t', \omega) \cdot \boldsymbol{\mu}\}\{U^0(t')\} \, dt' \qquad (3)$$

Here $\{U^0(t)\}$ is the time evolution operator of the molecule in the absence of an electromagnetic field. It is recalled in Appendix B, following an approach of Gordon (1968), how, in the framework of this linear response theory, the spectral density $\{I(\omega)\}$ characterizing the molecular transition may be related to the autocorrelation function $\{G(t)\}$ of the dipole moment operator of the molecule at time t, according to Eq. (B10):

$$\{I(\omega)\} = \frac{1}{\sqrt{2\pi}} \int_{-\infty}^{\infty} \exp(-i\omega t)\{G(t)\} \, dt \qquad (4)$$

with

$$\{G(t)\} = \langle \boldsymbol{\mu}^+(0) \cdot \boldsymbol{\mu}(t) \rangle_0 \qquad (5)$$

and where $\boldsymbol{\mu}(0)$ is the dipole moment operator at initial time and the same operator at time t is given in the Heisenberg picture by:

$$\{\boldsymbol{\mu}(t)\} = \exp(Ht/\hbar)\{\boldsymbol{\mu}(0)\} \exp(-Ht/\hbar) \qquad (6)$$

In the linear response theory, attention is focused on the time dependence of the autocorrelation function.

An important point is that no assumption is made in the obtainment of the autocorrelation function of the dipole moment operator about the random motion of the dipoles since it is shown in Appendix B that these dipoles move in time according to the Hamiltonian H of the system, whatever it is. Thus it is erroneous to think that a correlation function can be used only when a system is undergoing some kind of stochastic motion.

Of course, when there is random motion of the dipoles due to the influence of the medium, the correlation function $\{G(t)\}$ describes the decay of our knowledge about the system as it approaches equilibrium. If, for instance, we know the direction of the dipole moment at a given time, then after a long time and many collisions, it is equally likely to be pointing in any direction. The dipole correlation function is a quantitative statement of just how this loss of memory comes about.

C. The Purpose of the Present Study

A complete theory of band shapes of hydrogen-bonded systems has to cover the above phenomenology discussed in Section A. Obviously the main task is the explanation of the inherent band broadening, but the theory also has to provide for inclusion of the band shaping mechanism operating in the condensed phases and the effects of tunneling that enter the scene with low barrier proton potential functions. Clearly, the theory has to be quantitative and permit the reconstruction of spectra with the use of physically accessible parameters.

In the following we shall be dealing with theoretical studies of the shape of the high frequency stretching vibration of hydrogen bonds that are all performed in the framework of the linear response theory.

Instead of dealing with fitting experiments of the diverse, most fundamental theoretical approaches involving different mechanisms and formalisms, for which we refer to Hofacker, Y. Maréchal, and Ratner (1976), Hadzi and Bratos (1976), Sandorfi (1976), Y. Maréchal (1980, 1987a), and Sandorfi (1984), we have preferred here to emphasize the connection between these diverse treatments, in such a way as appears a matter of doctrine susceptible to give a consistent and unified theoretical tool built up on a clear physical basis; thus we have tried to perform unification in the notation and to clarify the links between the demonstrations.

For an overview dealing with this question, see Henri-Rousseau and Blaise (1997). In the present study, we shall omit the very important question of the infrared spectroscopy of water because of its specificity in the realm of hydrogen bond, and thus we shall not treat the corresponding theoretical studies of Rice et al. (1981, 1983), whose works remain the basic reference.

In the following, to avoid too extensive a development, we shall limit ourselves to the simplest situation, where it is possible to neglect Fermi resonance, the tunneling effect, or Davydov splitting. Of course, that leads to a restriction to weak or intermediate hydrogen bonds. Then the basic physical model is the strong anharmonic coupling between the high-frequency and low-frequency stretching mode that we shall now discuss.

II. THE STRONG COUPLING THEORY OF ANHARMONICITY

A. Physical Consideration

1. The Stepanov Idea (1946)

The quantum theory dealing with strong anharmonic coupling is known as *strong coupling theory*. The quantum approach is not without connection with the qualitative idea of Stepanov (1946), who assumed the adiabatic separability of the X–H and the H$\cdots Y$ movements of vibration in a linear hydrogen-bonded complex and observed that this enabled effective potential curves for the relatively slow motion to be defined for each quantum state of the phase vibration. Since the shape of these curves must depend on the vibrational states, the infrared spectrum associated with the $v_s(X$–H) transitions should have a substructure analogue of the Franck–Condon factors. These ideas have been defined by Sheppard (1959). Note that according to this author, the hydrogen bond should be stronger and shorter in the excited state of the fast mode.

2. Dynamical and Electrical Coupling

It is necessary, following Coulson and Robertson (1974) to distinguish kinematic coupling from electronic coupling: For describing the hydrogen bond as a linear triatomic system, it is possible, as performed by Coulson and Robertson (1974), to use in place of the coordinate of the slow mode, the Jacobi coordinates, that is, the distance between the center of mass of the X–H molecule and the Y atom. This Jacobi coordinate, which is the sum of the coordinate of the slow mode plus that of the fast mode times a ratio of masses, allows one to separate the center-of-mass motion of the hydrogen bond. In this frame, the dependence of the potential governing the slow mode appears to be dependent on the Jacobi coordinates minus the fast-mode coordinate times the above ratio of masses. As a consequence, the potential depending on the slow-mode coordinate depends therefore also on that of the fast mode. Since this coupling between the slow and fast mode is due to a mass effect, it may be viewed as a kinematic coupling. On the other hand, one may consider, as it is performed in the strong coupling theory, that there is an anharmonic coupling between the slow and fast mode, which has its origin in the electron-density redistribution accompanying hydrogen bond formation, that is, the electronic coupling.

3. Qualitative Considerations on the Strong Coupling Theory

In the anharmonicity theory, among the different anharmonic force constants, the most important is that coupling the high-frequency X–H and the slow-frequency H$\cdots Y$ movements. The second important anharmonic coupling is that between the high-frequency mode and the X–H$\cdots Y$ bending vibration, susceptible to Fermi resonance between the first excited state of the stretching mode and the second excited state of the bending mode. It is generally assumed that this anharmonic coupling is the unavoidable fundamental mechanism in an attempt to understand the spectral properties of the hydrogen bond. Note that this anharmonic coupling was first considered, 60 years ago, by Badger and Bauer (1937).

Let us consider more deeply the anharmonic coupling between the slow and fast modes. It is usually assumed to be quadratic with respect to the coordinate of the high-frequency mode and linear with respect to that of the low frequency. Note that there is also an approach of Borstnik (1976) dealing with anharmonic coupling that is both quadratic with respect to the slow and fast modes.

One may distinguish semiquantal approaches of this coupling, in which the fast mode is considered quantum mechanically, whereas the slow one is considered classical, from full quantum-mechanical approaches, in which both the fast and slow modes are considered to obey the uncertainty relation.

The first treatments dealing with the strong coupling have been performed quantum mechanically by Y. Maréchal and Witkowski (1968), by Fisher, Hofacker, and Ratner (1970), and also by Rösch and Ratner (1974). See also Singh and Wood (1968). Some years later have followed the semiquantal approaches of Bratos (1975) and of Robertson and Yarwood (1978) that are more simple and with which we shall begin.

B. The Basic Classical Equations of Anharmonicity

Here we follow the standard approach of Robertson and Yarwood (1978), which is similar to that of Bratos (1975) or of Sakun (1985). We start from the basis equations that are the same in the three approaches. Using mainly the notation of Robertson, the semiquantal Hamiltonian H of the hydrogen bond in the absence of a thermal bath is:

$$\{H_{sq}^0(Q, q)\} = [p^2/(2m) + \tfrac{1}{2}m\omega^{\circ 2}q^2] + [P^2/(2M) + \tfrac{1}{2}M\omega^{\circ\circ 2}Q^2]$$

$$+ (K_{112}\,Qq^2) \tag{7}$$

Here K_{112} is the anharmonic constant between the fast q and slow Q modes, p and P are the corresponding conjugate momentum, m and M the corresponding masses, and ω° and $\omega^{\circ\circ}$ the corresponding angular frequencies.

The full Hamiltonian (7) may be also written,

$$\{H_{sq}^0(Q, q)\} = [p^2/(2m) + (\tfrac{1}{2}m\omega^{\circ 2} + K_{112}Q)q^2] + [P^2/(2m) + \tfrac{1}{2}M\omega^{\circ\circ 2}Q^2]$$

(8)

or, after rearranging,

$$\{H_{sq}^0(Q, q)\} = [p^2/(2m) + \tfrac{1}{2}m\{\omega_{eff}^\circ(Q)\}^2 q^2] + [P^2/(2M) + \tfrac{1}{2}M\omega^{\circ\circ 2}Q^2] \quad (9)$$

ω_{eff}° is an effective angular frequency that is given by:

$$\{\omega_{eff}^\circ(Q)\} = \omega^\circ[1 + [2K_{112}/(m\omega^{\circ 2})]Q]^{1/2}$$

(10)

This effective angular frequency may be expanded to first order in Q, according to:

$$\{\omega_{eff}^\circ(Q)\} = \omega^\circ[1 + [K_{112}/(m\omega^{\circ 2})]Q]$$

(11)

This expansion may be written formally:

$$\{\omega_{eff}^\circ(Q)\} = \omega^\circ + a_{112}Q$$

(12)

$$a_{112} = K_{112}/(m\omega^{\circ 2})$$

(13)

Again, the eigenvalue equation of the Hamiltonian (9) is:

$$\{H_{sq}^0(Q, q)\}|\{k\}\rangle = k\hbar\{\omega_{eff}^\circ(Q)\}|\{k\}\rangle + [P^2/(2M) + \tfrac{1}{2}M\omega^{\circ\circ 2}Q^2]|\{k\}\rangle$$

(14)

which may be also written:

$$\langle\{k\}|\{H_{sq}^0(Q, q)\}|\{k\}\rangle = k\hbar\{\omega_{eff}^\circ(Q)\} + [P^2/(2M) + \tfrac{1}{2}M\omega^{\circ\circ 2}Q^2]$$

(15)

In order to simulate the influence of damping on the slow mode Q, Robertson and Yarwood (1978), and after them Sakun (1985), assumed that the $v_s(X-H\cdots Y)$ stretching is classical, and can be represented by a stochastic process that leads to a loss of the phase coherence of the $v_s(X-H)$ stretching because of the coupling between the two vibrational modes. Then they assumed the angular frequency involved in the Hamiltonian (9) as time dependent through the Q coordinate, that is, they wrote, in place of Eq. (12):

$$\{\omega^{\circ}_{\text{eff}}(Q, (t))\} = \omega^{\circ} + a_{112} Q(t) \qquad (16)$$

As a consequence, one may write the Hamiltonian in the same way as in Eqs. (8) and (9), but writing in these equations $Q(t)$ and $\omega(t)$ in place of Q and ω.

C.   The Quantum Approach of Anharmonicity

1.   The Witkowski and Maréchal Quantum Model of Anharmonic Coupling without Damping

Now we shall treat the quantum theory of Maréchal and Witkowski (1968) dealing with the anharmonic coupling between the slow and fast modes of hydrogen bonds that has been used extensively, and the spirit of which is near that of the canonical transformations approach of Fischer, Hofacker, and Ratner (1970).

The Hamiltonian decribing the two motions of q and Q in the absence of solvent is:

$$\{H^{0}(q, Q)\} = p^{2}/(2m) + P^{2}/(2M) + U(q, Q) \qquad (17)$$

with the commutators $[P, Q] = -i\hbar$, $[p, q] = -i\hbar$, and where $U(q, Q)$ is the potential that depends on q and Q. Note that the superscript 0 refers to the fact that there is no solvent.

In order to go further, the potential operator appearing in the expression of the Hamiltonian is considered the sum of two contributions:

$$\{U(q, Q)\} = \{U'(q, Q)\} + \{U''(Q)\}$$

$U'(q, Q)$, which governs the motion of the fast coordinate q, and which is assumed to depend weakly on Q, may be written:

$$\{U'(q, Q)\} = \tfrac{1}{2} \left(\frac{\partial^{2} U'(q, Q)}{\partial q^{2}} \right) q^{2}$$

with

$$\frac{\partial^{2} U'(q, Q)}{\partial q^{2}} = m\omega^{\circ}_{\text{eff}}(Q)^{2}$$

In this equation, since the potential of the fast mode also depends on the coordinate of the slow mode, the pulsation $\omega_{\text{eff}}^{\circ}$ of the fast mode must also depend on this slow mode as in the semiclassical approach. But here the situation is more complex since Q is not a scalar, as above, but an operator. On the other hand, $U''(Q)$ is the potential of the slow mode that, within the harmonic approximation is:

$$\{U''(Q)\} = (\tfrac{1}{2}M\omega^{\circ\circ2}Q^2)$$

In the above equations the symbols have the same meaning as in the preceding section dealing with the semiclassical approach of the anharmonic coupling. As a consequence, the full Hamiltonian becomes:

$$\{H^0(q, Q)\} = p^2/(2m) + \tfrac{1}{2}m\omega_{\text{eff}}^{\circ}(Q)^2 + \tfrac{1}{2}M\omega^{\circ\circ2}Q^2 + P^2/(2M) \qquad (18)$$

Let us perform the following partition of the full Hamiltonian according to:

$$\{H^0(q, Q)\} = \{h(q, Q)\} + \tfrac{1}{2}M\omega^{\circ\circ2}Q^2 + P^2/(2M) \qquad (19)$$

$h(q, Q)$ is the Hamiltonian of the fast mode given by:

$$\{h(q, Q)\} = [p^2/(2m) + \tfrac{1}{2}m\omega_{\text{eff}}^{\circ}(Q)^2] \qquad (20)$$

The eigenvalue equation of this Hamiltonian is:

$$\{h(q, Q)\} \,|\, \Phi_k(q, Q)\rangle = \{\varepsilon_k(Q)\} \,|\, \Phi_k(q, Q)\rangle \qquad (21)$$

$\{\varepsilon_k(Q)\}$ are the eigenvalues of the fast mode that depend on Q and $|\Phi_k(q, Q)\rangle$ are the corresponding kth eigenkets, which are assumed to depend parametrically on the coordinate Q of the slow mode and may define a basis according to:

$$\langle \Phi_g(q, Q) \,|\, \Phi_k(q, Q)\rangle = \delta_{g, k}$$

and

$$\sum |\Phi_k(q, Q)\rangle\langle\Phi_k(q, Q)| = 1$$

Besides, let us define k eigenvalue equations involving the Hamiltonian of the quantum harmonic oscillator describing the slow mode:

$$[P^2/(2M) + \tfrac{1}{2}M\omega^{\circ\circ2}Q^2] \,|\, \chi_n^{\{k\}}(Q)\rangle = \{E_n^{\{k\}}\} \,|\, \chi_n^{\{k\}}(Q)\rangle \qquad (22)$$

$\{E_n^{(k)}\}$ are the eigenvalues, whereas $|\chi_n^{(k)}(Q)\rangle$ are the corresponding eigenkets corresponding to different origins of the Q coordinate, and which define the k following different bases, according to:

$$\langle \chi_m^{(k)}(Q) | \chi_n^{(k)}(Q)\rangle = \delta_{mn}$$

$$\langle \chi_m^{(g)}(Q) | \chi_n^{(k)}(Q)\rangle = \{\Gamma_{mn}(g, k)\}$$

$$\sum |\chi_n^{(k)}(Q)\rangle\langle\chi_n^{(k)}(Q)| = 1$$

$\{\Gamma_{mn}(g, k)\}$ are the Franck–Condon factors.

In order to describe the full Hamiltonian, we must take the following basis built up from products of the two above bases:

$$\{|\Phi_k(q, Q)\rangle | \chi_n^{(k)}(Q)\rangle\}$$

As a consequence, the full Hamiltonian may be written:

$$
\begin{aligned}
\{H^0(q, Q)\} = \sum \sum \sum \sum & |\Phi_k(q, Q)\rangle | \chi_n^{(k)}(Q)\rangle \\
\times & \langle \chi_n^{(k)}(Q) | \langle \Phi_k(q, Q) | \{H^0(q, Q)\} | \Phi_g(q, Q)\rangle \\
\times & |\chi_m^{(g)}(Q)\rangle\langle\chi_m^{(g)}(Q) | \langle \Phi_g(q, Q) |
\end{aligned}
\tag{23}
$$

Besides, with Eq. (19), one obtains:

$$
\begin{aligned}
\{H^0(q, Q)\} = \sum \sum \sum \sum & \langle \chi_n^{(k)}(Q) | \langle \Phi_k(q, Q) | \\
\times & \{\{h(q, Q)\} + \tfrac{1}{2}M\omega^{\circ\circ 2}Q^2 + P^2/(2M)\} | \Phi_g(q, Q)\rangle \\
\times & |\chi_m^{(g)}(Q)\rangle | \Phi_k(q, Q)\rangle | \chi_n^{(k)}(Q)\rangle\langle\chi_m^{(g)}(Q) | \langle \Phi_g(q, Q) |
\end{aligned}
\tag{24}
$$

This Hamiltonian may be partitioned into:

$$\{H^0(q, Q)\} = \{H_{\text{diab}}^0(q, Q)\} + \{H_{\text{adiab}}^0(q, Q)\} \tag{25}$$

The first part, which is diagonal with respect to the eigenkets of the fast mode, is the adiabatic Hamiltonian:

$$
\begin{aligned}
\{H_{\text{adiab}}^0(q, Q)\} = \sum \sum & \langle \chi_n^{(k)}(Q) | \langle \Phi_k(q, Q) | \{\{h(q, Q)\} \\
& + \tfrac{1}{2}M\omega^{\circ\circ 2}Q^2 + P^2/(2M)\} | \Phi_k(q, Q)\rangle | \chi_n^{(k)}(Q)\rangle \\
\times & |\Phi_k(q, Q)\rangle | \chi_n^{(k)}(Q)\rangle\langle\chi_n^{(k)}(Q) | \langle \Phi_k(q, Q) |
\end{aligned}
\tag{26}
$$

On the other hand, the nondiagonal part of the Hamiltonian (i.e., the diabatic) is given by:

$$
\begin{aligned}
\{H_{\mathrm{diab}}^0(q, Q)\} = \sum \sum \sum \sum \langle \chi_n^{\{k\}}(Q)|\langle \Phi_k(q, Q)|\{\{h(q, Q)\} \\
+ \tfrac{1}{2}M\omega^{\circ\circ 2}Q^2 + P^2/(2M)\}|\Phi_g(q, Q)\rangle|\chi_m^{\{g\}}(Q)\rangle \\
\times |\Phi_k(q, Q)\rangle|\chi_n^{\{k\}}(Q)\rangle\langle\chi_m^{\{g\}}(Q)|\langle\Phi_g(q, Q)|, \qquad k \neq g
\end{aligned}
$$

This nondiagonal part reduces, owing to the orthogonality of the basis eigenfunctions, to:

$$
\begin{aligned}
\{H_{\mathrm{diab}}^0(q, Q)\} = \sum \sum \sum \sum \langle \chi_n^{\{k\}}(Q)|\chi_m^{\{g\}}(Q)\rangle\langle\Phi_k(q, Q)|P^2/(2M)|\Phi_g(q, Q)\rangle \\
+ [1/(2M)]\langle\chi_n^{\{k\}}(Q)|P|\chi_m^{\{g\}}(Q)\rangle\langle\Phi_k(q, Q)|P|\Phi_g(q, Q)\rangle \\
\times |\Phi_k(q, Q)\rangle|\chi_n^{\{k\}}(Q)\rangle\langle\chi_m^{\{g\}}(Q)|\langle\Phi_g(q, Q)|, \qquad k \neq g
\end{aligned}
$$

2. The Adiabatic Separation

Now, using a theorem given in Fong's book (1975), the above equation becomes:

$$
\begin{aligned}
\{H_{\mathrm{diab}}^0(q, Q)\} = \sum \sum \sum \sum \langle \chi_n^{\{k\}}(Q)|\chi_m^{\{g\}}(Q)\rangle\langle\Phi_k(q, Q)|P^2/(2M)|\Phi_g(q, Q)\rangle \\
+ \{[1/(2M)]\langle\chi_n^{\{k\}}(Q)|\partial U(q, Q)/\partial Q|\chi_m^{\{g\}}(Q)\rangle/\{\varepsilon_g(Q) - \varepsilon_k(Q)\}\} \\
\times \langle\Phi_k(q, Q)|P|\Phi_g(q, Q)\rangle|\Phi_k(q, Q)\rangle|\chi_n^{\{k\}}(Q)\rangle \\
\times \langle\chi_m^{\{g\}}(Q)|\langle\Phi_g(q, Q)|, \qquad k \neq g
\end{aligned}
\tag{27}
$$

In this equation it is well known that the first term is smaller than the last one. Besides, in the denominator of the last term appears the energy gap $\{\varepsilon_g(Q) - \varepsilon_k(Q)\}$ between the gth and kth levels of the fast mode, which allows simplifications when this gap is large.

Now it may be observed that the q mode is very fast with respect to the slow mode Q, so that it is possible to define a basis representation such as the fast q motion follows adiabatically the slow Q motion, just as electrons follow adiabatically the motion of the nuclei in a molecule.

Recall that Stepanov (1945) was the first to consider the adiabatic separation in the hydrogen bond. Later, Singh and Wood (1968), studying the vibrational dynamics of the stretching modes in the X–H$\cdots X$ systems, and assuming a Taylor expansion of the potential in the $X\cdots X$ and proton coordinate with strongly anharmonic coupling between them, have obtained accurate solutions for the low-lying states as expansion in harmonic oscillator basis states and have found on comparison that these solutions are well approximated using an adiabatic separation.

Maréchal in his thesis (1968) has shown, using the semiempirical treatment of Reid (1959) based on a slightly modified Lippincott and Schroeder potential (1956, 1957a and b) that the matrix elements neglected in the adiabatic separation are very small.

Sokolov and Saval'ev (1977) have presented a comprehensive discussion of the main features of the hydrogen spectra based on the Stepanov adiabatic approximation as the dominant concept. They treated asymmetric, weak to intermediate and strong, symmetric, hydrogen bonds, as distinct cases. For both types they have shown that a large amount of data such as isotope effects on frequencies, equilibrium distance versus frequencies, etc. follow theoretical relations derivable in an adiabatic description.

Later, Barton and Thorson (1979) have tested the validity of the Stepanov adiabatic approximation by including correct nonadiabatic couplings and solving the resultant eigenvalue problem for the FHF$^-$ model, by aid of ab initio potential energy surfaces. They have shown, by comparison with the exact results, that the adiabatic approximation gives transition frequencies within 1% and infrared relative intensities to about 10%. This systematic critical examination, which is in agreement with the above treatments, gives a good basis to the wide acceptance of the Stepanov approximation as a proper framework for interpreting vibrational dynamics in weak hydrogen bonds. As a consequence, in the following we shall neglect the diabatic part of the full Hamiltonian and write:

$$\{H^0(q, Q)\} \approx \sum \sum \langle \chi_n^{\{k\}}(Q) | \langle \Phi_k(q, Q) |$$
$$\times \{[p^2/(2M) + m\{\omega_{\text{eff}}^\circ(Q)\}^2] + [\tfrac{1}{2}M\omega^{\circ\circ2}Q^2] + P^2/(2M)\}$$
$$\times |\Phi_k(q, Q)\rangle |\chi_n^{\{k\}}(Q)\rangle |\Phi_k(q, Q)\rangle |\chi_n^{\{k\}}(Q)\rangle\langle\chi_n^{\{k\}}(Q) | \langle\Phi_k(q, Q) |$$

Moreover, given that the operators of the fast and slow modes are acting in different subspaces, that yields:

$$\{H^0(q, Q)\} \approx \sum \sum \langle \Phi_k(q, Q) | [p^2/(2m) + m\{\omega_{\text{eff}}^\circ(Q)\}^2] | \Phi_k(q, Q)\rangle$$
$$\times \langle\chi_n^{\{k\}}(Q) | [P^2/(2M) + \tfrac{1}{2}M\omega^{\circ\circ2}Q^2] | \chi_n^{\{k\}}(Q)\rangle$$
$$\times |\Phi_k(q, Q)\rangle |\chi_n^{\{k\}}(Q)\rangle\langle\chi_n^{\{k\}}(Q) | \langle\Phi_k(q, Q) |$$

Again, this simplifies to:

$$\{H^0(q, Q)\} \approx \sum |\Phi_k(q, Q)\rangle[(k + \tfrac{1}{2})\hbar\omega_{\text{eff}}^\circ(Q)]\langle\Phi_k(q, Q) |$$
$$\times \sum |\chi_n^{\{k\}}(Q)\rangle\langle\chi_n^{\{k\}}(Q) | [P^2/(2M)$$
$$+ \tfrac{1}{2}M\omega^{\circ\circ2}Q^2] | \chi_n^{\{k\}}(Q)\rangle\langle\chi_n^{\{k\}}(Q) |$$

At last, removing the closeness relation on the eigenkets of the slow mode, this gives the final result:

$$\{H^0(q, Q)\} \approx \sum |\Phi_k(q, Q)\rangle[(k + \tfrac{1}{2})\hbar\omega^\circ_{\text{eff}}(Q)]\langle\Phi_k(q, Q)|$$
$$\times \sum |\chi_n^{\{k\}}(Q)\rangle[P^2/(2M) + \tfrac{1}{2}M\omega^{\circ\circ2}Q^2]\langle\chi_n^{\{k\}}(Q)| \quad (28)$$

D. Effective Hamiltonians as a Consequence of Anharmonicity

Let us introduce the new notation:

$$|\{k\}\rangle = |\Phi_k(q, Q)\rangle, \qquad |n^{\{k\}}\rangle = |\chi_n^{\{k\}}(Q)\rangle$$

Of course we have the following orthonormality properties:

$$\langle\{g\}|\{k\}\rangle = \delta_{g, k}, \qquad \langle m^{\{k\}}|n^{\{k\}}\rangle = \delta_{mn}, \qquad \langle m^{\{g\}}|n^{\{k\}}\rangle = \{\Gamma_{mn}(g, k)\}$$
$$\sum |\{k\}\rangle\langle\{k\}| = 1, \qquad \sum |n^{\{k\}}\rangle\langle n^{\{k\}}| = 1$$

1. The Representation {I} of the Hydrogen Bond

Now we may observe that the Hamiltonian (28) we shall denote in the following with the subscript I, may be written in the new notation:

$$\{H_I^0(q, Q)\} = \sum |\{k\}\rangle\{H_I^{0\{k\}}(Q)\}\langle\{k\}| \quad (29)$$

$\{H_I^{0\{k\}}(Q)\}$ is an effective Hamiltonian, the nature of which depends on the quantum number k characterizing the excitation degree of the fast mode:

$$\{H_I^{0\{k\}}(Q)\} = \{H_I^{0\{k\}}\} = [(k + \tfrac{1}{2})\hbar\omega^\circ_{\text{eff}}(Q)] + [\tfrac{1}{2}M\omega^{\circ\circ2}Q^2 + P^2/(2M)] \quad (30)$$

We shall name this representation as the representation {I} of the hydrogen bond. Moreover, the angular frequency $\{\omega^\circ_{\text{eff}}(Q)\}$ depends on the slow coordinate, and may be expanded according to:

$$\{\omega^\circ_{\text{eff}}(Q)\} = \omega^\circ + a_{112}Q \quad (31)$$
$$a_{112} = [d\omega/dQ]^\circ \quad (32)$$

Then, using the above equations, the effective Hamiltonian becomes:

$$\{H_I^{0\{k\}}\} = [(k + \tfrac{1}{2})\hbar(\omega^\circ + a_{112}Q)] + [\tfrac{1}{2}M\omega^{\circ\circ2}Q^2 + P^2/(2M)] \quad (33)$$

As a consequence, the effective Hamiltonian of the slow mode corresponding to the ground state $k = 0$ and the first excited state $k = 1$ of the fast
mode are, respectively, after neglecting the zero-point energy of the fast
mode:

$$\{H_{\mathrm{I}}^{0\{0\}}\} = [\tfrac{1}{2}M\omega^{\circ\circ 2}Q^2 + P^2/(2M)] \tag{34}$$

$$\{H_{\mathrm{I}}^{0\{1\}}\} = [\tfrac{1}{2}M\omega^{\circ\circ 2}Q^2 + P^2/(2M)] + \hbar(\omega^\circ + a_{112}Q) \tag{35}$$

The transition from the ground state to the first excited one of the fast
mode q induces the occurring of a driven term in the Hamiltonian of the
slow mode Q, as a quantum consequence of the anharmonic coupling
between q and Q.

2.   The Boson Representation $\{\mathrm{II}\}$

a.   The Effective Hamiltonians (33) within the Boson Language. Now we
may express Eq. (33) by aid of the raising and lowering operators $a^\dagger$ and a,
obeying the commutation rule $[a, a^\dagger] = 1$ and characterizing the slow mode
Q according to the equations:

$$P = -iP^{\circ\circ}[a^\dagger - a], \qquad Q = Q^{\circ\circ}[a^\dagger + a]$$
$$P^{\circ\circ} = [M\hbar\omega^{\circ\circ}/2]^{1/2}, \qquad Q^{\circ\circ} = [\hbar/(2M\omega^{\circ\circ})]^{1/2}$$

In this representation, the effective Hamiltonians (33) becomes:

$$\{H_{\mathrm{I}}^{0\{k\}}\} = (k + \tfrac{1}{2})\hbar\omega^\circ + (k + \tfrac{1}{2})\alpha^\circ[a^\dagger + a]\hbar\omega^{\circ\circ} + (a^\dagger a + \tfrac{1}{2})\hbar\omega^{\circ\circ} \tag{36}$$

α° is an adimensional parameter characterizing the strength of the anharmonic coupling between the slow and fast mode and is given in view of Eq.
(32) by:

$$\alpha^\circ = a_{112}[Q^{\circ\circ}/\omega^{\circ\circ}] \tag{37}$$

$$\alpha^\circ = K_{112}[Q^{\circ\circ}/(m\omega^\circ\omega^{\circ\circ})] \tag{38}$$

$$\alpha^\circ = [d\omega/dQ][Q^{\circ\circ}/\omega^{\circ\circ}] \tag{39}$$

Of course, to simplify, one may substract the zero-point energy of the fast
mode, which leads to:

$$\{H_{\mathrm{I\,shift}}^{0\{k\}}\} = k\hbar\omega^\circ + k\alpha^\circ[a^\dagger + a]\hbar\omega^{\circ\circ} + \tfrac{1}{2}\alpha^\circ[a^\dagger + a]\hbar\omega^{\circ\circ} + (a^\dagger a + \tfrac{1}{2})\hbar\omega^{\circ\circ}$$

$$\tag{40}$$

b.   A Canonical Transformation Removing the Zero-Point Driven Term.
Note that in Eq. (40), there is a driven term $\frac{1}{2}\alpha°[a^{\dagger} + a]\hbar\omega°°$, linear in the raising and lowering operators, which is a consequence of the zero-point energy of the fast mode before expansion of the angular frequency of this mode in terms of the slow-mode coordinate. This driven operator, which will not play any role in the future, is $\frac{1}{2}\alpha°[a^{\dagger} + a]\hbar\omega°°$. In order to remove it, it is suitable to perform the following phase transformation on the above effective Hamiltonian:

$$\{H^{0\{k\}}_{\text{II shift}}\} = \{A(\tfrac{1}{2}\alpha°)\}\{H^{0\{k\}}_{\text{I shift}}\}\{A(\tfrac{1}{2}\alpha°)\}^{\dagger} \tag{41}$$

$\{A(\tfrac{1}{2}\alpha°)\}$ is the translation operator given by:

$$\{A(\tfrac{1}{2}\alpha°)\} = \exp\{\tfrac{1}{2}(\alpha°a^{\dagger} - \alpha°{*}a)\} = \{A(-\tfrac{1}{2}\alpha°)\}^{\dagger} \tag{42}$$

Then, by aid of theorem (A44) of Appendix A, the transformed effective Hamiltonian becomes after neglecting a scalar term $-(\tfrac{1}{2})\alpha°^{2}\hbar\omega°°$

$$\{H^{0\{k\}}_{\text{II shift}}\} = k\hbar\omega° + k\alpha°[a^{\dagger} + a]\hbar\omega°° - k\alpha°^{2}\hbar\omega°° + (a^{\dagger}a + \tfrac{1}{2})\hbar\omega°° \tag{43}$$

Again, neglecting the zero-point energy of the slow mode, one obtains in this new representation, which we shall name $\{II\}$:

$$\{H^{0\{k\}}_{\text{II}}\} = k\hbar\omega° + k\alpha°[a^{\dagger} + a]\hbar\omega°° - k\alpha°^{2}\hbar\omega°° + a^{\dagger}a\hbar\omega°° \tag{44}$$

Then, in view of Eq. (29), we may write in the new representation $\{II\}$ the full Hamiltonian of the hydrogen bond according to:

$$\{H^{0}_{\text{II}}\} = \sum | \{k\} \rangle \{H^{0\{k\}}_{\text{II}}\} \langle \{k\} | \tag{45}$$

Note that, in view of Eq. (44), the two first effective Hamiltonians of the slow mode corresponding, respectively, to the ground state and the first excited state of the fast mode, are, respectively:

$$\{H^{0\{0\}}_{\text{II}}\} = [(a^{\dagger}a)\hbar\omega°°] \tag{46}$$

$$\{H^{0\{1\}}_{\text{II}}\} = [(a^{\dagger}a)\hbar\omega°°] + \{\alpha°[a^{\dagger} + a] - \alpha°^{2}\}\hbar\omega°° + \hbar\omega° \tag{47}$$

c.   A Basis for the Representation of the Slow Mode. We may define the basis $\{|n\rangle\}$ built up from the eigenkets $|n\rangle$ of the number occupation operator according to:

$$(a^{\dagger}a)|n\rangle = n|n\rangle \tag{48}$$

$$\langle m \mid n \rangle = \delta_{mn}, \qquad \sum \mid n \rangle \langle n \mid = 1 \qquad (49)$$

Note that the eigenvalue equation of the effective Hamiltonian of the slow mode, when the first mode is not excited, is:

$$\{H_{\text{II}}^{0\{0\}}\} \mid n_{\text{II}}^{\{0\}} \rangle = n\hbar\omega^{\circ\circ} \mid n_{\text{II}}^{\{0\}} \rangle \qquad (50)$$

This shows that:

$$\mid n \rangle = \mid n_{\text{II}}^{\{0\}} \rangle \qquad (51)$$

Note that, in the coordinate representation of the slow mode, one has:

$$\langle Q \mid n_{\text{II}} \rangle = \{\chi_{\text{II}\,n}^{\{0\}}(Q)\} \qquad (52)$$

where the $\{\chi_{\text{II}\,n}^{\{0\}}(Q)\}$ are the wavefunctions of the slow mode appearing in the eigenvalue equation (22).

d. *The Density Operator.* Note that, just after the excitation of the fast mode, the density operator of the slow mode must remain the same as just before.

$$\{\rho_{\text{II}}^{\{1\}}(0)\} = \{\rho_{\text{II}}^{\{0\}}(0)\} \qquad (53)$$

As a consequence, if the slow mode is at thermal equilibrium before the excitation and thus is described by the Boltzmann density operator, it will also be described just after the excitation of the fast mode by the same density operator. Thus we may write the following equations characterizing these density operators:

$$\{\rho_{\text{II}}^{\{0\}}(0)\} = \varepsilon \exp\{-\lambda(a^{\dagger}a)\} \qquad (54)$$

$$\{\rho_{\text{II}}^{\{1\}}(0)\} = \varepsilon \exp\{-\lambda(a^{\dagger}a)\} \qquad (55)$$

$$\lambda = \hbar\omega^{\circ\circ}/(k_{\text{B}}\,T) \qquad (56)$$

$$\varepsilon = 1 - \exp\{-\lambda\} \qquad (57)$$

3. *The Boson Representation* {III}

In representation {II} only the effective Hamiltonian of the slow mode corresponding to the ground state of the fast mode is diagonal with respect to the number occupation operator. It is suitable to pass to a new representation we shall name {III} in which all the effective Hamiltonians become diagonal.

a. Canonical Transformation Allowing a Passage from Representation {II} *to* {III}. Consider an effective operator $\{B_{\mathrm{II}}^{\{k\}}\}$ characterizing a given excited state $|\{k\}\rangle$ of the fast mode in the representation {II}. We may transform it in the spirit of the work of Hofacker et al. (1976) according to the following canonical transformation:

$$|\{k\}\rangle\{B_{\mathrm{III}}^{\{k\}}\}\langle\{k\}| = |\{k\}\rangle\{A^{\{k\}}(k\alpha^\circ)\}\langle\{k\}||\{k\}\rangle\{B_{\mathrm{II}}^{\{k\}}\}\langle\{k\}|\{A^{\{k\}}(k\alpha^\circ)\}^\dagger\langle\{k\}| \tag{58}$$

$\{A(k\alpha^\circ)\}$ are the translation operators given by:

$$\{A^{\{k\}}(k\alpha^\circ)\} = \exp\{k(\alpha^\circ a^\dagger - \alpha^\circ * a)\} = \{A^{\{k\}}(-k\alpha^\circ)\}^\dagger \tag{59}$$

Note that for effective operators corresponding to the ground state of the fast mode (i.e., $k = 0$), the old and new representations are the same.

b. Diagonalization of the Effective Hamiltonians. The total Hamiltonian in representation {III} may be expressed as an expansion of effective Hamiltonians such as:

$$\{H_{\mathrm{III}}^0\} = \sum |\{k\}\rangle\{H_{\mathrm{III}}^{0\{k\}}\}\langle\{k\}| \tag{60}$$

The effective Hamiltonians in the new representation are:

$$\{H_{\mathrm{III}}^{0\{k\}}\} = \{A^{\{k\}}(k\alpha^\circ)\}\{H_{\mathrm{II}}^{0\{k\}}\}\{A^{\{k\}}(k\alpha^\circ)\}^\dagger \tag{61}$$

In view of Eq. (44) one obtains diagonal Hamiltonians, that is:

$$\{H_{\mathrm{III}}^{0\{k\}}\} = (a^\dagger a)\hbar\omega^{\circ\circ} + k\hbar\omega^\circ - k(k + 1)\alpha^{\circ 2}\hbar\omega^{\circ\circ} \tag{62}$$

Of course the first, which will be used extensively in the following, are, respectively:

$$\{H_{\mathrm{III}}^{0\{0\}}\} = (a^\dagger a)\hbar\omega^{\circ\circ} \tag{63}$$

$$\{H_{\mathrm{III}}^{0\{1\}}\} = (a^\dagger a)\hbar\omega^{\circ\circ} + \hbar\omega^\circ - 2\alpha^{\circ 2}\hbar\omega^{\circ\circ} \tag{64}$$

The eigenvalue equation of these effective Hamiltonians are:

$$\{H_{\mathrm{III}}^{0\{0\}}\}|n_{\mathrm{III}}^{\{0\}}\rangle = \{E_{\mathrm{III}\,n}^{\{0\}}\}|n_{\mathrm{III}}^{\{0\}}\rangle \tag{65}$$

$$\{H_{\mathrm{III}}^{0\{1\}}\}|n_{\mathrm{III}}^{\{1\}}\rangle = \{E_{\mathrm{III}\,n}^{\{1\}}\}|n_{\mathrm{III}}^{\{1\}}\rangle \tag{66}$$

with, respectively:

$$\{E^{\{0\}}_{\text{III}\,n}\} = n\hbar\omega^{\circ\circ} \tag{67}$$

$$\{E^{\{1\}}_{\text{III}\,n}\} = n\hbar\omega^{\circ\circ} + \hbar\omega^\circ - 2\alpha^{\circ 2}\hbar\omega^{\circ\circ} \tag{68}$$

This result shows that the ground state of the slow mode is not only raised in energy by the amount of the excitation energy of the fast mode $\hbar\omega^\circ$, as in the usual normal-modes approach, but also lowered by the amount $2\alpha^{\circ 2}\hbar\omega^{\circ\circ}$, because of the anharmonic coupling between the slow and fast modes, (see Fig. 1.1). We shall see later that this lowering is related to a displacement of the origin of the slow-mode coordinate.

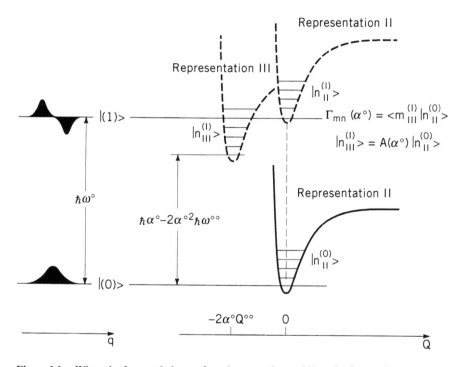

Figure 1.1. When the fast mode jumps from its ground state $|0\rangle$ to its first excited state $|1\rangle$, the effective Hamiltonian of the slow mode is modified and becomes driven. In representation {III}, the driven term is removed and the effective potential of the slow mode may be viewed as translated from $-2\alpha^\circ Q^{\circ\circ}$ and lowered in energy with respect to the excitation $\hbar\omega^\circ$ by $-2\alpha^\circ\hbar\omega^{\circ\circ}$. The displaced excited effective potential with respect to the ground effective potential of the slow mode leads to Franck–Condon factors.

c.   The Transformed Basis Describing the Slow Mode. Now consider the effective Hamiltonian in representation $\{II\}$. One may write:

$$\text{tr}\{|\,n_{II}^{\{k\}}\rangle\langle n_{II}^{\{k\}}\,|\,\{H_{II}^{\{k\}}\}\}$$

$$= \text{tr}\{|\,n_{II}^{\{k\}}\rangle\langle n_{II}^{\{k\}}\,|\,\{A^{\{k\}}(k\alpha^\circ)\}^\dagger\{A^{\{k\}}(k\alpha^\circ)\}\{H_{II}^{\{k\}}\}\{A^{\{k\}}(k\alpha^\circ)\}^\dagger\{A^{\{k\}}(k\alpha^\circ)\}\} \quad (69)$$

Then, using the invariance of the trace, one obtains:

$$\text{tr}\{|\,n_{II}^{\{k\}}\rangle\langle n_{II}^{\{k\}}\,|\,\{H_{II}^{\{k\}}\}\}$$

$$= \text{tr}\{\{A^{\{k\}}(k\alpha^\circ)\}\,|\,n_{II}^{\{k\}}\rangle\langle n_{II}^{\{k\}}\,|\,\{A^{\{k\}}(k\alpha^\circ)\}^\dagger\{A^{\{k\}}(k\alpha^\circ)\}\{H_{II}^{\{k\}}\}\{A^{\{k\}}(k\alpha^\circ)\}^\dagger\} \quad (70)$$

As a consequence, in view of Eq. (61), one obtains:

$$\text{tr}\{|\,n_{II}^{\{k\}}\rangle\langle n_{II}^{\{k\}}\,|\,\{H_{II}^{\{k\}}\}\} = \text{tr}\{|\,n_{III}^{\{k\}}\rangle\langle n_{III}^{\{k\}}\,|\,\{H_{III}^{\{k\}}\}\} \quad (71)$$

where the eigenkets in the new representation are:

$$|\,n_{III}^{\{k\}}\rangle = \{A^{\{k\}}(k\alpha^\circ)\}\,|\,n_{II}^{\{k\}}\rangle \quad (72)$$

Besides one has:

$$|\,n_{III}^{\{0\}}\rangle = |\,n_{II}^{\{0\}}\rangle = |\,n_{II}^{\{1\}}\rangle \quad (73)$$

For the first excited state of the fast mode, the transformed state of the slow mode is:

$$|\,n_{III}^{\{1\}}\rangle = \{A^{\{1\}}(\alpha^\circ)\}\,|\,n_{II}^{\{1\}}\rangle = \{A^{\{1\}}(\alpha^\circ)\}\,|\,n_{III}^{\{0\}}\rangle \quad (74)$$

The scalar products of the transformed states with the untransformed ones are therefore the matrix elements of the translation operator.

$$\langle m_{III}^{\{0\}}\,|\,n_{III}^{\{1\}}\rangle = \langle m_{III}^{\{0\}}\,|\,\{A^{\{1\}}(\alpha^\circ)\}\,|\,n_{III}^{\{1\}}\rangle \quad (75)$$

Besides, these matrix elements are nothing but the Franck–Condon factors $\Gamma_{mn}(\alpha^\circ)$.

$$\langle m_{III}^{\{0\}}\,|\,\{A^{\{1\}}(\alpha^\circ)\}\,|\,n_{III}^{\{1\}}\rangle = \Gamma_{mn}(\alpha^\circ) = \Gamma_{nm}(-\alpha^\circ) \quad (76)$$

Note that owing to Eq. (74), one has,

$$\langle Q \mid n_{\mathrm{III}}^{\{1\}} \rangle = \langle Q \mid \{ A^{\{1\}}(\alpha^{\circ}) \} \mid n_{\mathrm{II}}^{\{1\}} \rangle \tag{77}$$

or, in view of Eq. (A37) of Appendix A:

$$\langle Q \mid n_{\mathrm{III}}^{\{1\}} \rangle = \langle \chi_{\mathrm{II}\,m}^{\{0\}}(Q - 2\alpha^{\circ}Q^{\circ\circ}) \} \tag{78}$$

d. The Transformed Boltzmann Density Operators. Consider the transformed density operators characterizing the slow mode in representation {III}:

$$\{ \rho_{\mathrm{III}}^{\{k\}}(0) \} = \{ A^{\{k\}}(k\alpha^{\circ}) \} \{ \rho_{\mathrm{II}}^{\{k\}}(0) \} \{ A^{\{k\}}(k\alpha^{\circ}) \}^{\dagger} \tag{79}$$

Now, for the ground state of the fast mode, since $k = 0$ and owing to Eq. (54), we have:

$$\{ \rho_{\mathrm{III}}^{\{0\}}(0) \} = \varepsilon \exp\{ -\lambda a^{\dagger}a \} \tag{80}$$

For the first excited state of the fast mode, the transformed density operator of the slow mode is:

$$\{ \rho_{\mathrm{III}}^{\{1\}}(0) \} = \varepsilon \{ A^{\{1\}}(\alpha^{\circ}) \} \exp\{ -\lambda(a^{\dagger}a) \} \{ A^{\{1\}}(\alpha^{\circ}) \}^{\dagger} \tag{81}$$

Then, in view of theorem (A44) of Appendix A, one obtains

$$\{ \rho_{\mathrm{III}}^{\{1\}}(0) \} = \varepsilon \exp\{ -\lambda(a^{\dagger} - \alpha^{\circ})(a - \alpha^{\circ}) \} \tag{82}$$

e. The Transformed Q Coordinate and Dipole Moment Operator. Consider the Q coordinate of the slow mode. In representation {III} we have for $k = 1$:

$$\{ Q_{\mathrm{III}}^{\{1\}} \} = \{ A^{\{1\}}(\alpha^{\circ}) \} \{ Q_{\mathrm{II}} \} \{ A^{\{1\}}(\alpha^{\circ}) \}^{\dagger} \tag{83}$$

Recall that in representation {II} this coordinate is, according to Section 2a:

$$\{ Q_{\mathrm{II}} \} = Q^{\circ\circ}[a^{\dagger} + a] \tag{84}$$

Then, in view of theorem (A44) of Appendix A, we get:

$$\{Q_{III}^{\{1\}}\} = Q^{\circ\circ}[a^\dagger + a - 2\alpha^\circ] \tag{85}$$

As it appears, the unitary transform on the position coordinate operator of the slow mode, performed by aid of the translation operator, leads to a translation in the origin of this operator. As a consequence, this translation involves a displacement of the harmonic potential. More generally, that implies a displacement of the potential governing the dynamics of the slow mode: When the fast mode is excited, the effective potential of the slow mode is displaced towards short equilibrium distance of the slow mode, as is pictured in Fig. 1.1.

Finally, consider the dipole moment operator of the fast mode. In representation $\{II\}$ it is simply:

$$\{\mu_{II}\} = \sum \sum \{\mu_{II}^{\{k,\,l\}}\} = \sum \sum \mu_{kl} |\{k\}\rangle\langle\{l\}| \tag{86}$$

In the new representation $\{III\}$ it becomes:

$$\{\mu_{III}\} = \sum \sum \mu_{kl}\{A^{\{k\}}(k\alpha^\circ)\} |\{k\}\rangle\langle\{l\}| \{A^{\{l\}}(l\alpha^\circ)\}^\dagger \tag{87}$$

For the special situation of the transition from the ground state to the first excited state of the fast mode, that is, for $k = 1$ and $l = 0$, the above expression reduces to:

$$\{\mu_{III}^{\{1,\,0\}}\} = \{A^{\{1\}}(\alpha^\circ)\} |\{1\}\rangle\langle\{0\}| \mu_{10} \tag{88}$$

4. Isotope Effect

Equation (38) allows one to take into account the isotope effect. First we may observe that the ratio of the reduced masses μ_H/μ_D of the hydrogenated and deuterated hydrogen bond species concerning the fast mode is equal to 2. It is well known that the ratio of the angular frequencies is $\omega_H^\circ/\omega_D^\circ = \sqrt{2}$. Therefore, the ratio of the dimensionless coupling parameter is $\alpha_H^\circ/\alpha_D^\circ = \sqrt{2}$.

Within this formalism, Maréchal (1972) has shown that the $X \cdots Y$ equilibrium distance in a deuterated compounds $X-D \cdots Y$ differs from the corresponding equilibrium distance involved in the hydrogenated species $X-H \cdots Y$ by:

$$\Delta Q_{H,\,D} = -Q^{\circ\circ}\sqrt{2}\,\{\alpha_H^\circ - \alpha_D^\circ\} \tag{89}$$

The value of the parameter α° may be obtained from the infrared spectra by simulation of the spectrum within the strong anharmonic coupling theory (vide infra). On the other hand, $\Delta Q_{H, D}$ may be directly obtained, for instance by X-ray diffraction. Thus, as discussed by Maréchal, there is the possibility to compare the quantity obtained by the two ways that have led the authors to verify good agreement. In a similar way, Maréchal has shown that the equilibrium distance $X \cdots Y$, when the fast mode is in its first excited state, differs from the corresponding one when it is in its ground state by the negative quantity:

$$\{\Delta Q^{\{1, 0\}}\} = -2\alpha_H^{\circ} Q^{\circ\circ} \tag{90}$$

that is, the hydrogen bond is shortened and thus its strength enhanced when the fast mode is excited.

5.   The Strong Coupling Theory as a General Frame for Studying the Spectral Density

The strong coupling theory may be employed as above, using the position representation or in the framework of the double quantization language used in solid-state physics and in the quantum theory of light. Dealing with this representation, the works of Fischer, Hofacker, and Ratner (1970), Rösch and Ratner (1974), and Boulil et al. (1988, 1994a and b) should be mentioned.

The strong coupling method is both sufficiently general and elastic to include such other phenomena as Davydov coupling. Fermi resonances, tunneling effects, damping influence of the medium, and so forth. It seems to be the more general and applicable method for discussions on the dynamic properties of hydrogen-bonded systems.

The strong coupling method allows one to calculate the energy levels characterizing the effective Hamiltonians of the slow mode. The difference in energy between these levels leads, according to the Bohr principle, to a delta function spectrum. Of course, such a spectrum is never actually observed in practice since there are always such factors as instrumental resolution, rotational structure, and Doppler broadening, which can smear out such spectra. There is also, for hydrogen-bonded systems, a series of more important broadening and smoothing mechanisms, all of which can be also consistently included in a strong coupling picture.

III.   THE DYNAMICS OF THE SLOW MODE

A.   The Dynamics of the Slow Mode in the Absence of Damping

1.   The Time Evolution Operator of the Slow Mode in Representation {II}

In the absence of relaxation, it may be shown (Boulil et al., 1994a) that the time evolution operator $\{U_{II}^{0\{1\}}\}$ governing the dynamics of the slow mode just after the fast mode has been excited, that is, that of an undamped driven quantum harmonic oscillator generated by the Hamiltonian $\{H_{II}^{0\{1\}}\}$ given by Eq. (47), is:

$$i\hbar\ \partial\{U_{II}^{0\{1\}}(t)/\partial t\} = \{H_{II}^{0\{1\}}\}\{U_{II}^{0\{1\}}(t)\} \tag{91}$$

with the boundary condition for the initial situation:

$$\{U_{II}^{0\{1\}}(0)\} = 1$$

Substituting Eq. (47) into this equation, one obtains:

$$i\ \partial\{U_{II}^{0\{1\}}(t)\}/\partial t = \{(a^\dagger a)\omega^{\circ\circ} + \{\alpha^\circ[a^\dagger + a] - \alpha^{\circ 2}\}\omega^{\circ\circ} + \omega^\circ\}\{U_{II}^{0\{1\}}(t)\} \tag{92}$$

The resolution of the Schrödinger equation is given in Appendix C by Eq. (C30):

$$\{U_{II}^{0\{1\}}(t)\} = \exp\{-i(\omega^\circ - \alpha^{\circ 2}\omega^{\circ\circ})t\}\ \exp\{i\alpha^{\circ 2}\omega^{\circ\circ}t\}$$
$$\times \exp\{-i\alpha^{\circ 2}\sin(\omega^{\circ\circ}t)\}\ \exp\{-ia^\dagger a\omega^{\circ\circ}t\}\{A^{\{1\}}(-\Phi_{II}^0(t)*)\} \tag{93}$$

with:

$$\{\Phi_{II}^0(t)\} = \alpha^\circ[\exp\{-i\omega^{\circ\circ}t\} - 1] \tag{94}$$

The dynamics of the density operator of the slow mode in representation $\{II\}$ is given, according to Eq. (A3), by the unitary transform:

$$\{\rho_{II}^{0\{1\}}(t)\} = \{U_{II}^{0\{1\}}(t)\}\{\rho_{II}^{\{1\}}(0)\}\{U_{II}^{0\{1\}}(t)\}^\dagger$$

Now we may observe that, in view of Eq. (C29), the Eq. (93) may also be written:

$$\{U_{II}^{0\{1\}}(t)\} = \exp\{-i(\omega^\circ - \alpha^{\circ 2}\omega^{\circ\circ})t\}\ \exp\{i\alpha^{\circ 2}\omega^{\circ\circ}t\}$$
$$\times \exp\{-i\alpha^{\circ 2}\sin(\omega^{\circ\circ}t)\}\{A^{\{1\}}(\Phi_{II}^0(t))\}\ \exp\{-ia^\dagger a\omega^{\circ\circ}t\}$$

Then, owing to Eq. (55) and to the above equations, one finds, after canceling the scalar phase factors:

$$\{\rho_{II}^{0\{1\}}(t)\} = \varepsilon\{A^{\{1\}}(\Phi_{II}^0(t))\}\ \exp\{-ia^\dagger a\omega^{\circ\circ}t\}$$
$$\times \exp\{-\lambda a^\dagger a\}\ \exp\{ia^\dagger a\omega^{\circ\circ}t\}\{A^{\{1\}}(\Phi_{II}^0(t))\}^\dagger$$

which reduces to:

$$\{\rho_{\text{II}}^{0\{1\}}(t)\} = \varepsilon\{A^{\{1\}}(\Phi_{\text{II}}^0(t))\} \exp\{-\lambda a^\dagger a\}\{A^{\{1\}}(\Phi_{\text{II}}^0(t))\}^\dagger \tag{95}$$

Again, using theorem (A44) of Appendix A, one obtains:

$$\{\rho_{\text{II}}^{0\{1\}}(t)\} = \varepsilon \exp\{-\lambda(a^\dagger - \Phi_{\text{II}}^0(t))(a - \Phi_{\text{II}}^0(t)^*\} \tag{96}$$

It may be of interest to observe that the density operator of the driven damped quantum harmonic oscillator is that of a coherent state at the absolute temperature T. It may be shown, according to Eq. (A22), that when T approaches zero, the density operator given by the preceding equation narrows an operator projecting on the coherent state.

$$\{\rho_{\text{II}}^{0\{1\}}(t)\} \rightarrow |\Phi_{\text{II}}^{0\{1\}}(t)\rangle\langle\Phi_{\text{II}}^{0\{1\}}(t)|, \qquad \text{when } T \rightarrow 0 \tag{97}$$

A coherent state obeys the eigenvalue equation, according to property (A46) of Appendix A:

$$a|\Phi_{\text{II}}^{0\{1\}}(t)\rangle = \{\Phi_{\text{II}}^0(t)\}|\Phi_{\text{II}}^{0\{1\}}(t)\rangle \tag{98}$$

where a is the lowering operator. Recall that a coherent state minimizes the Heisenberg uncertainty relation, leading to a quasiclassical behavior. Such a coherent state may be obtained from the ground state $|0_{\text{II}}^{\{1\}}\rangle$ of the slow mode by aid of the time evolution operator (93) according to:

$$|\Phi_{\text{II}}^{0\{1\}}(t)\rangle = \{U_{\text{II}}^{0\{1\}}(t)\}|0_{\text{II}}^{\{1\}}\rangle \tag{99}$$

2. The Dynamics of the Density Operator of the Slow Mode in Representation {III}

The dynamics of the density operator of the slow mode in representation {III} is given by the unitary transform:

$$\{\rho_{\text{III}}^{0\{1\}}(t)\} = \exp\{-i\{H_{\text{III}}^0\}t/\hbar\}\{\rho_{\text{III}}^{0\{1\}}(0)\} \exp\{i\{H_{\text{III}}^0\}t/\hbar\} \tag{100}$$

Then the phase transformation acts inside the subspace corresponding to the first excited state of the fast mode, so that it may be expressed in terms of the corresponding effective Hamiltonians according to:

$$\{\rho_{\text{III}}^{0\{1\}}(t)\} = \exp\{-i\{H_{\text{III}}^{0\{1\}}\}t/\hbar\}\{\rho_{\text{III}}^{0\{1\}}(0)\} \exp\{i\{H_{\text{III}}^{0\{1\}}\}t/\hbar\} \tag{101}$$

Or, in view of Eqs. (64) and (82) and after simplification,

$$\{\rho_{\mathrm{III}}^{0\{1\}}(t)\} = \varepsilon \, \exp\{-ia^{\dagger}a\omega^{\circ\circ}t\} \, \exp\{-\lambda(a^{\dagger} - \alpha^{\circ})(a - \alpha^{\circ})\} \, \exp\{ia^{\dagger}a\omega^{\circ\circ}t\}$$

(102)

Moreover, by aid of theorem (A21) of Appendix A, one finds:

$$\{\rho_{\mathrm{III}}^{0\{1\}}(t)\} = \varepsilon \, \exp\{-\lambda(a^{\dagger} \exp\{-i\omega^{\circ\circ}t\} - \alpha^{\circ})(a \exp\{i\omega^{\circ\circ}t\} - \alpha^{\circ})\} \quad (103)$$

Again, inserting $\exp\{i\omega^{\circ\circ}t\} \, \exp\{-i\omega^{\circ\circ}t\} = 1$, one obtains:

$$\{\rho_{\mathrm{III}}^{0\{1\}}(t)\} = \varepsilon \, \exp\{-\lambda(a^{\dagger} - \alpha^{\circ} \exp\{i\omega^{\circ\circ}t\})(a - \alpha^{\circ} \exp\{-i\omega^{\circ\circ}t\})\} \quad (104)$$

Or

$$\{\rho_{\mathrm{III}}^{0\{1\}}(t)\} = \varepsilon \, \exp\{-\lambda(a^{\dagger} - \Phi_{\mathrm{III}}^{0}(t))(a - \Phi_{\mathrm{III}}^{0}(t)^{*})\}$$

(105)

with:

$$\{\Phi_{\mathrm{III}}^{0}(t)\} = \alpha^{\circ} \, \exp\{-i\omega^{\circ\circ}t\}$$

(106)

Note that the density operator may be obtained also owing to Eq. (A44) according to:

$$\{\rho_{\mathrm{III}}^{0\{1\}}(t)\} = \varepsilon\{A^{\{1\}}(\Phi_{\mathrm{III}}^{0}(t))\} \, \exp\{-\lambda(a^{\dagger}a)\}\{A^{\{1\}}(\Phi_{\mathrm{III}}^{0}(t))\}^{\dagger}$$

where, in view of Eq. (A32):

$$\{A^{\{1\}}(\Phi_{\mathrm{III}}^{0}(t))\} = \exp\{(\Phi_{\mathrm{III}}^{0}(t)a^{\dagger} - \Phi_{\mathrm{III}}^{0}(t)^{*}a)\}$$

(107)

By aid of the density operator (96), the time evolution of the average value of the slow-mode coordinate, when the fast mode is in its excited state, may be found, for instance, in representation $\{\mathrm{II}\}$ by aid of:

$$\langle Q^{\{1\}}(t)\rangle = \mathrm{tr}\{\{\rho_{\mathrm{II}}^{0\{1\}}(t)\}\{Q_{\mathrm{II}}\}\}$$

Note that, in representation $\{\mathrm{II}\}$, the Q coordinate is the same per se for all values of the quantum number characterizing the fast mode. In representation $\{\mathrm{III}\}$, the thermal average value of Q in the same conditions is:

$$\langle Q^{\{1\}}(t)\rangle = \mathrm{tr}\{\{\rho_{\mathrm{III}}^{0\{1\}}(t)\}\{Q_{\mathrm{III}}^{\{1\}}\}\}$$

(108)

where we have now defined that the Q coordinate has been subject to a phase transformation characterized by the quantum number of the fast mode $k = 1$, so that we have written $\{Q_{III}^{\{1\}}\}$.

In representation $\{III\}$, for instance, one has, in view of Eqs. (85), (105), and (108):

$$\langle Q^{\{1\}}(t) \rangle = \varepsilon Q^{\circ\circ}\, \mathrm{tr}\{\exp\{-\lambda(a^\dagger - \Phi_{III}^0(t))(a - \Phi_{III}^0(t)^*)\}[a^\dagger + a - 2\alpha^\circ]\} \quad (109)$$

Then, by aid of inference similar to that involved in Appendix F, one obtains:

$$\langle Q^{\{1\}}(t) \rangle = 2\alpha^\circ Q^{\circ\circ}[\cos(\omega^{\circ\circ}t) - 1] \quad (110)$$

This last result shows that just after the fast mode has jumped in its first excited state, the average value is zero, as before, and then, is going back and forth around its equilibrium position $-2\alpha^\circ Q^{\circ\circ}$.

B. The Dynamics of the Slow Mode in the Presence of Damping

In Section A we studied the dynamics of the slow mode, that is due to the quantum properties of the hydrogen bond system when there is no irreversible influence of the medium. Now, one has to look at the dynamics of the slow mode of the hydrogen bond when this slow mode is damped by the surroundings.

1.   The Dynamics in Representation $\{II\}$: The Driven Underdamped Quantum Harmonic Oscillator

a. The Hamiltonian of the Underdamped Slow Mode in Representation $\{II\}$. In order to treat the irreversible interaction of the hydrogen bond with an inert solvent, the following model is used: the total Hamiltonian $\{H_{II}\}$ is taken as the sum of the Hamiltonian (45), that of the thermal bath, and that of the interaction between it and the hydrogen bond. The thermal bath is considered as a set of quantum harmonic oscillators characterized by the bosons $b_k^\dagger$ and b_k involving quasicontinuous changes in their angular frequencies ω_k. If one assumes a weak coupling between the slow mode and the thermal bath, it is possible to consider this coupling as linear with respect to the bosons. That leads one to write the full Hamiltonian $\{H_{II}\}$ of the hydrogen bond embedded in a thermal bath as:

$$\{H_{II}\} = \{H_{II}^0\} + \{H_{II}^{damp}\} \quad (111)$$

$\{H_{II}^0\}$ is the Hamiltonian of the hydrogen bond in the absence of a thermal bath, whereas $\{H_{II}^{damp}\}$ is that of the thermal bath and of the coupling of the slow mode with it, that, in the linear coupling approximation, and neglecting the zero-point energies that do not play any role, may be written:

$$\{H_{\text{II}}^{\text{damp}}\} = \sum \{b_k^\dagger b_k \hbar\omega_k\} + \sum \{b_k^\dagger a\hbar\kappa_k + b_k a^\dagger \hbar\kappa_k^*\} \qquad (112)$$

Of course, since one assumes a weak coupling between the oscillator and the thermal bath, the coupling parameters κ_k must be small with respect to the characteristic angular frequencies, that is, $|\kappa_k| < \omega_k$. The above full Hamiltonian may also be written formally in terms of the effective Hamiltonians involved in the expression (45), that is:

$$\{H_{\text{II}}\} = \sum \{|\{k\}\rangle\{H_{\text{II}}^{0\{k\}}\}\langle\{k\}| + |\{k\}\rangle\{H_{\text{II}}^{\text{damp}}\}\langle\{k\}|\} \qquad (113)$$

or,

$$\{H_{\text{II}}\} = \sum |\{k\}\rangle\{H_{\text{II}}^{\{k\}}\}\langle\{k\}| \qquad (114)$$

with:

$$\{H_{\text{II}}^{\{k\}}\} = \{H_{\text{II}}^{0\{k\}}\} + \{H_{\text{II}}^{\text{damp}}\} \qquad (115)$$

The effective Hamiltonian of the slow mode embedded in the thermal bath, and corresponding to the ground state of the fast mode is:

$$\{H_{\text{II}}^{\{0\}}\} = [(a^\dagger a)\hbar\omega^{\circ\circ}] + \sum \{b_k^\dagger b_k \hbar\omega_k\} + \sum \{b_k^\dagger a\hbar\kappa_k + b_k a^\dagger \hbar\kappa_k^*\} \quad (116)$$

The effective Hamiltonian of the slow mode embedded in the thermal bath, and corresponding to the situation where the fast mode is excited, is:

$$\{H_{\text{II}}^{\{1\}}\} = (a^\dagger a)\hbar\omega^{\circ\circ} + \{\alpha^\circ[a^\dagger + a] - \alpha^{\circ 2}\}\hbar\omega^{\circ\circ}$$
$$+ \sum \{b_k^\dagger b_k \hbar\omega_k\} + \sum \{b_k^\dagger a\hbar\kappa_k + b_k a^\dagger \hbar\kappa_k^*\} \quad (117)$$

b. *The Dynamics of the Reduced Density Operator.* The reduced density operator characterizing the slow mode Q at time t may be found by solving the Liouville equation:

$$i\hbar\{\partial\rho_{\text{II}}^{\{1\}}(t)/\partial t\} = \text{tr}_\theta[\{H_{\text{II}}^{\{1\}}\}, \{\rho_{\text{II tot}}^{\{1\}}(t)\}] \qquad (118)$$

$\{\rho_{\text{II tot}}^{\{1\}}(t)\}$ is the full density operator at time t of the slow mode and of the thermal bath θ in representation $\{\text{II}\}$, tr_θ is the partial trace on the thermal bath. At initial time, the total density operator may be written as the product of the density operator of the slow mode times that of the thermal bath, that is:

$$\{\rho_{\mathrm{II}\,\mathrm{tot}}^{\{1\}}(0)\} = \{\rho_{\mathrm{II}}^{\{1\}}(0)\}\{\rho_\theta\} \tag{119}$$

with, in view of Eq. (55):

$$\{\rho_{\mathrm{II}}^{\{1\}}(0)\} = \varepsilon \exp\{-\lambda(a^\dagger a)\} \tag{120}$$

and where:

$$\{\rho_\theta\} = \prod \varepsilon_k \exp\{-\lambda_k(b_k^\dagger b_k)\} \tag{121}$$

with, respectively:

$$\lambda_k = \hbar\omega_k/(k_B\,T), \qquad \varepsilon_k = (1 - \exp\{-\lambda_k\})$$

According to Eq. (H83) of Appendix H, the solution of the Liouville equation (118), for situations where the damping is not too strong, is given according to:

$$\{\rho_{\mathrm{II}}^{\{1\}}(t)\} = \varepsilon \exp\{-\lambda(a^\dagger - \Phi_{\mathrm{II}}(t))(a - \Phi_{\mathrm{II}}^*(t))\} \tag{122}$$

$$\{\Phi_{\mathrm{II}}(t)\} = \beta[\exp\{-\gamma t/2\}\exp\{-i\omega^{\circ\circ}t\} - 1] \tag{123}$$

γ is the damping parameter related to the strength of the coupling of the slow mode with the thermal bath given by Eq. (H70),

$$\gamma = 2\pi\sum |\kappa_k|^2\,\delta(\omega_k - \omega^{\circ\circ}) \tag{124}$$

We must recall that, since the coupling of the oscillator with the bath is weak, one has $\gamma < \omega^{\circ\circ}$, which is the underdamped condition, whereas β is a nondimensional effective anharmonic coupling parameter given by:

$$\beta = \alpha^\circ(2\omega^{\circ\circ 2} + i\gamma\omega^{\circ\circ})/[2(\omega^{\circ\circ 2} + \gamma^2/4)] \tag{125}$$

Note that when the damping is missing (i.e., when $\gamma \to 0$),

$$\beta(\gamma \to 0) = \alpha^\circ \tag{126}$$

Then the argument of the translation operator reduces to that (94) involved in the absence of damping, that is:

$$\{\Phi_{II}(t, \gamma \to 0)\} = \{\Phi_{II}^0(t)\} \tag{127}$$

It may be observed that, in the same situation where there is no damping, the density operator (122) reduces to the corresponding one (96) in the absence of damping and in the same representation:

$$\{\rho_{II}^{\{1\}}(t, \gamma \to 0)\} = \{\rho_{II}^{0\{1\}}(t)\}$$

Note that the density operator (122) may be considered as the result of a phase transformation on the Boltzmann operator (55).

$$\{\rho_{II}^{\{1\}}(t)\} = \varepsilon\{A^{\{1\}}(\Phi_{II}(t))\} \exp\{-\lambda a^\dagger a\}\{A^{\{1\}}(\Phi_{II}(t))\}^\dagger \tag{128}$$

This phase transformation involves, in view of Eq. (A44) of Appendix A, the translation operator defined by:

$$\{A^{\{1\}}(\Phi_{II}(t))\} = \exp\{\Phi_{II}(t)a^\dagger - \Phi_{II}^*(t)a\} \tag{129}$$

It is shown in Appendix I that the reduced time evolution operator of the driven damped quantum harmonic oscillator may be obtained from the density operator (122) by aid of Eq. (16) according to the equations:

$$\{U_{II}^{\{1\}}(t)\} = \exp\{-i(\omega^\circ - \alpha^{\circ 2}\omega^{\circ\circ})t\} \exp\{i(|\beta|^2\omega^{\circ\circ})t\}$$
$$\times \exp\{-i|\beta|^2 \exp\{-\gamma t/2\} \sin(\omega^{\circ\circ}t)\}$$
$$\times \exp\{-ia^\dagger a\omega^{\circ\circ}t\}\{A^{\{1\}}(-\Phi_{II}(t)^*)\} \tag{130}$$
$$\{A^{\{1\}}(\Phi_{II}(t))\} = \exp\{\beta\{\exp\{-\gamma t/2\} \exp\{-i\omega^{\circ\circ}t\} - 1\}a^\dagger$$
$$- \beta^*\{\exp\{-\gamma t/2\} \exp\{i\omega^{\circ\circ}t\} - 1\}a\} \tag{131}$$

The reduced time evolution operator governing the eigenstates of the Hamiltonian of the slow mode may be viewed as a scattering operator inducing diffusion between the energy levels $|m_{II}^{\{1\}}\rangle$. It is this scattering that is at the origin of the broadened structure of the spectral density of the hydrogen bond.

Note that, when $\gamma \to 0$, owing to Eq. (127), the translation operator of the damped driven oscillator reduces to that of the corresponding undamped one.

$$\{A^{\{1\}}(\Phi_{II}(t, \gamma \to 0))\} = \{A^{\{1\}}(\Phi_{II}^0(t))\} \qquad (132)$$

Thus, as required, the above reduced time evolution operator reduces, in the absence of relaxation, to that of the undamped oscillator given by Eq. (93). At the opposite, when passing from the undamped driven situation to the damped one, the time evolution operators are transformed according to:

$$\{U_{II}^{0\{1\}}(t)\} \to \{U_{II}^{\{1\}}(t)\} \qquad (133)$$

c.   *The Quasiclassical Behavior of the Driven Damped Harmonic Oscillator.* It may be of interest to observe that the density operator of the driven damped quantum harmonic oscillator is that of a coherent state at the absolute temperature T. When T is approaching zero, then, according to Eq. (A22) of Appendix A, the density operator $\rho_{II}^{\{1\}}(t)$ given by Eq. (122) narrows an operator projecting on the coherent state $|\Phi_{II}(t)\rangle$, which may be written:

$$\{\rho_{II}^{\{1\}}(t)\} \to |\Phi_{II}^{\{1\}}(t)\rangle\langle\Phi_{II}^{\{1\}}(t)|, \qquad \text{when } T \to 0 \qquad (134)$$

Recall that a coherent state minimizes the Heisenberg uncertainty relation, leading to a quasiclassical behavior. That is the reason for which these states are also qualified as quasiclassical. In the present situation the coherent state defined by the above equation must also be obtained from the ground state $|0_{II}^{\{1\}}\rangle$ of the slow mode by aid of the time evolution operator (130), according to:

$$|\Phi_{II}^{\{1\}}(t)\rangle = \{U_{II}^{\{1\}}(t)\}|0_{II}^{\{1\}}\rangle \qquad (135)$$

By aid of the density operator (122), the average of the slow mode coordinate is, in the Schrödinger picture:

$$\langle Q^{\{1\}}(t)\rangle = \text{tr}\{\{\rho_{II}^{\{1\}}(t)\}\{Q_{II}^{\{1\}}\}\} \qquad (136)$$

Performing the trace, by aid of inferences similar to those involved in Appendix F, it may be shown that, for the damped situation:

$$\langle Q^{\{1\}}(t)\rangle = Q^{\circ\circ}[\omega^{\circ\circ}/(\omega^{\circ\circ2} + \gamma^2/4)]\{2\omega^{\circ\circ}\alpha^{\circ}[\exp\{-\gamma t/2\}$$
$$\times \cos(\omega^{\circ\circ}t) - 1] - \gamma \exp\{-\gamma t/2\} \sin(\omega^{\circ\circ}t)\} \qquad (137)$$

In the very underdamped situation, that reduces to:

$$\langle Q^{(1)}(t) \rangle \approx 2Q^{\circ\circ}\alpha^\circ[\exp\{-\gamma t/2\}\cos(\omega^{\circ\circ}t) - 1] \tag{138}$$

This last result shows that just after the fast mode has jumped in its first excited state, the average value is zero, as before. Then, it is performing a damped oscillation around an equilibrium value. Figure 1.2 depicts this dynamics.

2. The Dynamics in Representation {III}: The Underdamped Quantum Harmonic Oscillator

a. *Representation* {III} *in the Damped Situation.* In order to treat the interaction of the hydrogen bond with an inert solvent as above, but within representation {III}, we perform the phase transformation on the effective

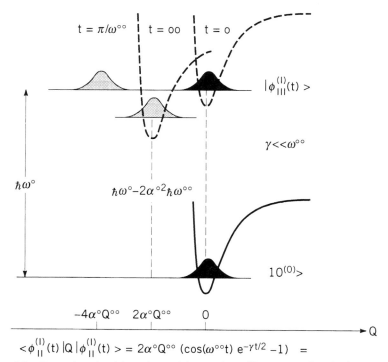

Figure 1.2. Dynamics of a driven damped harmonic oscillator. Just after the fast mode has jumped in its first excited state, the average value is zero. Then it leads to a damped oscillation around an equilibrium value that is decreasing with an enhancement of the damping parameter. As a consequence, for a very important damping there is no movement of the slow mode even just after the excitation of the fast mode.

Hamiltonian of the slow mode but not on that of the thermal bath and of that of its coupling with the slow mode. That leads to the full Hamiltonian of the damped slow mode:

$$\{H_{\mathrm{III}}\} = \{A^{\{k\}}(k\alpha^\circ)\}\{H_{\mathrm{II}}\}\{A^{\{k\}}(k\alpha^\circ)\}^\dagger \tag{139}$$

The transformed Hamiltonian of the hydrogen bond embedded in the thermal bath becomes:

$$\{H_{\mathrm{III}}\} = \{H_{\mathrm{III}}^0\} + \{A^{\{k\}}(k\alpha^\circ)\}\{H_{\mathrm{II}}^{\mathrm{damp}}\}\{A^{\{k\}}(k\alpha^\circ)\}^\dagger \tag{140}$$

$\{H_{\mathrm{III}}^0\}$ is given by Eq. (60).

Now, we may perform the approximation to neglect the phase transformation on the thermal bath. That leads to the simplified equation:

$$\{H_{\mathrm{III}}\} \approx \{H_{\mathrm{III}}^0\} + \{H_{\mathrm{II}}^{\mathrm{damp}}\} \tag{141}$$

This last equation may also be written formally in terms of effective Hamiltonians:

$$\{H_{\mathrm{III}}\} = \sum \{|\{k\}\rangle\{H_{\mathrm{III}}^{0\{k\}}\}\langle\{k\}| + |\{k\}\rangle\{H_{\mathrm{II}}^{\mathrm{damp}}\}\langle\{k\}|\} \tag{142}$$

or:

$$\{H_{\mathrm{III}}\} = \sum |\{k\}\rangle\{H_{\mathrm{III}}^{\{k\}}\}\langle\{k\}| \tag{143}$$

with:

$$\{H_{\mathrm{III}}^{\{k\}}\} = \{H_{\mathrm{III}}^{0\{k\}}\} + \{H_{\mathrm{II}}^{\mathrm{damp}}\} \tag{144}$$

b. *The Effective Hamiltonian of the Damped Slow Mode and the Corresponding Density Operator in the Ground and First Excited States of the Fast Mode.* As a consequence, neglecting the zero-point energies, and in view of Eqs. (63) and (112), one may write for the effective Hamiltonian of the slow mode before excitation of the fast mode:

$$\{H_{\mathrm{III}}^{\{0\}}\} = [a^\dagger a\hbar\omega^{\circ\circ}] + \sum \{b_k^\dagger b_k \hbar\omega_k\} + \sum \{b_k^\dagger a\hbar\kappa_k + b_k a^\dagger\hbar\kappa_k\} \tag{145}$$

Of course, as in representation $\{\mathrm{II}\}$, the coupling of the oscillator to the thermal bath is considered weak, and the coupling parameters must satisfy the condition $|\kappa_k| < \omega_k$. On the other hand, the corresponding density operator is, according to Eq. (80).

$$\{\rho_{\text{III}}^{\{0\}}(0)\} = \varepsilon \exp\{-\lambda a^\dagger a\} \tag{146}$$

Besides, according to Eqs. (64) and (112), one may write for the effective Hamiltonian of the slow mode just after excitation of the fast mode:

$$\{H_{\text{III}}^{\{1\}}\} = [a^\dagger a h \omega^{\circ\circ}] - 2\alpha^{\circ 2} \hbar \omega^{\circ\circ} + \hbar \omega^\circ + \sum \{b_k^\dagger b_k \hbar \omega_k\}$$
$$+ \sum \{b_k^\dagger a \hbar \kappa_k + b_k a^\dagger \hbar \kappa_k\} \tag{147}$$

Finally, the density operator at initial time is, according to Eq. (82)

$$\{\rho_{\text{III}}^{\{1\}}(0)\} = \varepsilon \exp\{-\lambda(a^\dagger - \alpha^\circ)(a - \alpha^\circ)\} \tag{148}$$

c. *The Dynamics of the Slow Mode in Representation* {III}. *The Damped Coherent State.* Besides, as at initial time, just after excitation of the fast mode, the slow mode is described by the density operator (148) of a coherent state, because of the nature of this density operator and of the effective Hamiltonian $\{H_{\text{III}}^{\{1\}}\}$ given by Eq. (147). The slow mode is thus a damped quasiclassical coherent state. The dynamics of its density operator is given by the Liouville equation:

$$i\hbar\{\partial\rho_{\text{III}}^{\{1\}}(t)/\partial t\} = \text{tr}_\theta[\{H_{\text{III}}^{\{1\}}\}, \{\rho_{\text{III}\,\text{tot}}^{\{1\}}(t)\}] \tag{149}$$

$\{\rho_{\text{III}\,\text{tot}}^{\{1\}}(t)\}$ is the full density operator of the slow mode and of the thermal bath:

$$\{\rho_{\text{III}\,\text{tot}}^{\{1\}}(0)\} = \{\rho_{\text{III}}^{\{1\}}(0)\}\{\rho_\theta\} \tag{150}$$

with

$$\{\rho_\theta\} = \prod \varepsilon_k \exp\{-\lambda_k(b_k^\dagger b_k)\} \tag{151}$$

Note that the symbols λ_k and ε_k used in this last equation are defined in Section B.1.b.

The solution of the Liouville equation (149) obtained by Louisell and Walker (1965) and given in Appendix H by Eq. (H83) is:

$$\{\rho_{\text{III}}^{\{1\}}(t)\} = \varepsilon \exp\{-\lambda(a^\dagger - \Phi_{\text{III}}(t))(a - \Phi_{\text{III}}(t)^*)\} \tag{152}$$

with, in view of Eq. (H75):

$$\{\Phi_{\text{III}}(t)\} = \alpha^\circ \exp\{-\gamma t/2\} \exp\{-i\omega^{\circ\circ}t\} \tag{153}$$

and where λ is given by Eq. (56). In the last equation, γ has the same expression as in representation $\{II\}$ and here must satisfy, as above, the same underdamped condition.

Note that when the damping is missing (i.e., when $\gamma \to 0$), the argument of the translation operator reduces to that (106) involved in the translation operator without damping:

$$\{\Phi_{III}(t, \gamma \to 0)\} = \{\Phi_{III}^0(t)\}$$

As a consequence and in the same situation when there is no damping, the density operator (152) reduces to the corresponding one (105) in the absence of damping and in the same representation:

$$\{\rho_{III}^{\{1\}}(t, \gamma \to 0)\} = \{\rho_{III}^{0\{1\}}(t)\} \tag{154}$$

On the other hand, it may be observed that the density operator may be viewed as the result of a phase transformation on the Boltzmann operator according to Eq. (A44):

$$\{\rho_{III}^{\{1\}}(t)\} = \varepsilon\{A^{\{1\}}(\Phi_{III}(t))\} \exp\{-\lambda a^\dagger a\}\{A^{\{1\}}(\Phi_{III}(t))\}^\dagger \tag{155}$$

where the translation operator involved in this equation is given by:

$$\{A^{\{1\}}(\Phi_{III}(t))\} = \exp\{\Phi_{III}(t)a^\dagger - \Phi_{III}(t)^* a\} \tag{156}$$

The following important result must be emphasized: In representation $\{III\}$ when one introduces the damping, the translation operator (107) in the absence of damping transforms simply into the damped one (156) according to:

$$\{A^{\{1\}}(\Phi_{III}^0(t))\} \to \{A^{\{1\}}(\Phi_{III}(t))\} \tag{157}$$

that is, the argument simply changes from Eq. (106) to Eq. (153).

3. The Connection between Representations $\{II\}$ and $\{III\}$

It must be emphasized that in the presence of damping, representation $\{II\}$ is more rigorous than $\{III\}$ because in representation $\{III\}$ the damping part $\{H_{II}^{damp}\}$ of the Hamiltonian has been assumed to be not transformed.

Now if we are in a very underdamped situation, it is possible to neglect γ with respect to $\omega^{\circ\circ}$ in expression (123) of the argument $\{\Phi_{II}(t)\}$ of the density operator (122) within representation $\{II\}$ [see Eq. (125)]. Then, in view of expression (153) of the argument $\{\Phi_{III}(t)\}$ appearing in representation $\{III\}$ of the density operator (152), the connection between the two arguments is:

$$\{\Phi_{II}(t)\} = \{\Phi_{III}(t)\} - \alpha^\circ$$

As a consequence, in view of Eq. (A45) of Appendix A, the connection between the damped translation operator is:

$$\{A^{\{1\}}(\Phi_{III}(0))\}\{A^{\{1\}}(\Phi_{III}(t))\}^\dagger = \exp\{-i\alpha^{\circ 2} \exp\{-\gamma t/2\}$$
$$\times \sin(\omega^{\circ\circ}t)\}\{A^{\{1\}}(-\Phi_{II}(t))\} \qquad (158)$$

where we have used the fact at initial time $\{\Phi_{III}(0)\} = \alpha^\circ$.

IV. THE SPECTRAL DENSITY WITHOUT DAMPING

Now, since we know the dynamics of the slow mode, which is anharmonically coupled to the fast mode, we are able to look at the spectral density of high-frequency vibrational mode of the hydrogen bond in the absence of influence of the medium, from a dynamical viewpoint within the linear response theory.

The spectral density will be given from the static viewpoint by a series of lines the angular frequencies of which will be the difference between the eigenvalues of the effective Hamiltonian:

$$\omega_{mn} = \{E^{\{1\}}_{III\,m} - E^{\{0\}}_{III\,n}\}/\hbar$$

That is, in view of Eqs. (67) and (68),

$$\omega_{mn} = (m - n)\omega^{\circ\circ} + (\omega^\circ - 2\alpha^{\circ 2}\omega^{\circ\circ})$$

The intensity of the lines will be given by:

$$I_{nm} = |\mu_{10}|^2 P_n \langle m^{\{1\}}_{III} | n^{\{0\}}_{III} \rangle |^2 \qquad (159)$$

P_n is the Boltzmann probability at temperature T:

$$P_n = \varepsilon \exp\{-\lambda n\}$$

The full spectral density is:

$$I^0(\omega) = \sum \sum I_{nm} \, \delta(\omega - \omega_{nm}) \tag{160}$$

That is, because of Eqs. (75) and (76):

$$I^0(\omega) = \varepsilon \, |\mu_{10}|^2$$
$$\times \sum \exp\{-\lambda n\} \sum |\Gamma_{mn}(\alpha^\circ)|^2 \, \delta(\omega - [(m-n)\omega^{\circ\circ} + (\omega^\circ - 2\alpha^{\circ 2}\omega^{\circ\circ})]) \tag{161}$$

where the Franck–Condon factors are given by Eqs. (A41) and (A42).

A. The Autocorrelation Function in Representation {II}

Now let us look at the dynamical viewpoint of the spectral density within the linear response theory. We first consider representation {II}.

1. The Starting Autocorrelation Function

Next let us consider the autocorrelation function of the dipole moment operator of the fast mode in representation {II}. By definition it is:

$$\{G_{II}^0(t)\} = \text{tr}'\{\{\rho_{II}^{\{0\}}(0)\}\{\mu_{II}^{\{0,\,1\}}(0)\}\{\mu_{II}^{\{1,\,0\}}(t)\}\}$$

where the trace must be performed on both the bases $\{|\{k\}\rangle\}$ and $\{|n\rangle\}$. The dipole moment operators involved in this last equation are, respectively, given by:

$$\{\mu_{II}^{\{0,\,1\}}(0)\} = \mu_{01} \, |\{0\}\rangle\langle\{1\}|$$

and

$$\{\mu_{II}^{\{1,\,0\}}(0)\} = \mu_{10} \, |\{1\}\rangle\langle\{0\}|$$

In the Heisenberg picture, the last operator is, according to Eqs. (A4) and (A10):

$$\{\mu_{II}^{\{1,\,0\}}(t)\} = \exp\{i\{H_{II}^0\}t/\hbar\}\{\mu_{II}^{\{1,\,0\}}(0)\} \, \exp\{-i\{H_{II}^0\}t/\hbar\}$$

$\{H_{II}^0\}$ is the full Hamiltonian in this representation and in the absence of damping, which is given by Eq. (45). The above result may be also written:

$$\{\mu_{II}^{\{1,\,0\}}(t)\} = \mu_{10} \, \exp\{i\{H_{II}^0\}t/\hbar\} \, |\{1\}\rangle\langle\{0\}| \, \exp\{-i\{H_{II}^0\}t/\hbar\} \tag{162}$$

As a consequence, the autocorrelation function becomes:

$$\{G_{II}^0(t)\} = \varepsilon|\mu_{10}|^2 \ \mathrm{tr}\{\sum \langle\{k\}| \exp\{-\lambda a^\dagger a\}|\{0\}\rangle$$
$$\times \langle\{1\}| \exp\{i\{H_{II}^0\}t/\hbar\}|\{1\}\rangle\langle\{0\}| \exp\{-i\{H_{II}^0\}t/\hbar\}|\{k\}\rangle\} \quad (163)$$

where the trace must now be performed only on the basis $\{|n\rangle\}$. The above equation may be written in terms of the effective Hamiltonians over which the Hamiltonian involved in the unitary transform is expanded. Thus, according to Eq. (45), the autocorrelation function becomes:

$$\{G_{II}^0(t)\} = \varepsilon|\mu_{10}|^2 \ \mathrm{tr}\{\sum \langle\{k\}| \exp\{-\lambda a^\dagger a\}|\{0\}\rangle\langle\{1\}|$$
$$\times \exp\{i\{H_{II}^{0\{1\}}\}t/\hbar\}\{1\}\rangle\langle\{0\}| \exp\{-i\{H_{II}^{0\{0\}}\}t/\hbar\}|\{k\}\rangle\} \quad (164)$$

The last result, because the exponential operators do not act on the eigenkets of the fast mode, may be written:

$$\{G_{II}^0(t)\} = \varepsilon|\mu_{10}|^2 \ \mathrm{tr}\{\sum \langle\{k\}|\{0\}\rangle \exp\{-\lambda a^\dagger a\}$$
$$\times \exp\{i\{H_{II}^{0\{1\}}\}t/\hbar\} \exp\{-i\{H_{II}^{0\{0\}}\}t/\hbar\}\langle\{0\}|\{k\}\rangle\}$$

2. The Autocorrelation function as Depending on the Time Evolution Operator of the Driven Harmonic Oscillator

Because of the orthonormality of the basis $\{|\{k\}\rangle\}$ characterizing the fast mode, the above result reduces to:

$$\{G_{II}^0(t)\} = \varepsilon|\mu_{10}|^2 \ \mathrm{tr}\{\exp\{-\lambda a^\dagger a\} \exp\{i\{H_{II}^{0\{1\}}\}t/\hbar\} \exp\{-i\{H_{II}^{0\{0\}}\}t/\hbar\}\}$$
$$(165)$$

Of course, in the last equation, the trace must be only performed on the basis $\{|n\rangle\}$. Again, this equation may also be written in view of Eq. (46):

$$\{G_{II}^0(t)\} = \varepsilon|\mu_{10}|^2 \ \mathrm{tr}\{\exp\{-\lambda a^\dagger a\} \exp\{i\{H_{II}^{0\{1\}}\}t/\hbar\} \exp\{-ia^\dagger a\omega^{\circ\circ}t\}\}$$
$$(166)$$

This last equation may be written formally:

$$\{G_{II}^0(t)\} = \varepsilon|\mu_{10}|^2 \ \mathrm{tr}\{\exp\{-\lambda a^\dagger a\}\{U_{II}^{0\{1\}}(t)\}^\dagger \exp\{-ia^\dagger a\omega^{\circ\circ}t\}\} \quad (167)$$

$\{U_{\text{II}}^{0\{1\}}(t)\}$ is the time evolution operator solution of the Schrödinger equation:

$$i\hbar\ \partial\{U_{\text{II}}^{0\{1\}}(t)\}/\partial t = \{H_{\text{II}}^{0\{1\}}\}\{U_{\text{II}}^{0\{1\}}(t)\} \tag{168}$$

The solution of this Schrödinger equation is given by Eq. (93). As a consequence, we have:

$$\begin{aligned}
\{U_{\text{II}}^{0\{1\}}(t)\}^{\dagger} = &\ \exp\{i(\omega^{\circ} - \alpha^{\circ 2}\omega^{\circ\circ})t\} \\
&\times \exp\{-i\alpha^{\circ 2}\omega^{\circ\circ}t\}\ \exp\{i\alpha^{\circ 2}\ \sin(\omega^{\circ\circ}t)\} \\
&\times \{A^{\{1\}}(\Phi_{\text{II}}^{0}(t))\}\ \exp\{ia^{\dagger}a\omega^{\circ\circ}t\}
\end{aligned} \tag{169}$$

Here appears the translation operator (A32) of Appendix A, the argument of which is:

$$\{\Phi_{\text{II}}^{0}(t)\} = \alpha^{\circ}[\exp\{-i\omega^{\circ\circ}t\} - 1] \tag{170}$$

3.   A Simplified Expression for the Autocorrelation Function

Owing to this equation, the autocorrelation function becomes, in view of Eq. (A13):

$$\{G_{\text{II}}^{0}(t)\} = \varepsilon\ |\mu_{10}|^{2}\ \text{tr}\{\exp\{-\lambda a^{\dagger}a\}\{U_{\text{II}}^{0\ \text{IP}}(t)\}^{\dagger}\} \tag{171}$$

where the right-hand side time evolution operator is the interaction picture of the time evolution operator $\{U_{\text{II}}^{0}(t)\}^{\dagger}$, which, in view of Eqs. (169) and (A13), is:

$$\begin{aligned}
\{U_{\text{II}}^{0\ \text{IP}}(t)\}^{\dagger} = &\ \exp\{i\alpha^{\circ 2}\ \sin(\omega^{\circ\circ}t)\}\ \exp\{-i\alpha^{\circ 2}\omega^{\circ\circ}t\} \\
&\times \exp\{i(\omega^{\circ} - \alpha^{\circ 2}\omega^{\circ\circ})t\}\{A^{\{1\}}(\Phi_{\text{II}}^{0}(t))\}
\end{aligned} \tag{172}$$

4.   The Autocorrelation Function as Depending on the Translation Operator

By aid of the two above equations, one obtains:

$$\begin{aligned}
\{G_{\text{II}}^{0}(t)\} = &\ \varepsilon\ |\mu_{10}|^{2}\ \exp\{i\alpha^{\circ 2}\ \sin(\omega^{\circ\circ}t)\} \\
&\times \exp\{-i\alpha^{\circ 2}\omega^{\circ\circ}t\}\ \exp\{i(\omega^{\circ} - \alpha^{\circ 2}\omega^{\circ\circ})t\} \\
&\times \text{tr}\{\exp\{-\lambda a^{\dagger}a\}\{A^{\{1\}}(\Phi_{\text{II}}^{0}(t))\}\}
\end{aligned} \tag{173}$$

Then, performing the thermal average on the translation operator with the aid of Appendix D, one obtains, according to Eq. (D3):

$$\varepsilon\ \text{tr}\{\exp\{-\lambda a^{\dagger}a\}\{A^{\{1\}}(\Phi_{\text{II}}^{0}(t))\}\} = \exp\{-\tfrac{1}{2}|\Phi_{\text{II}}^{0}(t)|^{2}[1 + 2\langle n\rangle]\}$$

$\langle n \rangle$ is the thermal average occupation number of a quantum harmonic oscillator given by Eq. (H90). In addition, the right-hand side term leads, owing to the expression of $\Phi_{\text{II}}^0(t)$ and after some rearrangements, to:

$$\exp\{-\tfrac{1}{2} \,|\, \Phi_{\text{II}}^0(t) \,|^2 [1 + 2\langle n \rangle]\} = \exp\{\alpha^{\circ 2}(1 + 2\langle n \rangle)[\cos(\omega^{\circ\circ} t) - 1]\} \quad (174)$$

As a consequence, the autocorrelation function takes the final form:

$$\{G_{\text{II}}^0(t)\} = |\,\mu_{10}\,|^2 \exp\{i\alpha^{\circ 2} \sin(\omega^{\circ\circ} t)\} \exp\{i(\omega^{\circ} - 2\alpha^{\circ 2}\omega^{\circ\circ})t\}$$
$$\times \exp\{\alpha^{\circ 2}(1 + 2\langle n \rangle)[\cos(\omega^{\circ\circ} t) - 1]\} \quad (175)$$

B. The Autocorrelation Function in Representation {III}

Next let us consider the autocorrelation function of the dipole moment operator of the fast mode in representation {III}. By definition it is:

$$\{G_{\text{III}}^0(t)\} = \varepsilon \ \text{tr}'\{\{\rho_{\text{III}}^{\{0\}}(0)\}\{\mu_{\text{III}}^{\{0, \ 1\}}(0)\}\{\mu_{\text{III}}^{\{1, \ 0\}}(t)\}\}$$

where the trace must be performed over both the bases $\{|\,\{k\}\rangle\}$ and $\{|\,n\rangle\}$. If we restrict ourselves to the transition from the ground state to the first excited state of the fast mode, the dipole moment operators in representation {III} appearing in the autocorrelation function are given by:

$$\{\mu_{\text{III}}^{\{0, \ 1\}}(0)\} = \mu_{01}\,|\,\{0\}\rangle\langle\{1\}\,|\,\{A^{\{1\}}(\alpha^{\circ})\}^{\dagger} \quad (176)$$
$$\{\mu_{\text{III}}^{\{1, \ 0\}}(t)\} = \mu_{10} \exp\{i\{H_{\text{III}}^0\}t/\hbar\}\{A^{\{1\}}(\alpha^{\circ})\}\,|\,\{1\}\rangle\langle\{0\}\,|\exp\{-i\{H_{\text{III}}^0\}t/\hbar\} \quad (177)$$

Therefore, after performing the trace on the basis $\{|\,\{k\}\rangle\}$, the autocorrelation function becomes:

$$\{G_{\text{III}}^0(t)\} = |\,\mu_{10}\,|^2 \ \text{tr}\{\sum \langle\{k\}\,|\,\{\rho_{\text{III}}^{\{0\}}(0)\}\,|\,\{0\}\rangle\langle\{1\}\,|\,\{A^{\{1\}}(\alpha^{\circ})\}^{\dagger}$$
$$\times \exp\{i\{H_{\text{III}}^0\}t/\hbar\}\{A^{\{1\}}(\alpha^{\circ})\}\,|\,\{1\}\rangle\langle\{0\}\,|$$
$$\times \exp\{-i\{H_{\text{III}}^0\}t/\hbar\}\,|\,\{k\}\rangle\}$$

Using the fact that the translation and density operators commute, respectively, with $|\,\{1\}\rangle$ and $|\,\{0\}\rangle$, we get:

$$\{G_{\text{III}}^0(t)\} = |\,\mu_{10}\,|^2 \ \text{tr}\{\sum \langle\{k\}\,|\,|\,\{0\}\rangle\{\rho_{\text{III}}^{\{0\}}(0)\}\{A^{\{1\}}(\alpha^{\circ})\}^{\dagger}\langle\{1\}\,|$$
$$\times \exp\{i\{H_{\text{III}}^0\}t/\hbar\}\,|\,\{1\}\rangle\{A^{\{1\}}(\alpha^{\circ})\}\langle\{0\}\,|$$
$$\times \exp\{-i\{H_{\text{III}}^0\}t/\hbar\}\,|\,\{k\}\rangle\} \quad (178)$$

Using again the expression of the full Hamiltonian in representation {III} in terms of effective Hamiltonians, we find:

$$\{G_{\text{III}}^0(t)\} = |\mu_{10}|^2 \, \text{tr}\{\sum \langle\{k\}|\{0\}\rangle\{\rho_{\text{III}}^{\{0\}}(0)\}\{A^{\{1\}}(\alpha^\circ)\}^\dagger\langle\{1\}| \\ \times \exp\{i\{H_{\text{III}}^{0\{1\}}\}t/\hbar\}|\{1\}\rangle\{A^{\{1\}}(\alpha^\circ)\} \\ \times \exp\{-i\{H_{\text{III}}^{0\{0\}}\}t/\hbar\}\langle\{0\}|\{k\}\rangle\} \tag{179}$$

Moreover, using the fact that the operator exponential commutes with the states of the basis $\{|\{k\}\rangle\}$ and the orthonormality properties of this basis, the above result reduces to:

$$\{G_{\text{III}}^0(t)\} = |\mu_{10}|^2 \, \text{tr}\{\{\rho_{\text{III}}^{\{0\}}(0)\}\{A^{\{1\}}(\alpha^\circ)\}^\dagger \\ \times \exp\{i\{H_{\text{III}}^{0\{1\}}\}t/\hbar\}\{A^{\{1\}}(\alpha^\circ)\} \exp\{-i\{H_{\text{III}}^{0\{0\}}\}t/\hbar\}\} \tag{180}$$

At last, using Eq. (80), giving the density operator, and Eqs. (63) and (64), giving the effective Hamiltonians in representation {III}, one finds:

$$\{G_{\text{III}}^0(t)\} = |\mu_{10}|^2\varepsilon \, \text{tr}\{\exp\{-\lambda a^\dagger a\}\{A^{\{1\}}(\alpha^\circ)\}^\dagger \\ \times \exp\{ia^\dagger a\omega^{\circ\circ}t\}\{A^{\{1\}}(\alpha^\circ)\} \\ \times \exp\{-ia^\dagger a\omega^{\circ\circ}t\}\} \exp\{i(\omega^\circ - 2\alpha^{\circ 2}\omega^{\circ\circ})t\} \tag{181}$$

C. Equivalence between the Autocorrelation Functions in Both Representations

Now we shall show that the two representations {II} and {III} lead, within the linear response theory, to autocorrelation functions that are equivalent. We may observe that, by aid of theorem (A21) of Appendix A, one has:

$$\exp\{ia^\dagger a\omega^{\circ\circ}t\}\{A^{\{1\}}(\alpha^\circ)\} \exp\{-ia^\dagger a\omega^{\circ\circ}t\} = \{A^{\{1\}}(\Phi_{\text{III}}^0(t)^*)\} \tag{182}$$

with:

$$\{\Phi_{\text{III}}^0(t)\} = \alpha^\circ \exp\{-i\omega^{\circ\circ}t\} \tag{183}$$

As a consequence, the autocorrelation function (181) in representation {III} takes the final form:

$$\{G_{\text{III}}^0(t)\} = \varepsilon|\mu_{10}|^2 \, \text{tr}\{\exp\{-\lambda a^\dagger a\}\{A^{\{1\}}(\alpha^\circ)\}^\dagger \\ \times \{A^{\{1\}}(\Phi_{\text{III}}^0(t)^*)\}\} \exp\{i(\omega^\circ - 2\alpha^{\circ 2}\omega^{\circ\circ})t\} \tag{184}$$

Besides, according to Eq. (A45) of Appendix A, the product of the translation operators appearing in this autocorrelation function is:

$$\{A^{\{1\}}(\alpha^\circ)\}^\dagger\{A^{\{1\}}(\Phi^0_{\mathrm{III}}(t)^*)\} = \exp\{i\alpha^{\circ 2}\,\sin(\omega^{\circ\circ}t)\}\{A^{\{1\}}(\Phi^0_{\mathrm{II}}(t)^*)\} \qquad (185)$$

As a consequence, the above autocorrelation function becomes:

$$\{G^0_{\mathrm{III}}(t)\} = \varepsilon\,|\mu_{10}|^2\,\exp\{i\alpha^{\circ 2}\,\sin(\omega^{\circ\circ}t)\}$$
$$\times\,\exp\{i(\omega^\circ - 2\alpha^{\circ 2}\omega^{\circ\circ})t\}\mathrm{tr}\{\exp\{-\lambda a^\dagger a\}\{A^{\{1\}}(\Phi^0_{\mathrm{II}}(t)^*)\}\} \qquad (186)$$

Then, by comparison with Eq. (173) giving the autocorrelation function in representation {II} and owing to Eq. (174), it is found that the autocorrelation functions are the same in both representations:

$$\{G^0_{\mathrm{III}}(t)\} = \{G^0_{\mathrm{II}}(t)\} = \{G^0(t)\} \qquad (187)$$

Because of the equivalence, we shall omit for the quantum autocorrelation function of the hydrogen bond in the absence of damping the subscripts {II} and {III} referring to the quantum representations.

It may be observed that it is possible to find the half-width $\Delta\omega$ of this spectral density. Then it is shown in the Appendix F, in a similar way as in Boulil (1988), that the half-width is given by Eq. (F4):

$$\Delta\omega = \alpha^\circ\omega^{\circ\circ}\,\coth^{1/2}[\hbar\omega^{\circ\circ}/(2k_{\mathrm{B}}T)]$$

D. Expressions of the Autocorrelation Function and of the Spectral Density

1. General Situation at a Given Temperature

Now we shall obtain two final expressions of the autocorrelation function that will be of interest in the following. Performing the trace in Eq. (186), one obtains, by aid of Eq. (D3) of Appendix D:

$$\{G^0(t)\} = |\mu_{10}|^2\,\exp\{i\alpha^{\circ 2}\,\sin(\omega^{\circ\circ}t)\}\,\exp\{i(\omega^\circ - 2\alpha^{\circ 2}\omega^{\circ\circ})t\}$$
$$\times\,\exp\{\alpha^{\circ 2}(1 + 2\langle n\rangle)[\cos(\omega^{\circ\circ}t) - 1]\} \qquad (188)$$

In this last equation $\langle n\rangle$ is the thermal average number occupation given by Eq. (H90).

Note that the autocorrelation function may also be written:

$$\{G^0(t)\} = \{G^{\circ\circ}(\alpha^\circ, t)\} |\mu_{10}|^2 \exp\{i\omega^\circ t\} \exp\{-i2\alpha^{\circ 2}\omega^{\circ\circ}t\} \quad (189)$$

with:

$$\{G^{\circ\circ}(\alpha^\circ, t)\} = \exp\{i\alpha^{\circ 2} \sin(\omega^{\circ\circ}t)\} \exp\{\alpha^{\circ 2}(1 + 2\langle n\rangle)[\cos(\omega^{\circ\circ}t) - 1]\} \quad (190)$$

On the other hand, using four times the closeness relation on the eigen-kets of $a^\dagger a$, and writing explicitly the trace performed on this same basis, the autocorrelation function (181) becomes:

$$\{G^0(t)\} = |\mu_{10}|^2 \sum\sum\sum\sum\sum \varepsilon\langle m| \exp\{-\lambda a^\dagger a\} |j\rangle$$
$$\times \langle j| \{A^{\{1\}}(\alpha^\circ)\}^\dagger |n\rangle\langle n| \exp\{ia^\dagger a\omega^{\circ\circ}t\} |k\rangle$$
$$\times \langle k| \{A^{\{1\}}(\alpha^\circ)\} |r\rangle\langle r| \exp\{-ia^\dagger a\omega^{\circ\circ}t\} |m\rangle$$
$$\times \exp\{i(\omega^\circ - 2\alpha^{\circ 2}\omega^{\circ\circ})t\} \quad (191)$$

Again, using the fact that the number occupation operator $a^\dagger a$ is diagonal in this basis, we get

$$\{G^0(t)\} = |\mu_{10}|^2 \sum\sum \varepsilon \exp\{-\lambda m\}\langle m| \{A^{\{1\}}(\alpha^\circ)\}^\dagger |n\rangle$$
$$\times \exp\{in\omega^{\circ\circ}t\}\langle n| \{A^{\{1\}}(\alpha^\circ)\} |m\rangle$$
$$\times \exp\{-im\omega^{\circ\circ}t\} \exp\{i(\omega^\circ - 2\alpha^{\circ 2}\omega^{\circ\circ})t\} \quad (192)$$

At last, using the fact that the matrix elements of the translation operator appearing in the above equation are the Franck–Condon factors defined in Eq. (A40) of Appendix A, we find:

$$\{G^0(t)\} = |\mu_{10}|^2 \sum\sum \varepsilon \exp\{-\lambda m\} |\Gamma_{mn}(\alpha^\circ)|^2$$
$$\times \exp\{-i(m - n)\omega^{\circ\circ}t\} \exp\{i(\omega^\circ - 2\alpha^{\circ 2}\omega^{\circ\circ})t\} \quad (193)$$

The spectral density $I^0(\omega)$ of the hydrogen bond is the Fourier transform of this autocorrelation function:

$$I^0(\omega) = [1/(\sqrt{2\pi})] \int \{G^0(t)\} \exp\{-i\omega t\} \, dt$$

Then, according to Eq. (A50), one obtains:

$$I^0(\omega) = \varepsilon |\mu_{10}|^2 \sum \exp\{-\lambda m\}$$
$$\times \sum |\Gamma_{mn}(\alpha^\circ)|^2 \, \delta(\omega - [\omega^\circ - 2\alpha^{\circ 2}\omega^{\circ\circ} - (m-n)\omega^{\circ\circ}])$$

$$(194)$$

which is the Franck–Condon progression obtained in Eq. (161).

Note that, as required, the autocorrelation function reduces simply, in the absence of damping:

$$\{G^0_{\text{harm}}(t)\} = \varepsilon |\mu_{10}|^2 \, \exp\{i\omega^\circ t\}$$

Therefore, the Fourier transform of the autocorrelation function is the simple delta peak centered on ω°, the angular frequency of the fast mode.

$$I^0(\omega) = [1/(\sqrt{2\pi})] \, \delta(\omega - \omega^\circ)$$

2. Special Situation at Zero Temperature

Now consider the special situation where the temperature is zero. Then, the spectral density (194) reduces to:

$$\{I^0_0(\omega)\} = |\mu_{10}|^2 \sum |\Gamma_{0n}(\alpha^\circ)|^2 \, \delta(\omega - [\omega^\circ - 2\alpha^{\circ 2}\omega^{\circ\circ} - n\omega^{\circ\circ}])$$

Again, in view of Eq. (A43) giving the Franck–Condon factor in this special situation, we get the Poisson distribution:

$$\{I^0_0(\omega)\} = |\mu_{10}|^2 \sum \exp\{-\alpha^{\circ 2}\}[\alpha^{\circ 2n}/n!] \, \delta(\omega - [\omega^\circ - 2\alpha^{\circ 2}\omega^{\circ\circ} - n\omega^{\circ\circ}])$$

$$(195)$$

Now we shall show that in the same zero-temperature condition, the autocorrelation function (188) leads, after taking the Fourier transform, to the same expression. At zero temperature, the thermal average of the number occupation is zero (i.e., $\langle n \rangle = 0$), so that the autocorrelation function reduces to:

$$\{G^0_0(t)\} = |\mu_{10}|^2 \, \exp\{\alpha^{\circ 2}[\cos(\omega^{\circ\circ}t) + i\sin(\omega^{\circ\circ}t)]\}$$
$$\times \exp\{-\alpha^{\circ 2}\} \, \exp\{i(\omega^\circ - 2\alpha^{\circ 2}\omega^{\circ\circ})t\}$$

Now let us look at the spectral density. It may be expanded in Fourier series according to:

$$\{I^0_0(\omega)\} = (1/T) \sum \int_{-T/2}^{T/2} \{G^0_0(t)\} \cos(-n\omega^{\circ\circ}t) \, dt$$

with $\omega^{\circ\circ}T = 2\pi$. Then, taking the real part of the exponential of $i \sin(\omega^{\circ\circ}t)$, one has:

$$\{I_0^0(\omega)\} = (1/T)|\mu_{10}|^2 \sum \int_{-T/2}^{T/2} \exp\{\alpha^{\circ 2} \cos(\omega^{\circ\circ}t)\}$$

$$\times \cos(\alpha^{\circ 2} \cos(\omega^{\circ\circ}t) - n\omega^{\circ\circ}t) \exp\{-\alpha^{\circ 2}\} \exp\{i(\omega^\circ - 2\alpha^{\circ 2}\omega^{\circ\circ})t\} \, dt$$

Now, according to the following integral:

$$\int_0^{2\pi} \exp\{p \cos(x)\} \cos[p \cos(x) - nx] \, dx = 2\pi[p^n/n!]$$

we obtain the spectral density (195). The importance of this subsection is to make clear the fact that the main factor that is at the origin of the asymmetry of the fine structure of the spectral density when one works in representation $\{II\}$ is:

$$\exp\{i\alpha^{\circ 2} \sin(\omega^{\circ\circ}t)\}$$

V. ABOUT BROADENING OF THE SPECTRAL DENSITY

We shall now consider the theories dealing with the shape of the spectral density of the high vibrational stretching modes in condensed medium. We shall consider the situation of bare hydrogen bonds for which the strong anharmonic coupling between the slow and fast mode plays the fundamental role, and thus for which it is possible to neglect subsidiary influences such as Fermi resonance or tunneling effects.

A. The Coulson and Robertson Approach and the Question of Predissociation

Coulson and Robertson have considered the following ideas of Stepanov (1946): The hydrogen bond dissociation energy is likely to be less than the energy of the high-frequency stretching mode vibrational quantum. In such cases, any excited state of the fast mode will be unstable against hydrogen bond dissociation and any spectral lines associated with transition to this quasistationary excited state will acquire a width that might account for the smooth contours of some of the bands found experimentally. In their fundamental study, published the year of Coulson's death, Coulson and Robertson (1974) used the theory of resonant scattering. They ignored the electronic anharmonic coupling between the slow and fast modes but took into account the kinematic coupling and considered anharmonicity of the slow mode. They found that predissociation broadening is negligible. In a subsequent paper (1975), they studied the structure of spectral density after improving their kinematic coupling model by also taking into account the anharmonicity of the fast mode. They found depth, curvature, and minimum position of each effective potential depending on the quantum state of the fast mode and calculated the Franck–Condon factors governing the excitation of the combination bands and the energy spread. In a subsequent paper, Robertson (1976) gave a didactic simplified approach to predissociation by aid of perturbation theory, which avoids the complicated mathematics of the original treatment but which is of course less rigorous since it is perturbative. In a last paper, Robertson (1977) has taken into account not only the kinematic coupling involved previously, but also following the ideas of Maréchal and Witkowski (1968), the anharmonic coupling between the slow and fast modes.

Ewing (1978, 1980), some years later, developed studies dealing with vibrational predissociation in hydrogen-bonded complexes, and probably without any knowledge of the Coulson and Robertson works (1974, 1975). He applied his theoretical model to the HF···HF complex and also obtained very long lifetimes incompatible with those required to explain the broadening of the spectrum by predissociation. He has shown that the lifetimes are exceedingly sensitive to the parameters that characterize the multidimensional surface. Related to the Ewing approach must be noted the theoretical works of Beswick and Jortner (1978) dealing with calculations of vibrational predissociation processes in non-hydrogen-bonded complexes.

Moreover, Mühle, Süsse, and Welsch (1978, 1980), at the same time as Ewing, have reconsidered predissociation by using a more realistic potential than that used by Coulson and Robertson (1974): The potential energy surface they used was derived by fitting the bond length, the dissociation energy of the hydrogen bond, the bond length shortening, and the frequency of the fast mode to their observed values. They have concluded some possibility of broadening of the spectral density by predissociation. This question seems therefore to be not completely answered.

B. About Broadening of the Spectral Density. General Considerations

Among possible origins of the broadening of the spectral density in the gas phase, there are the predissociation we have considered above and the rotational structure considered by Robertson (1977). There is also the possibility of centrifugal distortion effects as suggested by Lascombe (1973), which presupposes a strong dependence of the stretching frequency on the length of the hydrogen bridge.

In liquids, as a rule, as distinct from gaseous systems, combination bands are not observed in the proton infrared spectra, although the coupling of the high-frequency to the low-frequency mode, which is an intrinsic characteristic of the complex, cannot disappear on transition from gases to solutions.

More generally two main relaxation mechanisms leading to a broadening of the spectral density, have been proposed: direct and indirect mechanisms. In the direct mechanism, the relaxation of the fast mode is due to direct interaction of the dipole moment of the complex to the local fluctuating electric field. In the indirect mechanism, the relaxation is caused by the anharmonic coupling of the high-frequncy stretching mode with the low-frequency hydrogen-bridge motion.

1. Direct Mechanism of Damping

Following the idea of Janoschek, Weidemann, and Zundel (1973), Rösch and Ratner (1974) have modeled the direct mechanism of relaxation within the double quantization language. In their approach, the broadening of the spectral density is due to the direct dipolar interaction of the dipole moment of the fast mode to the local fluctuating electric field produced by the solvent dipoles. The spectral density is obtained by Fourier transforming an autocorrelation function involving a cumulant expansion of the dipole–dipole correlation function. Expressions for the shift and broadening of the individual line are given. The dynamics as well as the strength of the system–bath coupling are important for determining line shapes. The model proposed by Abramczyk (1985, 1987), who considered the direct dipolar interaction of the dipole moment of the proton donor and the dipole moment produced by the hydrogen bond in the complex regarded as a nonrigid supermolecule, should be mentioned.

2. Indirect Mechanism of Damping

In the indirect mechanism, because of the anharmonic coupling, attention is focused either on the stochastic or on the damping properties of the slow-mode coordinate. The studies performed within this mechanism can be classified according to the level of approximation dealing with the behavior of the slow mode.

Bratos (1975) used stochastic arguments that are related to general theory of the infrared band profile of liquids (Bratos 1970). The physical idea was that, in liquids, because the surrounding molecules are acting on the hydrogen bond, the motion associated with the low-frequency modes of the complex becomes stochastic. The Bratos theory was applicable to bands much wider in principle than the frequency of the slow mode. Bratos predicted that the band profile is essentially for weak hydrogen bonds, a broad, symmetric, distorted Gaussian line, produced by the anharmonic coupling between the high-frequency and low-frequency modes. However, experimentally, the bands of the fast mode exhibit visible deviation from Gaussian.

Robertson and Yarwood (1978) suggested that the slow modulation limit assumed by Bratos (1975) is not always realized and that the low-frequency motion of the hydrogen bond complex in solution is not fully stochastic. The major Robertson assumption was that the low-frequency stretching mode is susceptible to be represented by the Ornstein–Uhlenbeck stochastic processes and that the phase coherence of the fast mode is lost because of the coupling between the two vibrational modes. He analyzed his model in terms of a randomly modulated oscillator. At the basis of this random process of the slow mode, there is the Langevin approach, according to which the correlated random external force is delta Dirac-like so that the correlation time of the Langevin random force must be smaller than the relaxation time of the slow mode.

Sakun (1981) has improved the model of Robertson and Yarwood (1978) by taking in place of the classical anharmonic theory the strong coupling theory and by using the quasicrystalline model reviewed by Maradudin (1966). This approach enabled Sakun to avoid not only the restrictive conditions at the basis of the Langevin equation used by Robertson and Yarwood (1978), but also the restrictive assumption according to which the interaction of the slow-mode coordinate with the environment is weak as compared to $k_B T$, an assumption essentially in the Bratos approach and inconsistent with the requirement that the motion of the slow mode is stochastic.

Sakun (1985), in a later paper, has reconsidered the Robertson and Yarwood approach (1978) by assuming that the stochastic slow-mode coordinate Q obeys a generalized Langevin equation involving a memory function, in place of the Langevin equation used by Robertson. Kryachko (1994) in the third part of a recent paper has proposed another analysis, also using the generalized Langevin equation.

Boulil et al. (1988) have proposed a full quantum-statistical approach of the indirect relaxation. In a first study, they started from an autocorrelation function of the dipole moment of the fast mode that was in the same quantum representation as that used by Rösch and Ratner (1974). This autocorrelation function involves the product of a translation operator at the initial situation, and of the same translation operator at a further instant. The time dependence of this operator is a simple consequence of the fact that, just after the excitation of the high-frequency stretching mode, the slow mode becomes a quasiclassical coherent state (Carruthers, 1965) susceptible to oscillate back and forth that is damped by the solvent. Using a formalism employed in the quantum theory of light, they were able to find the expression of the damped translation operator, and thus to obtain the autocorrelation function of the fast stretching mode.

More recently, Boulil et al. (1994a) have improved their approach in the following way: They started from the strong anharmonic coupling theory that has been seen above to lead to different effective Hamiltonians for the slow mode according to the fact that the fast mode is in its ground state or in its first excited state; whereas the slow mode is a simple quantum harmonic oscillator, when the fast mode is in its ground state, it becomes a driven damped quantum harmonic oscillator when the fast mode has jumped in its first excited state. Because of the irreversible influence of the medium, this driven harmonic oscillator must become damped. This driven damped slow mode has the behavior of a damped quasiclassical coherent state that minimizes the uncertainty relation and acts in a similar way as that of a classical driven damped harmonic oscillator. The reduced time evolution operator governing the dynamics of this driven damped quantum harmonic oscillator may be viewed as a scattering source. As a consequence, the excitation of the fast mode induces diffusion between the energy levels of the slow mode and thus a broadening of the spectral density of the fast mode and a concomitant shift toward low frequencies.

VI.   CLASSICAL APPROACHES IN THE INDIRECT DAMPING MECHANISM

In this section, we shall consider more deeply the principal theories dealing with the indirect relaxation of the slow mode towards the medium via the anharmonic coupling between the fast mode considered quantum mechanically and the slow one considered as classical. The idea is that the slow mode is in close energetic contact with the Brownian motions of the solvent molecules, since its vibrational quantum is comparable with $k_B T$ at room temperature. As the vibrational mode exchanges energy with its surroundings, it will lose its phase coherence, which will in turn affect the phase coherence of the fast most. As a consequence, the phase coherence of the fast mode will be affected and a broadening of the corresponding bands will result.

A.   The Bratos Approach

Now let us look at the theory of Bratos (1975). We shall give a simplified version of his work in which will be neglected the highest orders in the expansion involved in the calculations, and where we shall modify the notation completely.

1. The Basis Physical Assumptions

Within the classical anharmonic coupling between the slow and fast modes, Bratos looks at the effective angular frequency ω_{eff} of the fast mode, which is given in view of Eq. (10) by:

$$\{\omega_{\text{eff}}^\circ(Q)\} = \omega^\circ[1 + [2K_{112}/(m\omega^{\circ 2})]Q]^{1/2} \tag{196}$$

Bratos expands the angular frequency of the fast mode up to second order in the slow mode Q.

$$\{\omega_{\text{eff}}^\circ(Q)\} = \omega^\circ + a_{112}Q + \tfrac{1}{2}b_{112}Q^2 \tag{197}$$

with, respectively:

$$a_{112} = K_{112}/(m\omega^{\circ 2}) \tag{198}$$

$$b_{112} = -\tfrac{1}{2}[K_{112}/(m\omega^{\circ 2})]^2 \tag{199}$$

Within the linear response theory Bratos starts from the autocorrelation function of the dipole moment operator of the fast mode:

$$\{G_{\text{Brat}}(t)\} = \left\langle \{\mu^{\{0,\,1\}}(0)\}\{\mu^{\{1,\,0\}}(t)\} \right\rangle_{q,\,Q} \tag{200}$$

Here the brackets denote the thermal average on the fast and slow modes. The dipole moment operator at initial and at time t, are, respectively, given by:

$$\{\mu^{\{0,\,1\}}(0)\} = |\{0\}\rangle\langle\{1\}|\mu_{01}$$

$$\{\mu^{\{1,\,0\}}(t)\} = \mu_{10}\exp\{i\{H^0(Q, q)\}t/\hbar\}|\{1\}\rangle\langle\{0\}|\exp\{-i\{H^0(Q, q)\}t/\hbar\}$$

Besides, using Eq. (8), the time-dependent dipole moment operator becomes:

$$\begin{aligned}
\{\mu^{\{1,\,0\}}(t)\} = \mu_{10}\exp\{ &i\{[p^2/(2m) + \tfrac{1}{2}m\{\omega_{\text{eff}}^\circ(Q)\}^2 q^2] \\
&+ [P^2/(2M) + \tfrac{1}{2}M\omega^{\circ\circ 2}Q^2]\}t/\hbar\} \\
\times\, |\{1\}\rangle\langle\{0\}|\exp\{ &-i\{[p^2/(2m) + \tfrac{1}{2}m\{\omega_{\text{eff}}^\circ(Q)\}^2 q^2] \\
&+ [P^2/(2M) + \tfrac{1}{2}M\omega^{\circ\circ 2}Q^2]\}t/\hbar\}
\end{aligned}$$

Moreover, using the fact that the slow-mode coordinate and conjugate momentum commutes with those of the fast mode and with $|\{1\}\rangle\langle\{0\}|$, the above expression reduces to:

$$
\begin{aligned}
\{\mu^{\{1,\,0\}}(t)\} = \mu_{10}\,\exp\{&i[p^2/(2m) + \tfrac{1}{2}m\{\omega_{\text{eff}}^\circ(Q)\}^2 q^2]t/\hbar\}\,|\{1\}\rangle\langle\{0\}| \\
&\times \exp\{-i[p^2/(2m) + \tfrac{1}{2}m\{\omega_{\text{eff}}^\circ(Q)\}^2 q^2]t/\hbar\}
\end{aligned}
\tag{201}
$$

The dipole moment operator at time t is therefore:

$$
\{\mu^{\{1,\,0\}}(t)\} = \mu_{10}\,|\{1\}\rangle\langle\{0\}|\,\exp\{i\omega_{\text{eff}}^\circ(Q)t\}
\tag{202}
$$

As a consequence, the autocorrelation function becomes, after using the fact that $\langle\{0\}|\{0\}\rangle = 1$, and after performing a partial trace on the basis $\{|\{k\}\rangle\}$:

$$
\begin{aligned}
\{G_{\text{Brat}}(t)\} = |\mu_{10}|^2 \sum \Big\langle &\langle\{k\}|\{0\}\rangle\langle\{1\}| \\
&\times \exp\{i\omega_{\text{eff}}^\circ(Q)t\}\,|\{1\}\rangle\langle\{0\}|\{k\}\rangle\Big\rangle_Q
\end{aligned}
\tag{203}
$$

Moreover, one obtains by aid of the orthogonality property $\langle\{0\}|\{k\}\rangle = 0$ with $k \neq 0$, the simplified results:

$$
\{G_{\text{Brat}}(t)\} = |\mu_{10}|^2\Big\langle \exp\{i\omega_{\text{eff}}^\circ(Q)t\} \Big\rangle_Q
\tag{204}
$$

Besides, the thermal average value involved in this autocorrelation function is the thermal average on the Boltzmann distribution of the potential energy of the slow mode:

$$
\Big\langle \exp\{i\omega_{\text{eff}}^\circ(Q)t\} \Big\rangle_Q = \int \{P_{\text{B}}(Q)\}\,\exp\{i\omega_{\text{eff}}^\circ(Q)t\}\,dQ
\tag{205}
$$

with

$$
\{P_{\text{B}}(Q)\} = \exp\{-M\omega^{\circ\circ 2}Q^2/(2k_{\text{B}}T)\}/Z
\tag{206}
$$

and where Z is the classical partition function of the slow mode given by:

$$
Z = \int \exp\{-M\omega^{\circ\circ 2}Q^2/(2k_{\text{B}}T)\}\,dQ
\tag{207}
$$

That is:

$$Z = \{2\pi k_B T/(M\omega^{\circ\circ 2})\}^{1/2} \tag{208}$$

Thus the autocorrelation function becomes:

$$\{G_{Brat}(t)\} = |\mu_{10}|^2 \int \{P_B(Q)\} \exp\{i\omega^{\circ}_{eff}(Q)t\} \, dQ \tag{209}$$

The spectral density that is the Fourier transform of this autocorrelation function is:

$$\{I_{Brat}(\omega)\} = [1/(2\pi)^{1/2}] \int \{G_{Brat}(t)\} \exp\{-i\omega t) \, dt \tag{210}$$

and thus

$$\{I_{Brat}(\omega)\} = |\mu_{10}|^2 [1/(2\pi)^{1/2}] \int \{P_B(Q)\}$$

$$\times \int \exp\{i\omega^{\circ}_{eff}(Q)t\} \exp\{-i\omega t\} \, dt \, dQ \tag{211}$$

2. Calculation of a Suitable Expression for the Spectral Density

The integration on time involved in the last equation is, according to Eq. (A50), the Dirac distribution:

$$\int \exp\{i[\omega^{\circ}_{eff}(Q) - \omega]t\} \, dt = \sqrt{2\pi} \, \delta(f(Q)) \tag{212}$$

$f(Q)$ is given by:

$$f(Q) = [\omega^{\circ}_{eff}(Q) - \omega] \tag{213}$$

Again, the Dirac distribution involved in the above equation is given according to Eq. (A54) by:

$$\{\delta(f(Q))\} = \sum [1/(df/dQ)_{Q=Q_i}]\{\delta(Q - Q_i)\} \tag{214}$$

where the Q_i are the solutions of $\omega^\circ_{\text{eff}}(Q) = \omega$. Then, in view of the two last equations, the spectral density becomes

$$\{I_{\text{Brat}}(\omega)\} = |\mu_{10}|^2\sqrt{2\pi} \sum \int dQ\ \{P_{\text{B}}(Q)\}\{\delta(Q - Q_i)\}/(df/dQ)_{Q=Q_i} \quad (215)$$

or, after simplification:

$$\{I_{\text{Brat}}(\omega)\} = |\mu_{10}|^2\sqrt{2\pi} \sum \{P_{\text{B}}(Q)/(df/dQ)\}_{Q=Q_i} \quad (216)$$

3.   The Spectral Density When Expansion of Eq. (196) Is up to First Order in Q

When we limit the expansion of the angular frequency to first order in Q, in view of Eq. (197), Eq. (213) reduces to:

$$f(Q) = a_{112}Q + (\omega^\circ - \omega) \quad (217)$$

The equation $f(Q) = 0$ has a single root given by:

$$Q_1 = (\omega - \omega^\circ)/a_{112} \quad (218)$$

Recall that according to Eq. (37), a_{112} is given by:

$$a_{112}(\alpha^\circ/Q^{\circ\circ})\omega^{\circ\circ} \quad (219)$$

As a consequence, owing to the general expression (206) of the Boltzmann distribution, the above spectral density becomes:

$$\{I_{\text{Brat}}(\omega)\} = |\mu_{10}|^2\sqrt{2\pi}\ \exp\{-(M\omega^{\circ\circ 2})(\omega - \omega^\circ)^2/(2k_{\text{B}}Ta^2_{112})\}/(Za_{112}) \quad (220)$$

Expressed in terms of the anharmonic coupling parameter, using Eq. (208) giving the partition function Z, this becomes, after using the fact that $M\omega^{\circ\circ 2}Q^{\circ\circ 2} = \hbar\omega^{\circ\circ}$:

$$\{I_{\text{Brat}}(\omega)\} = |\mu_{10}|^2(1/\alpha^\circ\omega^{\circ\circ})[\hbar\omega^{\circ\circ}/(k_{\text{B}}T)]^{1/2}$$
$$\times \exp\{-(1/\alpha^\circ)^2[(\omega - \omega^\circ)/\omega^{\circ\circ}]^2[\hbar\omega^{\circ\circ}/(2k_{\text{B}}T)]\}$$

or, in terms of λ appearing in the quantum approach:

$$\{I_{\text{Brat}}(\omega)\} = |\mu_{10}|^2[\lambda^{1/2}/(\alpha°\omega°°)] \exp\{-[\lambda/(\sqrt{2}\alpha°)^2][(\omega - \omega°)/\omega°°]^2\}$$

(221)

Figure 1.3 gives examples of spectral densities obtained with the aid of Eq. (221) for an angular frequency of the fast mode equal to 3000 cm^{-1}, an angular frequency of the slow mode equal to 150 cm^{-1} and for different values of the anharmonic coupling and of the temperature, the unities used for the spectral densities being unspecified.

Note that the autocorrelation function corresponding to the spectral density (221) is the inverse Fourier transform of this last equation:

$$\{G_{\text{Brat}}(t)\} = [1/(2\pi)^{1/2}] \int \{I_{\text{Brat}}(\omega)\} \exp\{i\omega t\} \, d\omega$$

(222)

The result is,

$$\{G_{\text{Brat}}(t)\} = |\mu_{10}|^2 \exp\{i\omega°t\} \exp\{-(\alpha°^2/2)(\omega°°t)^2[k_B T/(\hbar\omega°°)]\}$$ (223)

4.   The First Moment of the Spectrum

On the other hand, the average angular frequency of the spectrum is given by:

$$\langle\omega\rangle = \int \{P_B(Q)\}\{\omega°_{\text{eff}}(Q)\} \, dQ$$

(224)

Expanding $\{\omega°_{\text{eff}}(Q)\}$ up to second order, this gives:

$$\langle\omega\rangle = \int \exp\{-M\omega°°^2 Q^2/(2k_B T)\}$$

$$\times \omega°(1 + a_{112}Q + \tfrac{1}{2}b_{112}Q^2) \, dQ \bigg/ \int \exp\{-[M\omega°°^2 Q^2/(2k_B T)]\} \, dQ$$

(225)

and thus, after integration

$$\langle\omega\rangle = \omega°(1 + \tfrac{1}{2}b_{112}[k_B T/(M\omega°°^2)])$$

(226)

As it appears, up to first order the above equation reduces to:

$$\langle\omega\rangle = \omega°$$

(227)

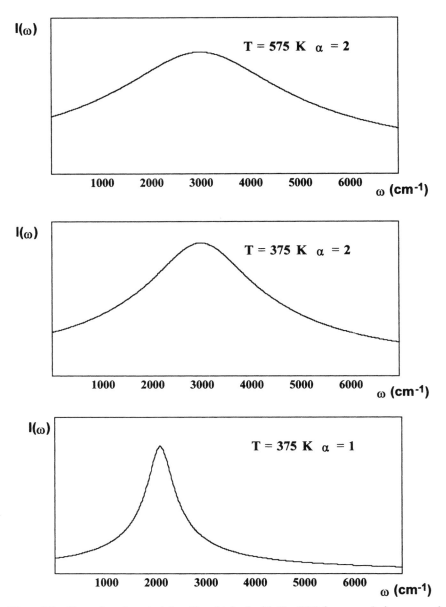

Figure 1.3. Examples of spectral densities obtained with Eq. (221) for an angle frequency of the fast mode equal to 3000 cm^{-1}, an angular frequency of the slow mode equal to 150 cm^{-1}, and for different values of the anharmonic coupling and of the temperature, the unities used for the spectral densities being unspecified.

Note that, according to Eq. (226), at zero temperature, and up to second order, there is no shift in the angular frequency with respect to the angular frequency of the fast mode.

5. The Spectral Density When the Expansion of the Pulsation (197) Is up to Second Order

Now consider the more general situation where the expansion of the angular frequency (197) is up to second order in Q. Then, in view of Eq. (197), the expression of $f(Q)$ involved in Eq. (213) is:

$$f(Q) = \tfrac{1}{2}b_{112} Q^2 + a_{112} Q + (\omega^\circ - \omega) \tag{228}$$

The roots of the equation $f(Q) = 0$ are:

$$Q_i = -(a_{112}/b_{112})\{1 \pm [1 - 2(b_{112}/a_{112}^2)(\omega^\circ - \omega)]^{1/2}\} \tag{29}$$

Bratos retains only the smallest of the two roots. Besides, one has, in view of these two equations, and permuting ω° and ω:

$$(df/dQ)_{Q=Q_i} = \pm a_{112}[1 + 2(b_{112}/a_{112}^2)(\omega - \omega^\circ)]^{1/2} \tag{230}$$

Finally, if one limits the expansion of the potentials of the slow mode to second order in Q,

$$\{P_B(Q = Q_i)\} = \exp\{-[M\omega^{\circ\circ 2}/(2k_B T)](a_{112}/b_{112})^2$$
$$\times \{1 - [1 + 2(b_{112}/a_{112}^2)(\omega - \omega^\circ)]^{1/2}\}^2\}/Z \tag{231}$$

Then, using the above equations,

$$\{I_{\mathrm{Brat}}(\omega)\} = \mathrm{const}\,|\mu_{10}|^2 \exp\{-[M\omega^{\circ\circ 2}/(2k_B T)](a_{112}/b_{112}^2)^2$$
$$\times \{1 - [1 + 2(b_{112}/a_{112}^2)(\omega - \omega^\circ)]^{1/2}\}^2\}$$
$$\times \{M\omega^{\circ\circ 2}/(2\pi k_B T)\}^{1/2} \tag{232}$$

Note that Bratos has introduced in his model the specific anharmonicity of the slow and fast modes that we do not give here.

B. The Robertson Approach

Now let us look at the very important classical approach of Robertson and Yarwood (1978).

1. The Starting Equations

The theory of Robertson is very similar in spirit to that of Bratos; however, the dynamics of the hydrogen bond relaxation are treated more precisely. The major assumption in that the stretching $v_s(XH\cdots Y)$ is classical and can be represented by the Ornstein–Uhlenbeck stochastic process that leads to a loss of the phase coherence of the $v_s(X–H)$ stretching mode because of the anharmonic coupling between the slow and fast modes. He analyzes his model in terms of the Kubo theory of a randomly modulated oscillator. For the slow coordinate, Robertson assumes that the behavior of this slow coordinate is classical and the same as that of a damped harmonic oscillator as considered in the Langevin theory. It must be emphasized that the Robertson model neglects the $v_s(X–H)$ anharmonicity.

Robertson considers the angular frequency of the fast mode within the same classical anharmonic coupling considered by Bratos. But he considers, in the spirit of Kubo, the angular frequency as a variable that is stochastically time dependent through the stochastic time dependence of the slow mode; that is, he writes:

$$\{\omega_{\text{eff}}^{\circ}(Q(t))\} = \omega^{\circ} + a_{112}\,Q(t) \tag{233}$$

The Robertson autocorrelation function has the same structure as that of Bratos, given by Eq. (200), but is now involving a thermal average performed on the stochastic time-dependent slow mode $Q(t)$, in place of the static slow mode Q, in the Bratos approach:

$$\{G_{\text{Rob}}(t)\} = \left\langle \{\mu^{\{0,\,1\}}(0)\}\{\mu^{\{1,\,0\}}(t)\} \right\rangle_{q,\,Q(t)} \tag{234}$$

where the brackets mean both a trace on the basis of the fast mode and a thermal average on the slow mode. The dipole moment operator is:

$$\{\mu^{\{0,\,1\}}(t)\} = |\{0\}\rangle\langle\{1\}|\,\mu_{01} \tag{235}$$

For the dynamics of the dipole moment operator in the absence of damping, we must look at the Heisenberg equation of motion.

$$i\hbar\,\partial\{\mu^{\{1,\,0\}}(t)\}/\partial t = [\{\mu^{\{1,\,0\}}(t)\},\,\{h(q,\,Q)\}] \tag{236}$$

where the Hamiltonian governing the $v_s(XH)$ motion is, owing to Eq. (9):

$$\{h(q,\,Q)\} = [p^2/(2m) + \tfrac{1}{2}m\{\omega_{\text{eff}}^{\circ}(Q)\}^2 q^2] \tag{237}$$

Developing the commutator, one finds:

$$i\hbar \ \partial\{\mu^{\langle 1, \ 0\rangle}(t)\}/\partial t = [\{\mu^{\langle 1, \ 0\rangle}(t)\}\{h(q, Q)\} - \{h(q, Q)\}\{\mu^{\langle 1, \ 0\rangle}(t)\}] \quad (238)$$

This yields:

$$i\hbar \ \partial\{\mu^{\langle 1, \ 0\rangle}(t)\}/\partial t = \mu_{10} |\{1\}\rangle\langle\{0\}| \{H_1\} - \{H_1\}|\{1\}\rangle\langle\{0\}| \mu_{10} \quad (239)$$

Now, owing to the eigenvalue equations:

$$\{h(q, Q)\}|\{0\}\rangle = \tfrac{1}{2}\hbar\{\omega^\circ_{\mathrm{eff}}(Q(t))\}|\{0\}\rangle \quad (240)$$

$$\{h(q, Q)\}|\{1\}\rangle = \tfrac{3}{4}\hbar\{\omega^\circ_{\mathrm{eff}}(Q(t))\}|\{1\}\rangle \quad (241)$$

the time dependence of the dipole moment operator is:

$$i \ \partial\{\mu^{\langle 1, \ 0\rangle}(t)\}/\partial t = -\{\omega^\circ_{\mathrm{eff}}(Q(t))\}\mu_{10}|\{1\}\rangle\langle\{0\}| \quad (242)$$

which leads after integration to:

$$\{\mu^{\langle 1, \ 0\rangle}(t)\} = \mu_{10}|\{1\}\rangle\langle\{0\}| \int_0^t \exp\{i\{\omega^\circ_{\mathrm{eff}}(Q(t'))\}\} \ dt' \quad (243)$$

Again, after performing a partial trace on the basis $\{|\{k\}\rangle\}$, this gives:

$$\{G_{\mathrm{Rob}}(t)\} = |\mu_{10}|^2 \sum \Big\langle \langle\{k\}|\{0\}\rangle\langle\{1\}|$$

$$\times \int_0^t dt' \ \exp\{i\{\omega^\circ_{\mathrm{eff}}(Q(t))\}\}|\{1\}\rangle\langle\{0\}|\{k\}\rangle \Big\rangle_{Q(t)} \quad (244)$$

Here, the symbols $\langle \ \rangle_Q$ denote simply an average on $Q(t)$ performed on the thermal bath, damping the movement of the slow mode. Moreover, by aid of the orthogonality property $\langle\{0\}|\{k\}\rangle = 0$ with $k \neq 0$, that reduces to the simplified result:

$$\{G_{\mathrm{Rob}}(t)\} = \varepsilon|\mu_{10}|^2 \Big\langle \exp\Big\{i \int_0^t \{\omega^\circ_{\mathrm{eff}}(Q(t'))\} \ dt'\Big\}\Big\rangle_{Q(t)} \quad (245)$$

Finally, owing to Eq. (233), the autocorrelation function becomes:

$$\{G_{\mathrm{Rob}}(t)\} = |\mu_{10}|^2 \ \exp\{i\omega^\circ t\}\Big\langle \exp\Big\{ia_{112} \int_0^t Q(t') \ dt'\Big\}\Big\rangle_{Q(t)} \quad (246)$$

2. Use of the Langevin Theory within the Kubo Cumulant Expansion

a. The Langevin Equation. In order to find the thermal average of the exponential involved in the autocorrelation function, Robertson and Yarwood (1978) use the cumulant expansion of Kubo and consider the statistical behavior of the slow-mode coordinate as that of a Brownian oscillator within the Langevin equation:

$$M \, d^2Q/dt^2 + M\gamma \, dQ/dt = F(t) - M\omega^2 Q$$

For an overview of this question, see Appendix H.

More precisely, Robertson and Yarwood (1978) have used the results of Uhlenbeck and Ornstein (1930). Owing to this crucial assumption, it is possible to perform a Kubo cumulant expansion in the above autocorrelation function, which can be truncated after two terms. It must be emphasized that the Uhlenbeck and Ornstein theory assumes, because it is a consequence of the Langevin equation, that the correlation time of the random force is infinitesimal. If necessary, this last assumption can be relaxed and the Langevin equation replaced by a generalized Langevin equation (vide infra the work of Sakun). Of course, the thermal average of the slow-mode coordinate viewed as a Brownian oscillator is:

$$\langle Q(t) \rangle = 0$$

b. The Second-Order Cumulant Expansion of the Autocorrelation Function of the Dipole Moment Operator. Let us perform a second-order cumulant expansion of the autocorrelation function (246):

$$\{G_{\text{Rob}}(t)\} = |\mu_{10}|^2 \exp\{i\omega^\circ t\}\left(\exp\left\{ia_{112}\int_0^t dt' \, \langle Q(t')\rangle\right\}\right.$$
$$\left. + \exp\left\{-\tfrac{1}{2}a_{112}^2 \int_0^t dt' \int_0^{t'} dt'' \, \langle Q(t')Q(t'')\rangle\right\}\right)$$

As the thermal average of the coordinate is zero for a Brownian oscillator, the autocorrelation function reduces to:

$$\{G_{\text{Rob}}(t)\} = |\mu_{10}|^2 \exp\{i\omega^\circ t\} \exp\left\{-\tfrac{1}{2}a_{112}^2 \int_0^t dt' \int_0^{t'} dt'' \, \langle Q(t')Q(t'')\rangle\right\}$$

$$(247)$$

Moreover, performing as usual the variable change $\tau = t' - t''$, the above equation becomes:

$$\{G_{\text{Rob}}(t)\} = |\mu_{10}|^2 \exp\{i\omega^\circ t\}$$

$$\times \exp\left\{ -\tfrac{1}{2}a_{112}^2 \int_0^t d\tau \; (t - \tau)\{\langle Q(\tau)Q(0)\rangle + \langle Q(-\tau)Q(0)\rangle\}\right\} \tag{248}$$

Uhlenbeck and Ornstein (1930) and Wang and Uhlenbeck (1945) have obtained by aid of the stochastic Langevin equation, which may generate a Fokker–Planck equation (see Appendix H, Section 1), the structure of the autocorrelation function of the position coordinate of the Brownian oscillator.

For the underdamped situation $\omega^{\circ\circ} > \gamma$, the expression of the autocorrelation function, which holds for positive time, is:

$$\langle Q(t)Q(0)\rangle = \langle Q^2\rangle \exp\{-\gamma t/2\}$$

$$\times \{\cos(\omega_{\text{und}}^{\circ\circ} t) + [\gamma/(2\omega_{\text{und}}^{\circ\circ})] \sin(\omega_{\text{und}}^{\circ\circ} t)\} \tag{249}$$

with

$$\omega_{\text{und}}^{\circ\circ 2} = [\omega^{\circ\circ 2} - \gamma^2/4] \tag{250}$$

For the overdamped situation $\gamma > \omega^{\circ\circ}$, the expression of the autocorrelation function, which holds for positive time, is:

$$\langle Q(t)Q(0)\rangle = \langle Q^2\rangle \exp\{-\gamma t/2\}\{\cos(\omega_{\text{over}}^{\circ\circ} t) + [\gamma/(2\omega_{\text{over}}^{\circ\circ})] \sin(\omega_{\text{over}}^{\circ\circ} t)\}$$

with

$$\omega_{\text{over}}^{\circ\circ 2} = -[\omega^{\circ\circ 2} - \gamma^2/4]$$

Note that the classical thermal average of the squared displacement appearing in these equations is by definition:

$$\langle Q_{\text{cl}}^2\rangle = \int \{P_{\text{B}}(Q)\}Q^2 \, dQ \tag{251}$$

By integration and using the expression for $Q^{\circ\circ}$, one finds:

$$\langle Q_{cl}^2 \rangle = Q^{\circ\circ 2}[2k_B T/(\hbar\omega^{\circ\circ})] = 2Q^{\circ\circ 2}/\lambda \tag{252}$$

Note that if the slow-mode oscillator is considered quantum mechanically, the corresponding expression is given by a sum in place of an integral, and the result is:

$$\langle Q_{qu}^2 \rangle = Q^{\circ\circ 2} \coth\{\hbar\omega^{\circ\circ}/(2k_B T)\}$$

which reduces for high temperature to the classical expression. In his study, Robertson considered, for the slow mode, the quantum expression in place of the classical one. But it is not in the spirit of his approach, since he considers for the autocorrelation function of the slow mode that of a classical Brownian oscillator. Thus it seems better to consider the classical expression, which we shall perform in the following.

We may observe that both the underdamped and overdamped autocorrelation function reduces, in the absence of damping, to:

$$\langle Q(t)Q(0) \rangle = Q^{\circ\circ 2}[2k_B T/(\hbar\omega^{\circ\circ})] \cos(\omega^{\circ\circ}t)$$

As it appears, it is the usual behavior of a harmonic oscillation the amplitude of which is vanishing when the temperature reduces to zero.

c. *Explicit Expression for the Robertson Autocorrelation Function.* In view of the above equations, the autocorrelation function of the dipole moment operator becomes, in the overdamped situation, after passing to the dimensionless anharmonic coupling given by Eq. (219):

$$\begin{aligned}
\{G_{Rob}(t)\} = |\mu_{10}|^2 \exp\{i\omega^\circ t\} \exp\{&-2[\alpha^{\circ 2}/\lambda] \\
\times \{\gamma t + [(\gamma^2 &- \omega^{\circ\circ 2})/\omega^{\circ\circ 2}][\exp\{-\gamma t/2\} \cos(\omega_{und}^{\circ\circ} t) - 1] \\
+ [\gamma/(2\omega_{und}^{\circ\circ})]&[(\gamma^2 - 3\omega^{\circ\circ 2})/\omega^{\circ\circ 2}] \\
\times \exp\{-\gamma t/2\} &\sin(\omega_{und}^{\circ\circ} t)\}\}
\end{aligned} \tag{253}$$

Note that this result may be compared to that (223) obtained by Bratos for the corresponding situation. The spectral density of the Robertson approach is the Fourier transform:

$$\{I_{Rob}(\omega)\} = 1/\sqrt{2\pi} \int \{G_{Rob}(t)\} \exp(-i\omega t)\, dt \tag{254}$$

d. The Very Underdamped Situation. Now let us look at the spectral situation involving very small damping, $\gamma° \ll \omega°°$. Then we may perform, in view of Eq. (250), the approximation: $\omega°°_{und} \approx \omega°°$. The autocorrelation function becomes:

$$\{G_{Rob}(t)\} = |\mu_{10}|^2 \exp\{i\omega°t\} \exp\{-2[\alpha°^2/\lambda]$$
$$\times \{\gamma t - [\exp\{-\gamma t/2\} \cos(\omega°°t) - 1]$$
$$- 3(\gamma/(2\omega°°)) \exp\{-\gamma t/2\} \sin(\omega°°t)\}\} \tag{255}$$

e. The Undamped and Zero-Temperature Limits. When the damping is missing, the Robertson autocorrelation function reduces to the very simple form:

$$\{G_{Rob}(t)\} = |\mu_{10}|^2 \exp\{i\omega°t\} \exp\{2[\alpha°^2/\lambda][\cos(\omega°°t) - 1]\}$$

Finally, note that the Robertson autocorrelation function reduces, at zero temperature with and without damping, to the simple form:

$$\{G^0_{Rob}(\omega)\} = |\mu_{10}|^2 \exp\{i\omega°t\}$$

Note that for zero temperature and in the presence of damping, the autocorrelation function would be the same. Of course, the Fourier transform of this function is the simple delta peak centered on the angular frequency $\omega°$ of the fast mode.

$$\{I_{Rob}(\omega)\} = [1/\sqrt{2\pi}] \, \delta(\omega - \omega°)$$

3. The Amplitude of Modulation and the Correlation Time of Kubo

In his study of the random properties of classical harmonic oscillators involving stochastic time-dependent angular frequencies, Kubo (1961) distinguished two important parameters, the amplitude of modulation and the correlation time of modulation τ_c, which have been applied by Robertson to his model.

According to Kubo, the amplitude of modulation Δ (i.e., the fluctuation of the angular frequency of the fast mode in the present situation) and the correlation time of modulation τ_c are given, respectively, by:

$$\Delta^2 = \langle\{\omega°_{eff}(Q(t))\}^2\rangle \tag{256}$$

$$\tau_c = (1/\Delta^2) \int_0^\infty \langle\{\omega°_{eff}(Q(t))\}\{\omega°_{eff}(Q(0))\}\rangle \, dt \tag{257}$$

Then, in view of Eq. (233), the amplitude of modulation appears to be:

$$\Delta = a_{112}\langle Q^2\rangle^{1/2} \tag{258}$$

Owing to Eqs. (37) and (252), it becomes:

$$\Delta = \alpha^\circ\omega^{\circ\circ}(2/\lambda)^{1/2} \tag{259}$$

On the other hand, in the theory of Robertson, the Kubo correlation time of modulation appears, according to Eq. (233) to be:

$$\tau_c = \int_0^\infty dt\ \langle Q(t)Q(0)\rangle/\langle Q^2\rangle \tag{260}$$

Thus it becomes, respectively, in the under- and overdamped situations:

$$\tau_c = \int_0^\infty \exp\{-\gamma t/2\}[\cos(\omega_{und}\,t) + \gamma/(2\omega_{und})\,\sin(\omega_{und}\,t)]\,dt \tag{261}$$

$$\tau_c = \int_0^\infty \exp\{-\gamma t/2\}[\cos(\omega_{over}\,t) + \gamma/(2\omega_{over})\,\sin(\omega_{over}\,t)]\,dt \tag{262}$$

After integration, one obtains in both situations:

$$\tau_c = \gamma/\omega^{\circ\circ 2} \tag{263}$$

4. The Kubo Slow and Fast Modulation Limits Applied to the Robertson Theory

Now, following Kubo, consider the nondimensional product $\tau_c\Delta$, which may be written:

$$\begin{aligned}
\tau_c\Delta &= [\gamma/\omega^{\circ\circ}][\Delta/\omega^{\circ\circ}] \\
\tau_c\Delta &= [\alpha^\circ\gamma/\omega^{\circ\circ}][2k_B\,T/(\hbar\omega^{\circ\circ})]^{1/2}
\end{aligned} \tag{264}$$

Then, according to Kubo, one may distinguish two limiting situations:

- The slow-modulation limit corresponding to $\tau_c\Delta \gg 1$;
- The rapid-modulation limit corresponding to $\tau_c\Delta \ll 1$.

5.   The Autocorrelation Function of Robertson for Large Time

Consider the situation where the correlation time τ_c is very small with respect to the time of interest, which is practically large times. Then the upper limit time t of integration may be extended to infinite in the expression (248) of the autocorrelation function, so that, in view of Eq. (258), it takes the form:

$$\{G_{\text{Rob}}(t)\} \approx |\mu_{10}|^2 \exp\{i\omega^\circ t\}$$

$$\times \exp\left\{-\tfrac{1}{2}\Delta^2 \int_0^\infty d\tau \, (t - \tau)\{\langle Q(\tau)Q(0)\rangle + \langle Q(-\tau)Q(0)\rangle\}/\langle Q^2\rangle\right\}$$

(265)

The result of the integration is then, owing to Eq. (260), given by the equations:

$$\{G_{\text{Rob}}(t)\} \approx |\mu_{10}|^2 \exp\{i\omega^\circ t\} \exp\{-\gamma_L t\}$$ (266)

$$\gamma_L = \Delta^2 \tau_c$$ (267)

The Fourier transform of this autocorrelation function is a Lorentzian of the form:

$$\{I_{\text{Rob}}(\omega)\} \approx |\mu_{10}|^2 (2/\pi)\gamma_L/[(\omega - \omega^\circ)^2 + \gamma_L^2]$$ (268)

with, in view of Eqs. (259) and (263):

$$\gamma_L = 2\gamma[\alpha^{\circ 2}/\lambda]$$ (269)

6.   The Robertson Autocorrelation Function for Short Times

On the other hand, for very short times with respect to the correlation time, one may consider that the autocorrelation function at time t is practically the same as at initial time:

$$\langle Q(\tau)Q(0)\rangle = \langle Q(-\tau)Q(0)\rangle \approx \langle Q^2\rangle$$ (270)

As a consequence the general autocorrelation function (248) in view of Eq. (258), giving Δ leads to

$$\{G_{\text{Rob}}(t)\} \approx |\mu_{10}|^2 \exp\{i\omega^\circ t\} \exp\left\{-\Delta^2 \int_0^t d\tau \, (t - \tau)\right\}$$ (271)

The result is

$$\{G_{\text{Rob}}(t)\} \approx |\mu_{10}|^2 \exp\{i\omega°t\} \exp\{-\tfrac{1}{2}\Delta^2 t^2\} \tag{272}$$

Or, in view of Eq. (259):

$$\{G_{\text{Rob}}(t)\} = |\mu_{10}|^2 \exp\{i\omega°t\} \exp\{-[\alpha°\omega°°/\sqrt{\lambda}]^2 t^2\} \tag{273}$$

The Fourier transform of this autocorrelation function is, of course, a Gaussian:

$$\{I_{\text{Rob}}(\omega)\} = |\mu_{10}|^2 [\lambda^{1/2}/(\sqrt{2}\alpha°\omega°°)] \exp\{-[\lambda/(2\alpha°)^2][(\omega - \omega°)/\omega°°]^2\} \tag{274}$$

As it appears, this result is very near that of Bratos [Eq. (221)].

7.   The Line Shape

One may perform, according to Kubo, the following previsions about the line shape of the spectral density involving hydrogen bond in solution.

a.   *Slow- and Fast-Modulation Limit.*   There are two possible situations: (1) The slow-modulation limit corresponding to $\tau_c \Delta > 1$ for which the shape is Gaussian, and that may be in connection with the result obtained for large times. According to Eq. (264), this condition is:

$$\alpha°[2k_{\text{B}} T/(\hbar\omega°°)]^{1/2} > [\omega°°/\gamma] \tag{275a}$$

(2) The fast-modulation limit corresponding to $\tau_c \Delta < 1$ for which the shape is Lorentzian and which may be in connection with the result we have obtained for short times. In view of Eq. (264), this condition is:

$$\alpha°[2k_{\text{B}} T/(\hbar\omega°°)]^{1/2} < [\omega°°/\gamma] \tag{275b}$$

b.   *Discussion.*   In the overdamped situation, where $\omega°°/\gamma < 1$, the slow-modulation limit is very easy to realize, whereas in the underdamped cases, where $\omega°°/\gamma > 1$, the slow-modulation limit is only realized for very high temperature and very strong anharmonic coupling. At the opposite, in the overdamped situation, where $\omega°°/\gamma < 1$, the fast-modulation limit is only favored by very small anharmonic coupling and very low temperature, whereas the slow-modulation limit is not difficult to realize. It appears that an increase in the temperature, or in the anharmonic coupling, or in the damping parameter favors the passage from a Lorentzian shape to a Gaussian one.

8. *Limit of Validity of the Theory*

The semiclassical theory of Robertson has the merit to convey the fact that, in solution, the profile of the infrared spectra of the high-frequency mode often diverges from a Gaussian shape. It gives the possibility to reproduce some features of the band shape. However, it should be noted that the Robertson model is not satisfactory for low temperatures or when the system becomes very underdamped, as might be the case in the gas phase since the basic assumption that the character of the slow mode is determined largely by its interaction with the environment is no longer valid. In the very underdamped situation, each excited state of the high-frequency vibrational mode is probably sufficiently long-lived to define for the low-frequency mode an effective potential, in which a number of complete oscillations can take place, so that vibrational Franck–Condon transitions need to be considered. More generally, the asymmetry of the band shape, everywhere it appears, cannot be taken into account from the semiclassical model, which neglects the possibility of the effective potentials appearing in the quantum approach of anharmonicity.

Dealing with the Robertson model, and applying it with success to experimental results, the papers of Yarwood and Robertson (1975), Yarwood, Ackroyd, and Robertson (1978), and Bournay and Robertson (1978) should be cited.

9. *Numerical Simulation of Spectral Densities within the Quantum Indirect Relaxation*

We shall now give some calculations of spectral densities obtained with Eq. (253) for an angular frequency of the fast mode equal to 3000 cm^{-1}, and an angular frequency of the slow mode equal to 150 cm^{-1}. Two series of calculations have been chosen, one in which the nondimensional anharmonic constant α° is $\sqrt{2}$ and another one corresponding to isotopic substitution of the X–H bond by deuterium and leading to $\alpha^\circ = 1$ and to $\omega^\circ = 2120$ cm^{-1}. For each series has been considered room temperature (300 K), and three different values of the damping parameter γ have been selected, all being compatible with the underdamped situation. The results, which are within the fast-modulation limit, are given, respectively, in Figs. 1.4a and 1.4b without giving the units of the intensities. Inspection of the figures manifests:

1. A decrease in the half-width of the spectrum with deuterium substitution.

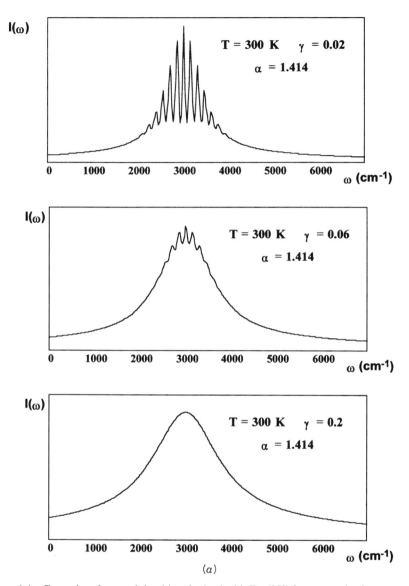

Figure 1.4. Examples of spectral densities obtained with Eq. (253) for an angular frequency of the fast mode equal to 3000 cm^{-1}, and an angular frequency of the slow mode equal to 150 cm^{-1}. Two series of calculations have been chosen, one in which the nondimensional anharmonic constant α° is $\sqrt{2}$ and another one corresponding to isotopic substitution of the X–H bond by deuterium, leading to $\alpha^\circ = 1$. For each series has been considered the room temperature (300 K), and three different values of the damping parameter γ have been selected, all being compatible with the underdamped situation. The results, which are within the fast-modulation limit, are given, respectively, in (a) and (b) without giving the units of the intensities.

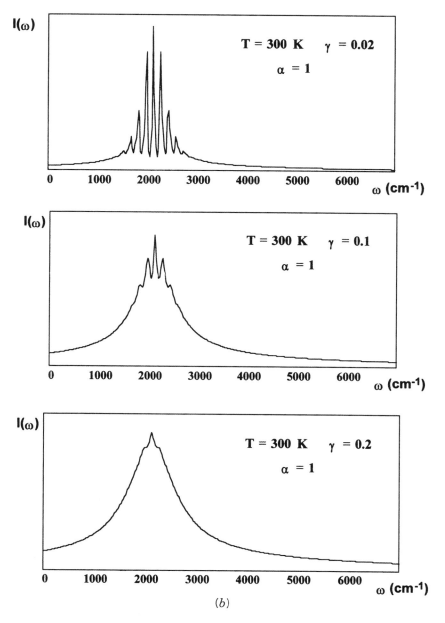

(b)

Figure 1.4—continued

2. An increase of the half-width with temperature.
3. An influence of the damping parameter γ on the coalescence of the sub-bands.
4. A deviation of the line shape from the Gaussian shape corresponding to the slow-modulation limit and to the Bratos model, where there is coalescence of the sub-bands.

C. The Sakun Approach

It may be now of interest to look at the generalization of the Robertson model by Sakun (1985).

1. The Basis Idea of Sakun

As we have seen, one of the assumptions used in the Robertson approach is that the correlation time of the random force characterizing the medium and which induces the damping of the slow mode is infinitesimal. When this assumption is relaxed, the Langevin equation used by Robertson must be replaced by the generalized Langevin equation:

$$M[d^2Q/dt^2]_t + (M/k_B T) \int_0^t \{F(t-t')F(0)\}[dQ/dt]_{t'}\, dt' = F(t) - M\omega^{\circ\circ 2}Q$$

For this question see Appendix H dealing with the question of relaxation. That is at the basis of the work of Sakun (1985), who generalizes the approach of Robertson by removing the approximation dealing with the assumption that the autocorrelation function of the force characterizing the thermal bath may be considered as Dirac delta peaked.

Sakun starts from the autocorrelation function of Robertson (247):

$$\{G_{\text{Sak}}(t)\} = |\mu_{10}|^2 \exp\{i\omega^\circ t\}$$
$$\times \exp\left\{-\tfrac{1}{2}a_{112}^2 \int_0^t dt' \int_0^{t'} dt'' \langle Q(t')Q(t'')\rangle\right\} \tag{276}$$

Then, in order to find the average value of the Q coordinate and the corresponding autocorrelation function, he uses in place of the Langevin equation used by Robertson the generalized Langevin equation involving a memory function (see Appendix H).

2. The Autocorrelation Functions of the Q Coordinate and of the Dipole Moment Operator

Sakun has shown that, within the generalized Langevin formalism, the autocorrelation function of the slow-mode coordinate may be written according to the equations:

$$\langle Q(t)Q(0)\rangle = \langle Q^2\rangle\{\lambda_\mu \exp\{\omega_\mu t\} + 2R_e \lambda_\nu \exp\{\omega_\nu t\}\} \tag{277}$$

$$\omega_\mu = -[1 - \zeta]/\tau_c^S, \qquad \omega_\nu = i\omega_{\text{eff}}^{\circ\circ} - \gamma/2$$

$$\zeta = \gamma\tau_c^S \qquad \omega_{\text{eff}}^{\circ\circ} = \omega^{\circ\circ}/[1 - \zeta]^{1/2}$$

τ_χ^S is the correlation time of the stochastic generalized Langevin force, whereas λ_μ and λ_ν are dimensionless parameters given in terms of $\langle F(0)^2\rangle$ [which may be obtained from Eq. (H29) at initial time], by:

$$\lambda_\pi = \{1 - (\omega_\pi^2/\omega^{\circ\circ 2})$$

$$+ (\omega^{\circ\circ}2Mk_B T/\langle F(0)^2\rangle)^2[1 + \omega_\pi^2/\omega^{\circ\circ 2}]^2\}^{-1} \tag{278}$$

Sakun has shown that:

$$\lambda_\mu = \zeta^3\xi/[(\xi + 2)\zeta^2 - 3\zeta + 1]$$

$$\lambda_\nu = \tfrac{1}{2}[1 - \lambda_\mu] - \tfrac{1}{2}i[4\xi - 1]^{-1/2}\{1 - \lambda_\mu + 2(1 - \zeta)\lambda_\mu/\zeta\}$$

with:

$$\xi = [\tfrac{1}{4} + \omega_{\text{eff}}^{\circ\circ 2}/\gamma^2]$$

3.   The Autocorrelation Function within the Generalized Langevin Equation

The autocorrelation function (276) of the dipole moment operator of the hydrogen bond appears to be given in the Sakun approach by:

$$\{G_{\text{Sak}}(t)\} = |\mu_{01}|^2 \exp\{-i\omega^\circ t\}\{G_\mu(t)\}\,|\,\{G_\nu(t)\}\,|^2 \tag{279}$$

where the two time-dependent terms appearing in the right-hand side are autocorrelation functions given, respectively, by:

$$\{G_\mu(t)\} = \exp\{-\tfrac{1}{2}S_\mu[\exp\{\omega_\mu t\} - \omega_\mu t - 1]\} \tag{280}$$

$$\{G_\nu(t)\} = \exp\{-\tfrac{1}{2}S_\nu[\exp\{\omega_\nu t\} - \omega_\nu t - 1]\} \tag{281}$$

with:

$$S_\pi = 2\alpha^{\circ 2}\{\langle Q^2(T)\rangle/Q^{\circ\circ 2}\}[\omega_{\text{eff}}^{\circ\circ}/\omega_\pi]^2\lambda_\pi \tag{282}$$

where π stands for μ and ν. According to expression (252), the quantities may be written:

$$S_\pi = 2\alpha^{\circ 2}[2k_B T/(\hbar\omega^{\circ\circ})][\omega^{\circ\circ}_{\text{eff}}/\omega_\pi]^2 \lambda_\pi \tag{283}$$

By aid of the expansion:

$$\exp\{u + u^*\} = \sum [u/|u|]^n\{I_n(|2u|)\} \tag{284}$$

where $I_n(x)$ is the modified Bessel function, the autocorrelation function becomes:

$$
\begin{aligned}
\{G_{\text{Sak}}(t)\} = &|\mu_{01}|^2 \exp\{-i\omega^\circ t\} \exp\{-\tfrac{1}{2}S_\mu[\exp\{\omega_\mu t\} - \omega_\mu t - 1]\} \\
&\times \exp\{R_e(S_\nu \omega_\nu)t\} \sum \exp\{in\omega^{\circ\circ}_{\text{eff}} t\} \\
&\times [S_\nu/|S_\nu|]^n\{I_n(|S_\nu| \exp\{-\tfrac{1}{2}\gamma t\})\}
\end{aligned}
\tag{285}
$$

4. The Robertson Autocorrelation Function as a Special Situation When the Time Correlation of the Generalized Langevin Equation Approaches Zero

Note that for the Langevin equation, that is, when the time correlation of the generalized Langevin equation reduces to zero (i.e., $\tau^S_c \to 0$), the above parameters reduce in the following way:

$$\omega^{\circ\circ}_{\text{eff}} \to \omega^{\circ\circ}, \qquad \zeta \to 0, \qquad \lambda_\mu \to 0, \qquad \lambda_\nu \to \tfrac{1}{2}[1 - i\gamma/(2\omega^{\circ\circ})]$$

$$S_\mu \to 0, \qquad S_\nu \to i\alpha^{\circ 2}\omega^{\circ\circ}[2k_B T/(\hbar\omega^{\circ\circ})][(i\omega^{\circ\circ} - \gamma/2)]/[(\omega^{\circ\circ 2} + \gamma^2/4)]$$

$$\{G_\mu(t)\} \to 1, \qquad \{G_{\text{Sak}}(t)\} \to \{G_{\text{Rob}}(t)\} \tag{286}$$

The autocorrelation function of Sakun reduces therefore to that of Robertson that is Eq. (253).

It must be observed that, owing to the limit of validity of the Robertson approach, which neglects the quantum effects related to the existence of effective potentials for the slow mode, the introduction of the memory function by Sakun into the classical behavior of the slow mode appears to be perhaps somewhat superfluous, the quantum damping generally smoothing the classical details.

D. Generalization of the Robertson Model by Abramczyk: Direct and Indirect Relaxation

Abramczyk (1985) has generalized the model of Robertson taking into account the indirect relaxation within the anharmonic coupling between the slow and fast mode, by proposing a model that includes both the indirect relaxation and the dipole–dipole interaction between the molecular dipole moment directed along the X–H axis and the dipole moment generated by the hydrogen bond formation, and treated as a stochastic process due to thermal motion. This model assumes that both the indirect relaxation and the rotation–diffusion of the dipole moment are statistically independent.

1. Basic Equations

The Hamiltonian H of the hydrogen bond in the absence of a thermal bath is:

$$\{H^0_{\text{Abram}}(Q, q, \theta)\} = [p^2/(2m) + \tfrac{1}{2}m\omega^{\circ 2}q^2] + [P^2/(2M) + \tfrac{1}{2}M\omega^{\circ\circ 2}Q^2]$$
$$+ [K_{112}Qq^2] + \{V_D(q, \theta)\} \tag{287}$$

where the potential V_D is the Hamiltonian of interaction between the dipole moment μ of the X–H bond and the dipole moment produced by the hydrogen bond formation (see Fig. 1.5) given by:

$$\{V_D(q, \theta)\} = [\boldsymbol{\mu}(q) \cdot \Delta\boldsymbol{\mu}(q)r^2 - 3(\boldsymbol{\mu}(q) \cdot \mathbf{r}\,\Delta\boldsymbol{\mu}(q) \cdot \mathbf{r})]/[4\pi\varepsilon^\circ r^5] \tag{288}$$

$\Delta\boldsymbol{\mu}(q)$ is the change in the dipole moment produced by the hydrogen bond formation and $\mathbf{r}$ is the distance between the centers of the X–H and H$\cdots$Y bonds. The potential may be written as a function of angles:

$$\{V_D(q, \theta)\} = (3\cos(\theta_1)\cos(\theta_2) - \cos(\theta))\mu(q)\,\Delta\mu(q)/[4\pi\varepsilon^\circ r^3] \tag{289}$$

where θ is the angle between the X–H and the H$\cdots$Y axis, whereas θ_1 and θ_2 are, respectively, the angle between $\mathbf{r}$ and the X–H and H$\cdots$Y axes.

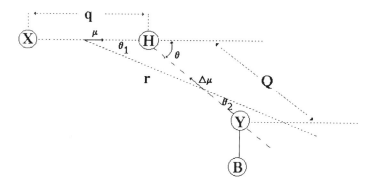

Figure 1.5. The normal coordinates and angles used by Abramczyk in the calculation of the potential V_D.

Besides, this potential may be written as a function of angles: Due to thermal motions in liquid, the angle θ deviates from its average value, which is zero for a linear hydrogen bond. Since the slow-mode Q distance is much greater than that q of the fast mode, it appears that one has $\theta \approx \theta_1$ and $\theta_2 \approx 0$. As a consequence, the dipole–dipole interaction potential appears to be a function of the single angle θ and of the fast-mode coordinate q.

$$\{V_D(q, \theta)\} = 2 \cos(\theta)\mu(q) \, \Delta\mu(q)/[4\pi\varepsilon°r^3] \qquad (290)$$

The dipole moments may be expanded up to first order with respect to the fast-mode coordinate:

$$\mu(q) = \mu° + (\partial\mu/\partial q)_0 \, q \qquad (291)$$

$$\Delta\mu(q) = \Delta\mu° + (\partial \, \Delta\mu/\partial q)_0 \, q \qquad (292)$$

As a consequence, the potential may be written:

$$\{V_D(q, \theta)\} = B(t) + 2Aq^2 \cos \theta/[4\pi\varepsilon°r^3] \qquad (293)$$

with:

$$A = (\partial\mu/\partial q)_0(\partial \, \Delta\mu/\partial q)_0 \qquad (294)$$

$B(t)$ is a function that up to first order in q will give zero diagonal matrix elements with respect to the eigenkets of the fast-mode Hamiltonian, and will therefore disappear in the calculation of the dipole moment operator. Note that the Abramczyk Hamiltonian may be written also:

$$\{H^0_{Abram}(Q, q, \theta)\} = \{H°(Q, q)\} + \{V_D(q, \theta)\} \qquad (295)$$

or, after rearranging,

$$\{H^0_{Abram}(Q, q, \theta)\} = [p^2/(2m) + \tfrac{1}{2}m(\omega°_{Abr}(Q, \theta))^2q^2]$$
$$+ [P^2/(2M) + \tfrac{1}{2}M\omega°°^2Q^2] \qquad (296)$$

Expanding up to first order the binomial expression of the angular frequency, we get:

$$\{\omega°_{Abram}(Q, \theta)\} = \omega° + a_{112}Q + b \cos \theta \qquad (297)$$

with:

$$b = 2A/[4\pi\varepsilon^\circ mr^3\omega^\circ]$$

Abramczyk, generalizing the approach of Robertson, considers not only Q as time dependent but also θ because of the thermal movements of the molecule. That leads to writing up to first order:

$$\{\omega^\circ_{\text{Abram}}(Q(t), \theta(t))\} = [\omega^\circ + a_{112}Q(t) + b\cos[\theta(t)]] \qquad (298)$$

2.   The Autocorrelation Function of the Hydrogen Bond

By aid of an inference similar to that used by Robertson to pass from Eq. (234) to Eq. (245), the autocorrelation function of the hydrogen bond in the presence of indirect and dipole–dipole relaxation appears to be given by:

$$\{G_{\text{Abram}}(t)\} = \varepsilon|\mu_{10}|^2\left\langle\exp\left\{i\int_0^t\{\omega^\circ_{\text{Abr}}(Q(t), \theta(t))\}\,dt'\right\}\right\rangle_{Q(t),\,\theta(t)} \qquad (299)$$

Again, using Eq. (298), one obtains:

$$\{G_{\text{Abram}}(t)\} = \varepsilon|\mu_{10}|^2\left\langle\exp\left\{i\int_0^t[\{\omega^\circ_{\text{eff}}(Q(t'))\} + b\{\cos(\theta(t'))\}]\,dt'\right\}\right\rangle_{Q(t),\,\theta(t)} \tag{300}$$

Assuming that the indirect relaxation and the dipole–dipole relaxation are statistically independent, one may write the autocorrelation function by aid of the equations:

$$\{G_{\text{Abram}}(t)\} \approx \varepsilon|\mu_{10}|^2\left\langle\exp\left\{i\int_0^t\{\omega^\circ_{\text{eff}}(Q(t'))\}\,dt'\right\}\right\rangle_{Q(t)}$$
$$\times\left\langle\exp\left\{ib\left\{\int_0^t\cos(\theta(t'))\right\}dt'\right\}\right\rangle_{\theta(t)} \tag{301}$$

$$\{\omega^\circ_{\text{eff}}(Q(t))\} = \omega^\circ + a_{112}Q(t) \qquad (302)$$

Moreover, owing to Eq. (246), the above autocorrelation function may be expressed as the Robertson autocorrelation function times an angular auto-correlation function:

$$\{G_{\text{Abram}}(t)\} = \{G_{\text{Rob}}(t)\}\{G_{\text{ang}}(t)\} \qquad (303)$$

$$\{G_{ang}(t)\} = \left\langle \exp\left\{ib \int_0^t \{\cos(\theta(t'))\}\, dt'\right\}\right\rangle_{\theta(t)} \tag{304}$$

Here the subscript refers to the thermal average on the angular distribution.

3. The Angular Autocorrelation Function

Assuming that the thermal average of cos $\theta(t)$ is zero,

$$\langle \cos(\theta(t))\rangle_{\theta(t)} = 0 \tag{305}$$

and has a Gaussian distribution, one obtains after expansion in cumulant series:

$$\{G_{ang}(t)\} = \exp\left\{-\tfrac{1}{2}b^2 \int_0^t dt' \int_0^{t'} dt'' \langle \cos(\theta(t'))\cos(\theta(t''))\rangle\right\} \tag{306}$$

Moreover, one may perform as usual the change of variables $\tau = t' - t''$. Then the above equation becomes:

$$\{G_{ang}(t)\} = \exp\left\{-\tfrac{1}{2}b^2 \int_0^t d\tau\, (t - \tau)\right.$$

$$\left. \times \{\langle \cos(\theta(\tau))\cos(\theta(0))\rangle + \langle \cos(\theta(-\tau))\cos(\theta(0))\rangle\}\right\} \tag{307}$$

Besides, the stochastic function $\theta(t)$, which deviates from its average value $\langle\theta\rangle_{\theta(t)}$ by $\Delta\theta(t)$, may be written:

$$\theta(t) = \langle\theta\rangle_{\theta(t)} + \Delta\theta(t) \tag{308}$$

Again, using this equation, one may write:

$$\cos(\langle\theta\rangle_{\theta(t)} + \Delta\theta(t)) = \{D_{00}^1(\langle\theta\rangle)\}\{D_{00}^1(\Delta\theta(t))\} \tag{309}$$

$\{D_{00}^1(\langle\theta\rangle)\}$ and $\{D_{00}^1(\Delta\theta(t))\}$ are the rotation matrices. Then the autocorrelation function appearing in the autocorrelation function may be written:

$$\langle \cos(\theta(t))\cos(\theta(0))\rangle = \cos^2\langle\theta\rangle\langle\{D_{00}^1(\Delta\theta(t))\}\{D_{00}^1(\Delta\theta(0))\}\rangle \tag{310}$$

Within the rotation–diffusion model, one has in the molecular frame:

$$\langle\{D_{00}^1(\Delta\theta(t))\}\{D_{00}^1(\Delta\theta(0))\}\rangle = \tfrac{1}{3}\exp\{-2D_{aa}\,t\} \tag{311}$$

D_{aa} is the rotational diffusion coefficient of proton–acceptor molecule describing a rotation around the axis perpendicular to the z axis. In the laboratory fixed system, Abramczyk has shown that the above equation becomes:

$$\langle \cos(\theta(t))\, \cos(\theta(0))\rangle = \tfrac{1}{3}[\cos^2\langle\theta\rangle]\, \exp\{-Dt\} \tag{312}$$

with:

$$D = 2[D_{aa} + D_{bb}] \tag{313}$$

D_{bb} is the rotational–diffusion coefficient of proton–donor molecule describing a rotation around the axis perpendicular to the z axis.

Finally, introducing this equation into the angular dipole–dipole auto-correlation function (307), one obtains, after integration:

$$\{G_{ang}(t)\} = \exp\{-b^2\, \cos^2\langle\theta\rangle\{Dt + [\exp\{-Dt\} - 1]\}/(3D^2)\} \tag{314}$$

Note that, since the time scale of vibrational relaxation for the H bond is very short in comparison with the time constant of external motion, one may write approximately $\exp\{-Dt\} \approx 1$, and the autocorrelation function reduces to the simple form:

$$\{G_{ang}(t)\} \approx \exp\{-b^2 t\, \cos^2\langle\theta\rangle/(3D)\} \tag{315}$$

4. The Spectral Density

Now let us consider the spectral density of the hydrogen bond involving both indirect and angular dipole–dipole relaxation. It is the Fourier transform of the autocorrelation function (303):

$$\{I_{Abram}(\omega)\} = [1/\sqrt{2\pi}] \int \{G_{Rob}(t)\}\{G_{ang}(t)\}\, \exp\{-i\omega t\}\, dt \tag{316}$$

Again, by aid of a theorem dealing with Fourier transform theory, one obtains the convolution product of the spectral density of the Robertson theory as defined by the Fourier transform of Eq. (253) times the line shape resulting from the angular dipole–dipole rotational diffusion:

$$\{I_{Abram}(\omega)\} = [1/(2\pi)] \int \{I_{Rob}(\omega)\}\{I_{ang}(\omega - \omega')\}\, d\omega' \tag{317}$$

with:

$$\{I_{ang}(\omega)\} = \int \{G_{ang}(t)\} \exp\{-i\omega t\} \, dt \qquad (318)$$

Note that this line shape is a Lorentzian:

$$\{I_{ang}(\omega)\} = \sqrt{2/\pi} \, \gamma_{ang}/\{\gamma_{ang}^2 + \omega^2\}$$

with for the situation corresponding to Eq. (315)

$$\gamma_{ang} = b^2 \cos^2\langle\theta\rangle/3D$$

VII. QUANTUM APPROACH IN THE INDIRECT DAMPING MECHANISM

After discussing the classical approaches of the spectral density of the high-frequency vibrational mode of the hydrogen bond within the indirect damping mechanism, we now have to consider the corresponding quantum approach.

A. Approach in Representation {III}

In the first place we shall consider the approach of indirect damping within representation {III}.

1. The Basis Autocorrelation Function

In this representation, the starting autocorrelation function is

$$\{G_{III}(t)\} = \mathrm{tr}'\{\{\rho_{III}^{\{0\}}(0)\}\{\mu_{III}^{\{0,\,1\}}(0)\}\{\mu_{III}^{\{1,\,0\}}(t)\}\}$$

where the trace must be performed on both the basis $\{|\{k\}\rangle\}$ and $\{|n\rangle\}$. The dipole moment operator at initial time and at time t are given, respectively, by:

$$\{\mu_{III}^{\{0,\,1\}}(0)\} = \mu_{01} |\{0\}\rangle\langle\{1\}| \{A^{\{1\}}(\alpha)\}^\dagger$$
$$\{\mu_{III}^{\{1,\,0\}}(t)\} = \mu_{10}\langle\exp\{i\{H_{III}\}t/\hbar\}\{A^{\{1\}}(\alpha^\circ)\} |\{1\}\rangle\langle\{0\}| \exp\{-i\{H_{III}\}t/\hbar\}\rangle_\theta$$

The phase transformation appearing in the last equation involves the full Hamiltonian of the coherent state characterizing the slow mode and damped by the thermal bath. Therefore, implicit in this phase transformation is a partial trace on the thermal bath, which is the reason for the brackets $\langle \ \rangle_\theta$. This leads, owing to Eq. (143), to:

$$\{\mu_{\text{III}}^{(1,\ 0)}(t)\} = \mu_{10}\left\langle \exp\{i\{H_{\text{III}}^{(1)}\}t/\hbar\}\{A^{(1)}(\alpha^\circ)\} \mid \{1\}\rangle\langle\{0\} \mid \exp\{-i\{H_{\text{III}}^{(0)}\}t/\hbar\}\right\rangle_\theta$$

Again, with the expression of the effective Hamiltonians (145) and (147), that leads to:

$$\{\mu_{\text{III}}^{(1,\ 0)}(t)\} = \mu_{10}\left\langle \exp\{i\{[a^\dagger a\omega^{\circ\circ}] - 2\alpha^{\circ 2}\omega^{\circ\circ} + \omega^\circ + \sum\ \{b_k^\dagger b_k\ \omega_k\}\right.$$
$$+ \sum\ \{b_k^\dagger a\kappa_k + b_k\ a^\dagger\kappa_k\}\}t\}\{A^{(1)}(\alpha^\circ)\} \mid \{1\}\rangle\langle\{0\} \mid$$
$$\times \exp\{-i\{[a^\dagger a\omega^{\circ\circ}] + \sum\ \{b_k^\dagger b_k\ \omega_k\} + \sum\ \{b_k^\dagger a\kappa_k + b_k\ a^\dagger\kappa_k\}\}t\}\right\rangle_\theta$$

That may simplify to:

$$\{\mu_{\text{III}}^{(1,\ 0)}(t)\} = \mu_{10}\left\langle \exp\{i\{[a^\dagger a\omega^{\circ\circ}] + \sum\ \{b_k^\dagger b_k\ \omega_k\}\right.$$
$$+ \sum\ \{b_k^\dagger a\kappa_k + b_k\ a^\dagger\kappa_k\}\}t\} \mid \{1\}\rangle\{A^{(1)}(\alpha^\circ)\}\langle\{0\} \mid$$
$$\times \exp\{-i\{[a^\dagger a\omega^{\circ\circ}] + \sum\ \{b_k^\dagger b_k\ \omega_k\}$$
$$\left. + \sum\ \{b_k^\dagger a\kappa_k + b_k\ a^\dagger\kappa_k\}\}t\}\right\rangle_\theta \exp\{i(\omega^\circ - 2\alpha^{\circ 2}\omega^{\circ\circ})t\}$$

The phase transformation appearing here and involving the partial trace over the thermal bath may be viewed as a unitary transform involving a reduced time evolution operator. Practically, the solution is to solve the Liouville equation (149) allowing to find the damped reduced density operator (152), according to considerations of III.B.2. Then, from this density operator, may be extracted the damped translation operator (156) that leads to write:

$$\{\mu_{\text{III}}^{(1,\ 0)}(t)\} = \mu_{10}\mid\{1\}\rangle\{A^{(1)}(\Phi_{\text{III}}(t)^*))\}\langle\{0\} \mid \exp\{i(\omega^\circ - 2\alpha^{\circ 2}\omega^{\circ\circ})t\}$$

Thus, proceeding as in Section III.C, and after performing the partial trace on the basis $\{\mid\{k\}\rangle\}$ the autocorrelation function becomes:

$$\{G_{\text{III}}(t)\} = \varepsilon\mid\mu_{10}\mid^2 \text{tr}\{\exp\{-\lambda a^\dagger a\}$$
$$\times \{A^{(1)}(\alpha^\circ)\}^\dagger\{A^{(1)}(\Phi_{\text{III}}(t)^*)\}\} \exp\{i(\omega^\circ - 2\alpha^{\circ 2}\omega^{\circ\circ})t\} \quad (319)$$

Note the analogy with Eq. (184), the only difference being the change in the argument of the translation operator. This result may also be written, in view of Eq. (A45) of Appendix A:

$$\{G_{III}(t)\} = \varepsilon|\mu_{10}|^2 \ \text{tr}\{\exp\{-\lambda a^\dagger a\}\{A^{\{1\}}(\Phi_{II}(t)^*)\}\}$$
$$\times \exp\{i(\omega^\circ - 2\alpha^{\circ 2}\omega^{\circ\circ})t\} \ \exp\{i\alpha^{\circ 2} \ \exp\{-\gamma t/2\} \ \sin\{\omega^{\circ\circ}t\}\} \quad (320)$$

Performing the trace, according to Eq. (D3) of Appendix D, one obtains:

$$\{G_{III}(t)\} = |\mu_{10}|^2 \ \exp\{i\omega^\circ t\} \ \exp\{\tfrac{1}{2}\alpha^{\circ 2}(1 + 2\langle n\rangle)$$
$$\times [2 \ \exp\{-\gamma t/2\} \ \cos(\omega^{\circ\circ}t) - 1 - \exp\{-\gamma t\}]\}$$
$$\times \exp\{i\alpha^{\circ 2} \ \exp\{-\gamma t/2\} \ \sin(\omega^{\circ\circ}t)\} \ \exp\{-i2\alpha^{\circ 2}\omega^{\circ\circ}t\} \quad (321)$$

$\langle n\rangle$ is the thermal average occupation number of a quantum harmonic oscillator given by Eq. (H90).

2. Expansions of the Underdamped Autocorrelation Function and Its Fourier Transform

a. *Expansion of Autocorrelation Function (320).* Note that the autocorrelation function (320) may be written with the closeness relation $\sum |n\rangle\langle n| = 1$:

$$\{G_{III}(t)\} = \varepsilon|\mu_{10}|^2 \sum\sum \langle m|\exp\{-\lambda a^\dagger a\}|n\rangle$$
$$\times \langle n|\{A^{\{1\}}(\Phi_{II}(t)^*)\}|m\rangle \ \exp\{i(\omega^\circ - 2\alpha^{\circ 2}\omega^{\circ\circ})t\}$$
$$\times \exp\{i\alpha^{\circ 2} \ \exp\{-\gamma t/2\} \ \sin(\omega^{\circ\circ}t)\} \quad (322)$$

Again, since the operator $a^\dagger a$ is diagonal in the basis used, that leads, after simplification, and with the aid of the definition of the Franck–Condon factors, to the result:

$$\{G_{III}(t)\} = \varepsilon|\mu_{10}|^2 \sum\sum \exp\{-\lambda m\}\{\Gamma_{mn}(\Phi_{II}(t)^*)\}$$
$$\times \exp\{i(\omega^\circ - 2\alpha^{\circ 2}\omega^{\circ\circ})t\} \ \exp\{i\alpha^{\circ 2} \ \exp\{-\gamma t/2\} \ \sin(\omega^{\circ\circ}t)\} \quad (323)$$

b. *Expansion of Autocorrelation Function (319).* Another possibility is to start from Eq. (319) to perform formally the trace and to use twice the closeness relation. That leads to:

$$\{G_{III}(t)\} = \varepsilon|\mu_{10}|^2 \sum\sum\sum \langle m|\exp\{-\lambda a^\dagger a\}|n\rangle\langle n|\{A^{\{1\}}(\alpha^\circ)\}^\dagger|k\rangle$$
$$\times \langle k|\{A^{\{1\}}(\Phi_{III}(t)^*)\}|m\rangle \ \exp\{i(\omega^\circ - 2\alpha^{\circ 2}\omega^{\circ\circ})t\} \quad (324)$$

Then, after simplification and in view of the expressions (A40) and (A41) of the matrix elements of the translation operator, one obtains the following expansion:

$$\{G_{III}(t)\} = \varepsilon |\mu_{10}|^2 \sum_{m=0}^{\infty} \exp\{-\lambda m\}$$

$$\times \sum_{l=0}^{m} \sum_{p=0}^{l} \sum_{r=0}^{l} \sum_{u=0}^{\infty} \sum_{v=0}^{\infty} A_{m,l,p,r,u,v} \, C_{m,l,p,r,u,v}(\alpha^{\circ})$$

$$\times \exp\{-i\omega_{m,l,r}^{\circ\circ} t\} \exp\{\gamma_{m,l,r,v} t/2\} \tag{325}$$

In this equation, the indices m and n must be permuted in the terms where they appear according to the fact that $n > m$, or $n \leq m$, and:

$$\gamma_{m,l,r,v} = (m - l + 2r + 2v)\gamma$$

$$\omega_{m,l,r}^{\circ\circ} = \omega^{\circ} - 2\alpha^{\circ 2}\omega^{\circ\circ} - (m - l + 2r)\omega^{\circ\circ}$$

$$A_{m,l,p,r,u,v} = (-1)^{p+r+u+v} l! \, m! / [2^{u+v}(l-p)!$$

$$\times (l-r)!(m-l+p)!(m-l+r)! \, p! \, r! \, u! \, v!]$$

$$C_{m,l,p,r,u,v}(\alpha^{\circ}) = \alpha^{\circ 2(m-l+r+p+u+v)}$$

Now let us look at the Fourier transform of the autocorrelation function:

$$\{I(\omega)\} = [1/\sqrt{2\pi}] \int \{G_{III}(t)\} \exp\{-i\omega t\} \, dt \tag{326}$$

Performing the integral, one obtains:

$$\{I(\omega)\} = \sum \sum \{I_{ml}(\omega)\} \tag{327}$$

$$\{I_{ml}(\omega)\} = \varepsilon |\mu_{10}|^2 \exp\{-\lambda m\} \sum_{p=0}^{l} \sum_{r=0}^{l} \sum_{u=0}^{\infty} \sum_{v=0}^{\infty} A_{m,l,p,r,u,v} \, C_{m,l,p,r,u,v}(\alpha^{\circ})$$

$$\times \{\gamma_{mnqj}/[(\omega - \omega_{m,l,r})^2 + (\gamma_{m,l,r,v}/2)^2]\} \tag{328}$$

Each line is the superposition of a quadruple sum of Lorentzians, each of them characterized by specific half-width. Note that this expression is valid only for the underdamped situation owing to the approximation (141).

3. The Quantum Highly Underdamped Autocorrelation Function at High Temperature and Its Comparison with the Corresponding Robertson Autocorrelation Function

Consider the highly underdamped situation where $\omega^{\circ\circ} \gg \gamma$ at high temperature. Of course, the quantum autocorrelation function (321) that was obtained for underdamped situations also holds for highly underdamped situations. Note that the thermal average number occupation appearing in it is:

$$\langle n \rangle = [\exp\{\hbar\omega^{\circ\circ}/(k_B T)\} - 1]^{-1} \tag{329}$$

Besides, when the temperature is high, that is, when $k_B T > \hbar\omega^{\circ\circ}$, one may expand the exponential involved in the thermal average occupation number to first order so that one obtains the well-known high-temperature result:

$$\langle n \rangle = k_B T/(\hbar\omega^{\circ\circ}) \tag{330}$$

As a consequence, in this special situation, if we denote:

$$\{\alpha_{qu}^{\circ}\} = \alpha^{\circ}[1 + 2k_B T/(\hbar\omega^{\circ\circ})]^{1/2} \tag{331}$$

the quantum autocorrelation function (321) takes the form:

$$\begin{aligned}
\{G_{III}(t)\} = &|\mu_{10}|^2 \exp\{i\omega^{\circ}t\} \exp\{-i2\alpha^{\circ 2}\omega^{\circ\circ}t\} \exp\{-\tfrac{1}{2}\alpha_{qu}^{\circ 2} \exp\{-\gamma t\}\} \\
&\times \exp\{\{\alpha_{qu}^{\circ}\}^2[2 \exp\{-\gamma t/2\} \cos(\omega^{\circ\circ}t) - 1] \\
&\times \exp\{i\alpha^{\circ 2} \exp\{-\gamma t/2\} \sin(\omega^{\circ\circ}t)\}
\end{aligned} \tag{332}$$

This last result, which is fully quantum mechanical, may be compared to the classical one of Robertson, under the same conditions. Recall that in the highly underdamped situation, the Robertson autocorrelation function is given by Eq. (255), which, in order to make clear the comparison, may also be written according to:

$$\begin{aligned}
\{G_{Rob}(t)\} = &|\mu_{10}|^2 \exp\{i\omega^{\circ}t\} \exp\{-\gamma_{eff} t\} \\
&\times \exp\{\{\alpha_{cl\,1}^{\circ}\}^2[\exp\{-\gamma t/2\} \cos(\omega^{\circ\circ}t) - 1]\} \\
&\times \exp\{\{\alpha_{cl\,2}^{\circ}\}^2 \exp\{-\gamma t/2\} \sin(\omega^{\circ\circ}t)\}
\end{aligned} \tag{333}$$

$$\gamma_{eff} = \gamma\alpha_{cl\,1}^{\circ 2}; \quad \{\alpha_{cl\,1}^{\circ}\} = \alpha^{\circ}[2k_B T/(\hbar\omega^{\circ\circ})]^{1/2};$$

$$\{\alpha_{cl\,2}^{\circ}\} = \alpha_{cl\,1}^{\circ}[3\gamma/(2\omega^{\circ\circ})]^{1/2}$$

Note the analogy between the quantum and the classical autocorrelation function, although the effective parameters involved in both situations do not have exactly the same structure.

An important difference is the absence of the phase shift appearing in the quantum autocorrelation function, which is deeply connected to the existence of effective Hamiltonians in the quantum approach.

B. Approach in Representation {II}

1. The Basis Autocorrelation Function in Representation {II}

Next let us consider the autocorrelation function of the dipole moment operator of the fast mode in representation $\{II\}$. It is:

$$\{G_{II}(t)\} = \mathrm{tr}'\{\{\rho_{II}^{\{0\}}(0)\}\{\mu_{II}^{\{0,\,1\}}(0)\}\{\mu_{II}^{\{1,\,0\}}(t)\}\}$$

where the trace must be performed on both the basis $\{|\{k\}\rangle\}$ and $\{|n\rangle\}$. The dipole moment operators are, respectively, given by:

$$\{\mu_{II}^{\{0,\,1\}}(0)\} = \mu_{01}|\{0\}\rangle\langle\{1\}|$$

$$\{\mu_{II}^{\{1,\,0\}}(t)\} = \mu_{10}\langle\exp\{i\{H_{II}\}t/\hbar\}\rangle_\theta|\{1\}\rangle\langle\{0\}|\langle\exp\{-i\{H_{II}\}t/\hbar\}\rangle_\theta$$

$\{H_{II}\}$ is the full Hamiltonian in representation $\{II\}$ involving the slow mode, the thermal bath, and the coupling between them, and where the meaning of the brackets is that we must perform, at least formally, a partial trace over the thermal bath, that is, damping the slow mode.

The autocorrelation function in the presence of damping is after performing the trace over the basis $\{|\{k\}\rangle\}$:

$$\{G_{II}(t)\} = \varepsilon|\mu_{10}|^2\,\mathrm{tr}\{\sum\langle\{k\}|\exp\{-\lambda a^\dagger a\}|\{0\}\rangle\langle\{1\}|$$
$$\times\,\langle\exp\{i\{H_{II}\}t/\hbar\}\rangle_\theta|\{1\}\rangle\langle\{0\}|\exp\{-i\{H_{II}\}t/\hbar\}|\{k\}\rangle\}$$

where the trace must be performed on basis $\{|n\rangle\}$. Again, using the expression (114) of the full Hamiltonian in terms of the effective Hamiltonians, this becomes after simplification:

$$\{G_{II}(t)\} = \varepsilon|\mu_{10}|^2\,\mathrm{tr}\{\sum\langle\{k\}|\exp\{-\lambda a^\dagger a\}|\{0\}\rangle\langle\{1\}|$$
$$\times\,\langle\exp\{i\{H_{II}^{\{1\}}\}t/\hbar\}\rangle_\theta|\{1\}\rangle\langle\{0\}|\exp\{-i\{H_{II}^{\{0\}}\}t/\hbar\}|\{k\}\rangle\}$$

Because of the orthonormality of the basis $\{|\{k\}\rangle\}$, this reduces to:

$$\{G_{II}(t)\} = \varepsilon |\mu_{10}|^2 \, \mathrm{tr}\{\exp\{-\lambda a^\dagger a\}\langle\exp\{i\{H_{II}^{\{1\}}\}t/\hbar\}\rangle_\theta \exp\{-i\{H_{II}^{\{0\}}\}t/\hbar\}\}$$

$$(334)$$

According to Eq. (117), the thermal average of the first phase transformation may be written formally:

$$\langle\exp\{i\{H_{II}^{\{1\}}\}t/\hbar\}\rangle_\theta = \langle\exp\{i\{(a^\dagger a)\omega^{\circ\circ} + \{\alpha^\circ[a^\dagger + a] - \alpha^{\circ 2}\}\omega^{\circ\circ}$$
$$+ \sum \{b_k^\dagger b_k \omega_k\} + \sum \{b_k^\dagger a\kappa_k + b_k a^\dagger \kappa_k^*\}\}\rangle_\theta$$

According to the results of Section III.B.1, the reduced time evolution operator corresponding to the partial trace performed on the thermal bath may be viewed as given by:

$$\langle\exp\{i\{H_{II}^{\{1\}}\}t/\hbar\}\rangle_\theta = \{U_{II}^{\{1\}}(t)\}^\dagger \qquad (335)$$

where the time evolution operator appearing on the right-hand side is given by Eq. (130).

On the other hand, the second reduced phase transformation appearing in Eq. (334) is the same as that without the thermal bath, because for this subspace corresponding to the ground state of the fast mode, the thermal bath does not play any role, so that we may write for the ground state of the high-frequency mode:

$$\langle\exp\{-i\{H_{II}^{\{0\}}\}t/\hbar\}\rangle_\theta = \exp\{-i\{H_{II}^{0\{0\}}\}t/\hbar\} = \exp\{-ia^\dagger a\omega^{\circ\circ}t\} \qquad (336)$$

$\{H_{II}^{\{0\}}\}$ is the effective Hamiltonian in the absence of damping, and we have used Eq. (46).

2. The Autocorrelation Function Expressed in Terms of the Translation Operator

As a consequence of the above equations, the autocorrelation function of the damped hydrogen bond in representation {II} becomes:

$$\{G_{II}(t)\} = \varepsilon |\mu_{10}|^2 \, \mathrm{tr}\{\exp\{-\lambda a^\dagger a\}\{U_{II}^{\{1\}}(t)\}^\dagger \exp\{-ia^\dagger a\omega^{\circ\circ}t\}\} \qquad (337)$$

with, in view of Eq. (130)

$$\{U_{II}^{\{1\}}(t)\}^\dagger = \exp\{i(\omega^\circ - \alpha^{\circ 2}\omega^{\circ\circ})t\} \exp\{-i|\beta|^2\omega^{\circ\circ}t\}$$
$$\times \exp\{i|\beta|^2 \exp\{-\gamma t/2\} \sin(\omega^{\circ\circ}t)\}$$
$$\times \{A^{\{1\}}(\Phi_{II}(t))\} \exp\{ia^\dagger a\omega^{\circ\circ}t\} \qquad (338)$$

These two equations reduce, respectively, in the absence of relaxation to Eqs. (167) and (169). Owing to Eqs (337) and (338), the autocorrelation function reduces to:

$$\{G_{II}(t)\} = \varepsilon |\mu_{10}|^2 \ \text{tr}\{\exp\{-\lambda a^\dagger a\}\{U_{II}^{IP}(t)\}^\dagger\} \qquad (339)$$

with,

$$\{U_{II}^{IP}(t)\}^\dagger = \exp\{i(\omega^\circ - \alpha^{\circ 2}\omega^{\circ\circ})t\} \ \exp\{-i(|\beta|^2\omega^{\circ\circ})t\}$$
$$\times \exp\{i|\beta|^2 \ \exp\{-\gamma t/2\}$$
$$\times \sin(\omega^{\circ\circ}t)\}\{A^{(1)}(\Phi_{II}(t))\} \qquad (340)$$

This equation (339) corresponds for the damped situation to Eq. (171) in the absence of relaxation, whereas Eq. (340) reduces, in the absence of damping, as required, to (172)

Moreover, the above autocorrelation function leads to:

$$\{G_{II}(t)\} = \varepsilon |\mu_{10}|^2 \ \exp\{i(\omega^\circ - \alpha^{\circ 2}\omega^{\circ\circ})t\} \ \exp\{-i(|\beta|^2\omega^{\circ\circ})t\}$$
$$\times \exp\{i|\beta|^2 \ \exp\{-\gamma t/2\} \ \sin(\omega^{\circ\circ}t)\}$$
$$\times \text{tr}\{\exp\{-\lambda a^\dagger a\}\{A^{(1)}(\Phi_{II}(t))\}\} \qquad (341)$$

Then, performing the trace, one obtains by aid of Eq. (D3) of Appendix D:

$$\{G_{II}(t)\} = |\mu_{10}|^2 \ \exp\{-\tfrac{1}{2}|\beta|^2(1 + 2\langle n\rangle)\} \ \exp\{\tfrac{1}{2}|\beta|^2(1 + 2\langle n\rangle)$$
$$\times [2 \ \exp\{-\gamma t/2\} \ \cos(\omega^{\circ\circ}t) - \exp\{-\gamma t\}]\}$$
$$\times \exp\{i|\beta|^2 \ \exp\{-\gamma t/2\} \ \sin(\omega^{\circ\circ}t)\}$$
$$\times \exp\{-i|\beta|^2\omega^{\circ\circ}t\} \ \exp\{i(\omega^\circ - \alpha^{\circ 2}\omega^{\circ\circ})t\} \qquad (342)$$

3.   The Final Autocorrelation Function

At last, using Eq. (H90) of the thermal average of the mean number occupation,

$$\langle n\rangle = 1/[\exp\{\hbar\omega/(k_B T)\} - 1]$$

the autocorrelation function takes the final form:

$$\{G_{\text{II}}(t)\} = |\mu_{10}|^2 \exp\{-\tfrac{1}{2}|\beta|^2(1 + 2[\exp\{\hbar\omega^{\circ\circ}/(k_B T)\} - 1]^{-1})\}$$
$$\times \exp\{\tfrac{1}{2}|\beta|^2(1 + 2[\exp\{\hbar\omega^{\circ\circ}/(k_B T)\} - 1]^{-1})$$
$$\times [2 \exp\{-\gamma t/2\} \cos(\omega^{\circ\circ}t) - \exp\{-\gamma t\}]\}$$
$$\times \exp\{i|\beta|^2 \exp\{-\gamma t/2\} \sin(\omega^{\circ\circ}t)\}$$
$$\times \exp\{-i|\beta|^2\omega^{\circ\circ}t\} \exp\{i(\omega^{\circ} - \alpha^{\circ 2}\omega^{\circ\circ})t\} \qquad (343)$$

with:

$$|\beta| = \alpha^{\circ}[4\omega^{\circ\circ 4} + \gamma^2\omega^{\circ\circ 2}]^{1/2}/[2(\omega^{\circ\circ 2} + \gamma^2/4)] \qquad (344)$$

The final expression for the autocorrelation function of the hydrogen bond, damped according to the indirect mechanism, depends on the following physical parameters:

The absolute temperature T, the angular frequencies ω° and $\omega^{\circ\circ}$ of the fast and slow modes, the damping parameter γ of the slow mode and the effective anharmonic coupling parameter β between the slow and fast modes, which, in turn, depends through $|\beta|$ on the anharmonic coupling parameter α° given by Eq. (38), on the damping parameter γ and on the angular frequency $\omega^{\circ\circ}$ of the slow mode.

4. Half-width Temperature Dependence and Isotope Effect

It may be observed that it is possible to find the dependence of the half-width $\Delta\omega$ of this spectral density on temperature. It is shown in the appendix that owing to Eq. (F4) it is possible to write:

$$\Delta\omega(T_1)/\Delta\omega(T_2) = \coth^{1/2}[\hbar\omega^{\circ\circ}/(2k_B T_1)]/\coth^{1/2}[\hbar\omega^{\circ\circ}/(2k_B T_2)] \quad (345)$$

5. Limit Situation When the Damping is Missing

When the indirect damping is missing (i.e., for $\gamma \to 0$), the effective dimensionless parameter reduces to $\beta = \alpha^{\circ}$. As a consequence, the autocorrelation function (342) reduces to:

$$\{G_{\text{II}}(t, \gamma \to 0)\} = |\mu_{10}|^2 \exp\{i\alpha^{\circ 2} \sin(\omega^{\circ\circ}t)\} \exp\{i(\omega^{\circ} - 2\alpha^{\circ 2}\omega^{\circ\circ})t\}$$
$$\times \exp\{\alpha^{\circ 2}(1 + 2\langle n\rangle)[\cos(\omega^{\circ\circ}t) - 1]\} \qquad (346)$$

As it appears, this autocorrelation function is nothing but Eq. (188), as obtained in the absence of damping:

$$\{G_{\text{II}}(t, \gamma \to 0)\} = \{G_{\text{II}}^{\circ}(t)\}$$

6. Comparison between Representations {II} and {III} in the Highly Underdamped Situations

Note that when the indirect damping is decreasing, β reduces to α°, so that the autocorrelation function (342) reduces to:

$$\{G_{II}(t)\} = |\mu_{10}|^2 \exp\{-\tfrac{1}{2}\alpha^{\circ 2}(1 + 2\langle n\rangle)\}$$
$$\times \exp\{\tfrac{1}{2}\alpha^{\circ 2}(1 + 2\langle n\rangle)[2\exp\{-\gamma t/2\}\cos(\omega^{\circ\circ}t) - \exp\{-\gamma t\}]\}$$
$$\times \exp\{i\alpha^{\circ 2}\exp\{-\gamma t/2\}\sin(\omega^{\circ\circ}t)\}$$
$$\times \exp\{-i\alpha^{\circ 2}\omega^{\circ\circ}t\}\exp\{i(\omega^\circ - \alpha^{\circ 2}\omega^{\circ\circ})t\} \tag{347}$$

As it appears, by inspection of Eq. (321) giving the damped autocorrelation function in the indirect mechanism within representation {III}, the above autocorrelation function in representation {II} is equal in the underdamped situation to that in representation {III}:

$$\{G_{II}(t)\} = \{G_{III}(t)\}, \qquad \text{for } \gamma \ll \omega^{\circ\circ} \tag{348}$$

7. Numerical Simulation of Spectral Densities within the Quantum Indirect Relaxation

It may be of interest to give some numerical simulations of spectral densities obtained with the aid of Eq. (343). Have been considered an H-bonded compound for which the angular frequency ω° of the fast mode is 3000 cm^{-1} and the angular frequency $\omega^{\circ\circ}$ of the slow mode equal to 150 cm^{-1}, and its deuterated parent for which ω° is reduced by $1/\sqrt{2}$ and becomes therefore 2120 cm^{-1} whereas $\omega^{\circ\circ}$ remains equal to 150 cm^{-1}. The non-dimensional anharmonic constant α° has been taken equal to $\sqrt{2}$ for the H-bonded species, and therefore equal to 1 for the D-bonded isomer owing to the considerations of Section II.D.4. For both the H and D compounds, two values of temperature have been chosen, one corresponding to low temperatures (30 K) and the other one around the room temperature (300 K). For each temperature, three different values of the damping parameter γ have been selected, all being compatible with the underdamped situation. In order to stabilize at zero value for infinite time the autocorrelation function (343), this function has been multiplied by a damped exponential factor, the parameter of which has been chosen to be the fourth of γ. Figures 1.6a and 1.6b give the spectral density for the H-bonded species, the first one at room temperature and the last one at 30 K, whereas Figs. 1.6c and 1.6d give the corresponding spectral densities for the deuterated isomer, the damping parameter increasing from the top to the bottom of the figures.

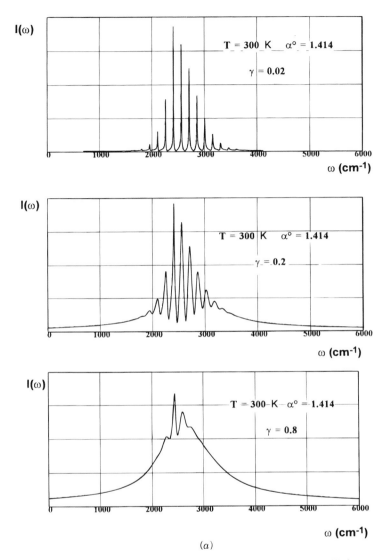

(a)

Figure 1.6. Examples of spectral densities obtained with the use of Eq. (343) for an angular frequency of the slow mode equal to 150 cm^{-1}. Two series of calculations have been considered. A first one in which the nondimensional anharmonic constant α° takes some suitable value, i.e. $\sqrt{2}$ and the angular frequency of the fast mode $\omega^\circ = 3000$ cm^{-1} and another one corresponding to isotopic substitution of the X–H bond by deuterium, leading for α° to the value 1 and for ω° to the value 2120 cm^{-1}. For each series, two values of temperature have been chosen, one corresponding to low temperatures (30 K) and the other one around room temperature (300 K). For each temperature, three different values of the damping parameter γ have been selected, all being compatible with the underdamped situation. The results are given in (a), (b), (c), and (d) without defining the intensity units.

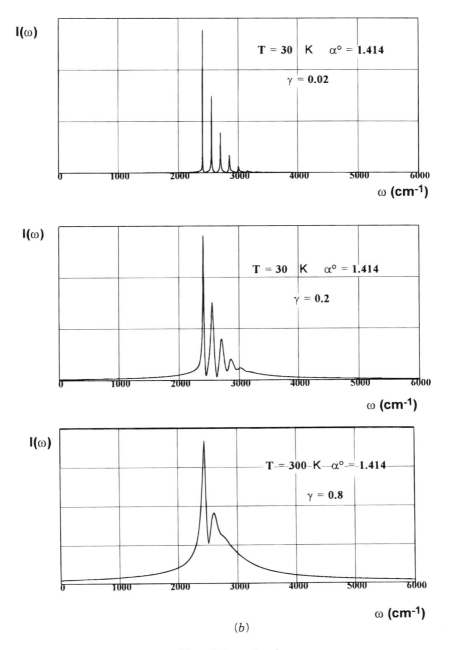

(b)

Figure 1.6—*continued*

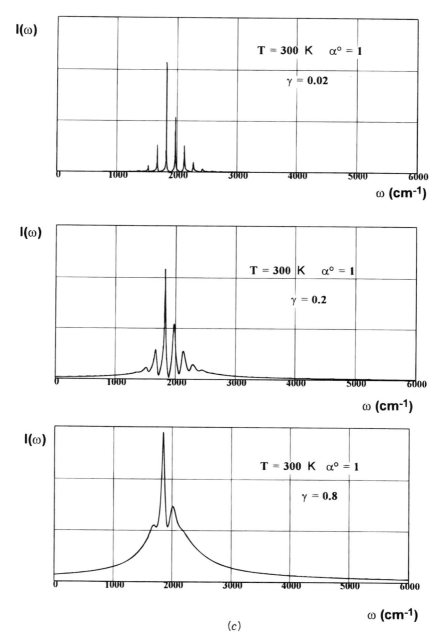

(c)

Figure 1.6—*continued*

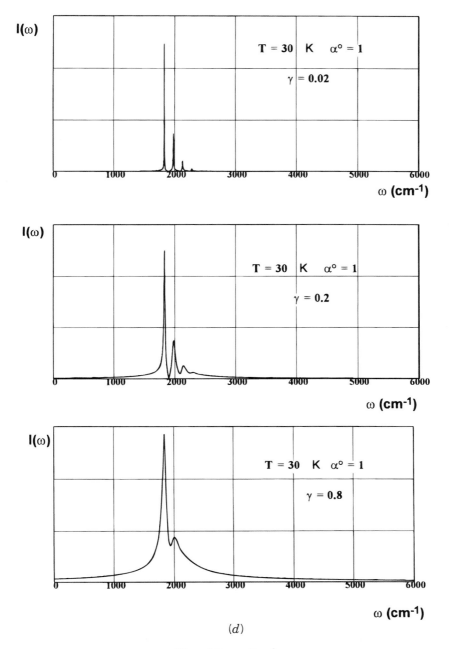

(d)

Figure 1.6—*continued*

Inspection of the figures makes clear the fine structure and the asymmetry of the spectrum, that is a specificity of the quantum approach. Must be quoted:

1. The shift with respect to $\omega°$ of the barycentre of the spectrum towards low frequencies.
2. The decrease of the angular frequency shift with deuterium substitution (compare Figs. 1.6a with 1.6c and 1.6b with 1.6d).
3. The decrease in the half-width of the spectrum with deuterium substitution (compare Figs. 1.6a with 1.6c and 1.6b with 1.6d).
4. The increase of the half-width with temperature (compare Figs. 1.6a with 1.6b and 1.6c with 1.6d).
5. The increase of the asymmetry of the global spectrum with a lowering of the temperature (compare Figs. 1.6a with 1.6c and 1.6b with 1.6d).
6. The increase of the asymmetry with a decrease of the anharmonic constant.
7. The relatively weak influence of the increase of the damping parameter on the width of the sub-bands.
8. The influence of the damping parameter on the relative intensities of the sub-bands.

Comparison with the classical approach of indirect damping (Figs. 1.4 and 1.6) shows that the quantum approach has the merit to introduce an asymmetry of the global band shape, the possibility of shoulders of inequal height and a shift of the barycentre towards slow frequencies. The comparison shows also that all the other physical parameters being the same in both the classical and quantum approaches, the damping parameter must be greater in the classical than in the quantum in order to produce the same coalescence of the sub-bands.

VIII. QUANTUM APPROACH WITHIN THE DIRECT DAMPING MECHANISM

In the preceding section we have seen the quantum approaches of the spectral density that were within the indirect damping mechanism. We now have to look at the theory of spectral density that is within the direct damping mechanism. This theory is in the very important work of Rösch and Ratner (1974).

A. Basic Equations in Representation {III}

Now consider more precisely the direct damping mechanism where the relaxation is due to the coupling between the fluctuating local electric field and the dipole moment of the complex and that may well contribute particularly when polar solvents are involved so that the local electrical field fluctuations are correspondingly large. Owing to vibrational anharmonicity the expectation value of the interaction will differ according as the fast mode is in its ground state or in its first excited state, and thus the energy difference between these states will depend on the local field. Fluctuations in this field will result in a broadening of the spectrum. The magnitude of the broadening produced by this mechanism must be proportional to the derivative of the molecular dipole moment with respect to the $X \cdots H$ separation, and since this derivative is known to increase substantially on hydrogen bond formation, the broadening produced by this direct relaxation mechanism should also increase. We shall now give the theory of the band broadening of the stretching fast mode as developed by Rösch and Ratner (1974) and simplify their approach by considering only the transition from the ground state to the first excited state of the fast mode.

1. The Hamiltonian of the Hydrogen Bond Interacting with the Charged Dipoles

Rösch and Ratner consider the direct relaxation of the dipole moment μ of the fast mode of the hydrogen bond, with the thermal bath formed by the time-dependent local electric field $E^0(t)$ produced by the oscillations of charged particles. They write the total Hamiltonian of the hydrogen bond coupled to the thermal bath in representation $\{III\}$:

$$\{H_{Rat}\} = \{H^0_{III}\} + \{H_{III\,damp}\} \tag{349}$$

Here, the first right-hand term is the Hamiltonian of the hydrogen bond with no damping, written in representation $\{III\}$, which may be written as an expansion in terms of the effective Hamiltonians $\{H^{0\{k\}}_{III}\}$:

$$\{H^0_{III}\} = \sum |\{k\}\rangle \{H^{0\{k\}}_{III}\} \langle\{k\}| \tag{350}$$

Recall that, according to Eq. (62), these effective Hamiltonians are:

$$\{H^{0\{k\}}_{III}\} = [(a^\dagger a) - k(k+1)\alpha^{\circ 2}]\hbar\omega^{\circ\circ} + k\hbar\omega^\circ \tag{351}$$

On the other hand, the last right-hand term is the Hamiltonian coupling the chemical bond to the electric field generated by all the other hydrogen bonds. More precisely, it is in representation $\{III\}$, the product of the dipole moment operator $\{\mu_{III}\}$ of the hydrogen bond in this representation by the local electric field $\{E^0(t)\}$.

$$\{H_{\text{III damp}}\} = \{\mu_{\text{III}}\}\{E^0(t)\} \tag{352}$$

Next, owing to the expression (87) of the dipole moment operator in this representation, the above Hamiltonian becomes:

$$\{H_{\text{III damp}}\} = \sum \sum \mu_{kl}\{E^0(t)\}\{A^{\{k\}}(k\alpha^\circ)\} \, | \, \{k\}\rangle\langle\{l\}| \, \{A^{\{l\}}(l\alpha^\circ)^\dagger\} \tag{353}$$

2. Assumption of the Local Electrical Field Generated by the Charged Dipoles of the Hydrogen Bond

$\{E^0(t)\}$ in the preceding equation is the local electrical field at the hydrogen bond, due to oscillations of charged particles of dipoles that make up the thermal bath. The time dependence of $\{E^0(t)\}$ is stochastic, and its oscillations occur on a broad range of time scale, ranging from rotational reorientation down to solvent vibrations. It is assumed by Rösch and Ratner that the thermal mean value of the electric field and its autocorrelation function obey an expression similar to that considered for the stochastic force appearing in the generalized Langevin equation in the framework of the Doob theorem [see Appendix H, Eqs. (H27) and (H28)].

$$\langle E^0(\tau)\rangle = 0 \tag{354}$$

$$\langle E^0(0)E^0(\tau)\rangle = \langle E^0(0)^2\rangle \, \exp\{-\kappa\tau\} \tag{355}$$

3. The Autocorrelation Function

The autocorrelation function of the hydrogen bond is, in representation $\{\text{III}\}$:

$$\{G_{\text{Rat}}(t)\} = \text{tr}'\{\rho_{\text{III}}^{\{1\}}(0)\{\mu_{\text{III}}^{\{0,\,1\}}(0)\}\langle\{\mu_{\text{III}}^{\{1,\,0\}}(t)\}\rangle\}$$

Here a thermal average on the stochastic properties of the electrical field must be performed, and the trace must be performed on the basis $\{|\{k\}\rangle\}$ and $\{|n\rangle\}$. The dipole moment operator at initial time and at time t are given, respectively, in view of Eq. (88) by:

$$\{\mu_{\text{III}}^{\{0,\,1\}}(0)\} = \mu_{01} \, | \, \{0\}\rangle\langle\{1\}| \, \{A^{\{1\}}(\alpha^\circ)\}^\dagger$$

$$\langle\{\mu_{\text{III}}^{\{1,\,0\}}(t)\}\rangle = \mu_{10}\langle\{U_{\text{Rat}}(t)\}^\dagger\rangle\{A^{\{1\}}(\alpha^\circ)\} \, | \, \{1\}\rangle\langle\{0\}| \, \langle\{U_{\text{Rat}}(t)\}\rangle$$

where:

$$i\hbar(\partial U_{\text{Rat}}(t)/\partial t) = \{H_{\text{Rat}}\}\{U_{\text{Rat}}(t)\}$$

Then the autocorrelation function becomes:

$$\{G_{\mathrm{Rat}}(t)\} = |\mu_{10}|^2 \ \mathrm{tr}'\{\rho_\theta | \{0\}\rangle\langle\{1\} | \{A^{\{1\}}(\alpha^\circ)\}^\dagger$$

$$\times \langle\{U_{\mathrm{Rat}}(t)\}^\dagger\rangle | \{1\}\rangle\langle\{0\} | \{A^{\{1\}}(\alpha^\circ)\}\langle\{U_{\mathrm{Rat}}(t)\}\rangle\} \quad (356)$$

In Appendix G it is shown that, according to Eq. (G5), this autocorrelation function appears to be:

$$\{G_{\mathrm{Rat}}(t)\} = |\mu_{10}|^2 \ \exp\{i(\omega^\circ - 2\alpha^{\circ 2}\omega^{\circ\circ})t\}$$

$$\times \varepsilon \ \mathrm{tr}\{\exp\{-\lambda a^\dagger a\}\{A^{\{1\}}(\alpha^\circ)\}^\dagger\{A^{\{1\}}(\Phi_{\mathrm{III}}^0(t)^*)\}$$

$$\times \langle\{1\} | \langle\{U_{\mathrm{Rat}}^{\mathrm{IP}}(t)\}^\dagger\rangle | \{1\}\rangle\langle\{0\} | \langle\{U_{\mathrm{Rat}}^{\mathrm{IP}}(t)\}\rangle | \{0\}\rangle\} \quad (357)$$

The IP time evolution operator appearing in this equation is given by the Schrödinger equation:

$$i\hbar(\partial U_{\mathrm{Rat}}^{\mathrm{IP}}(t)/\partial t) = \{H_{\mathrm{III\ damp}}^{\mathrm{IP}}\}\{U_{\mathrm{Rat}}^{\mathrm{IP}}(t)\}$$

where the perturbation Hamiltonian in the interaction picture is given by:

$$\{H_{\mathrm{III\ damp}}^{\mathrm{IP}}\} = \exp\{i\{H_{\mathrm{III}}^0\}t/\hbar\}\{H_{\mathrm{III\ damp}}\} \exp\{-i\{H_{\mathrm{III}}^0\}t/\hbar\}$$

The autocorrelation function (357) may also be written:

$$\{G_{\mathrm{Rat}}(t)\} \approx |\mu_{10}|^2 \ \exp\{i(\omega^\circ - 2\alpha^{\circ 2}\omega^{\circ\circ})t\}$$

$$\times \varepsilon \ \mathrm{tr}\{\exp\{-\lambda a^\dagger a\}\{A^{\{1\}}(\alpha^\circ)\}^\dagger$$

$$\times \{A^{\{1\}}(\Phi_{\mathrm{III}}^0(t)^*)\}\}\langle\{U^{\mathrm{IP}\{1,\ 1\}}(t)\}\rangle\langle\{U^{\mathrm{IP}\{0,\ 0\}}(t)\}\rangle \quad (358)$$

with:

$$\langle\{U^{\mathrm{IP}\{r,\ r\}}(t)\}\rangle = \langle\{r\} | \langle\{U_{\mathrm{Rat}}^{\mathrm{IP}}(t)\}\rangle | \{r\}\rangle \quad (359)$$

The thermal average involved in Eq. (359) has to take into account the statistical properties involving the fluctuating electrical field and given by Eqs. (354) and (355).

In view of Eq. (184), giving the expression of the autocorrelation function in the absence of damping, the autocorrelation function of the hydrogen bond, damped by its direct interaction with the dipole moments generated by the other hydrogen bonds, may be written:

$$\{G_{\mathrm{Rat}}(t)\} = \{G^0(t)\}\langle\{U^{\mathrm{IP}\{1,\ 1\}}(t)\}^\dagger\rangle\langle\{U^{\mathrm{IP}\{0,\ 0\}}(t)\}\rangle \quad (360)$$

B. Obtaining the Thermal Average Involved in Eq. (360)

1. The Time Evolution Operator in the Interaction Picture

To find a closed expression for the correlation function (360), it is required to know the time evolution operator in the interaction picture that is shown in Appendix G to be given by Eq. (G7):

$$\{U^{\mathrm{IP}}_{\mathrm{Rat}}(t)\} = \exp\left\{-i \sum \sum \mu_{kl} \int_0^t \{E^0(\tau)\} \exp\{i\omega_{kl}\tau\}\right.$$

$$\left. \times \{A^{\{k\}}(k\Phi^0_{\mathrm{III}}(\tau)^*)\} \,|\, \{k\}\rangle\langle\{l\}\,|\, \{A^{\{l\}}(l\Phi^0_{\mathrm{III}}(\tau)^*)\}^\dagger \, d\tau\right\} \quad (361)$$

$$\omega_{kl} = (k - l)\omega^\circ - [k(k+1) - l(l+1)]\alpha^{\circ 2}\omega^{\circ\circ} \quad (362)$$

Now consider the thermal average of the diagonal matrix elements of this IP evolution operator appearing in the expression (360) of the autocorrelation function. Owing to the above equations (359) and (361) they are:

$$\langle\{U^{\mathrm{IP}\{r,\,r\}}(t)\}\rangle = \left\langle \langle\{r\}\,|\, \exp\left\{-i \sum \sum \mu_{kl} \int_0^t d\tau \, \{E^0(\tau)\} \exp\{i\omega_{kl}\tau\}\right.\right.$$

$$\left.\left. \times \{A^{\{k\}}(k\Phi^0_{\mathrm{III}}(\tau)^*)\} \,|\, \{k\}\rangle\langle\{l\}\,|\, \{A^{\{l\}}(l\Phi^0_{\mathrm{III}}(\tau)^*)\}^\dagger\right\} |\,\{r\}\rangle \right\rangle \quad (363)$$

2. Cumulant Expansion of the Thermal Average Value of the Exponential Operator

Then, expanding the average of the exponential in cumulant cluster series, retaining only the first two terms and using the fact that, according to Eq. (354), the average value of the electric field is zero, one obtains:

$$\langle\{U^{\mathrm{IP}\{r,\,r\}}(t)\}\rangle = \langle\{r\}\,|\, \exp\left\{-\sum \sum \sum \sum \mu_{kl}\,\mu_{pq} \int_0^t \int_0^{\tau_1} d\tau_1 \, d\tau_2\right.$$

$$\times \exp\{i[\omega_{kl}\tau_1 + \omega_{pq}\tau_2]\}\langle E^0(\tau_1)E^0(\tau_2)\rangle$$

$$\times \varepsilon \, \mathrm{tr}\{\exp\{-\lambda a^\dagger a\}\{A^{\{k\}}(k\Phi^0_{\mathrm{III}}(\tau_1)^*)\} \,|\, \{k\}\rangle$$

$$\times \langle\{l\}\,|\, \{A^{\{l\}}(l\Phi^0_{\mathrm{III}}(\tau_1)^*)\}^\dagger\{A^{\{p\}}(p\Phi^0_{\mathrm{III}}(\tau_2)^*)\} \,|\, \{p\}\rangle$$

$$\left. \times \langle\{q\}\,|\, \{A^{\{q\}}(q\Phi^0_{\mathrm{III}}(\tau_2)^*)\}^\dagger\}\right\} |\,\{r\}\rangle \quad (364)$$

In Appendix G it is shown that performing the thermal average of the translation operators by aid of a trace on the Boltzmann operator involving the free Bosons, changing the time coordinates, and performing one integration, one finds, according to Eq. (G9):

$$
\langle\{U^{\text{IP}\{r,\ r\}}(t)\}\rangle = \exp\Bigg\{-\sum|\mu_{rl}|^2\int_0^t d\tau\,(t-\tau)\exp\{i\omega_{rl}(\tau_1-\tau_2)\}
$$
$$
\times\{\langle E^0(\tau)E^0(0)\rangle+\langle E^0(-\tau)E^0(0)\rangle\}
$$
$$
\times\exp\{[r-l]^2|\alpha^\circ|^2(1+2\langle n\rangle)[\cos(\omega^{\circ\circ}\tau)-1]\}
$$
$$
\times\exp\{i(r-l)^2|\alpha^\circ|^2\sin(\omega^{\circ\circ}\tau)\}\Bigg\}
\tag{365}
$$

where $\tau=\tau_1-\tau_2$, whereas $\langle n\rangle$ is the thermal average of the number occupation operator given by Eq. (H90).

3. The IP Time Evolution Operator within the Doob Approximation for the Electric Field Autocorrelation Function

Then using Eq. (355), one obtains:

$$
\langle\{U^{\text{IP}\{r,\ r\}}(t)\}\rangle = \exp\Bigg\{-\sum|\mu_{rl}|^2 2\langle E^0(0)^2\rangle
$$
$$
\times\int_0^t d\tau\,(t-\tau)\exp\{-\kappa\tau\}\exp\{-i\omega_{rl}\tau\}
$$
$$
\times\exp\{(r-l)^2|\alpha^\circ|^2(1+2\langle n\rangle)[\cos(\omega^{\circ\circ}\tau)-1]\}
$$
$$
\times\exp\{i(r-l)^2|\alpha^\circ|^2\sin(\omega^{\circ\circ}\tau)\}\Bigg\}
\tag{366}
$$

Again, extending the upper limit of integration to infinity, as it is usually performed, one may write:

$$
\langle\{U^{\text{IP}\{r,\ r\}}(t)\}\rangle \approx \exp\{-\{\gamma^{\{r\}}+i\,\Delta^{\{r\}}\}t\}
\tag{367}
$$

with:

$$
\{\gamma^{\{r\}}+i\,\Delta^{\{r\}}\} = \sum|\mu_{rl}|^2 2\langle E^0(0)^2\rangle\int_0^\infty dt\,\exp\{-[i\omega_{rl}+\kappa]t\}
$$
$$
\times\exp\{(r-l)^2|\alpha^\circ|^2(1+2\langle n\rangle)[\cos(\omega^{\circ\circ}t)-1]\}
$$
$$
\times\exp\{i(r-l)^2|\alpha^\circ|^2\sin\omega^{\circ\circ}t\}
\tag{368}
$$

C.   The Final Autocorrelation Function and the Spectral Density

1.   *The Final Autocorrelation Function*

Owing to the above equations, the autocorrelation function (360) may be written:

$$\{G_{\text{Rat}}(t)\} = \{G^0(t)\} \exp\{-\gamma^{\{1,\,0\}}t\} \exp\{-i\,\Delta^{\{1,\,0\}}t\} \tag{369}$$

with, respectively, $\gamma^{\{1,\,0\}} = \gamma^{\{1\}} + \gamma^{\{0\}}$ and $\Delta^{\{1,\,0\}} = \Delta^{\{1\}} - \Delta^{\{0\}}$.

In view of Eq. (188), the autocorrelation function may be written after neglecting the angular frequency shift:

$$\{G_{\text{Rat}}(t)\} = |\mu_{10}|^2 \exp\{i\alpha^{\circ 2} \sin(\omega^{\circ\circ}t)\} \exp\{i(\omega^\circ - 2\alpha^{\circ 2}\omega^{\circ\circ})t\}$$
$$\times \exp\{\alpha^{\circ 2}(1 + 2\langle n\rangle)[\cos(\omega^{\circ\circ}t) - 1]\} \exp\{-\gamma^{\{1,\,0\}}t\} \tag{370}$$

2.   *Expansion of the Autocorrelation Function and the Spectral Density*

On the other hand, in view of Eq. (193), which gives the expansion of $\{G^0(t)\}$, the above autocorrelation function yields:

$$\{G_{\text{Rat}}(t)\} = \varepsilon|\mu_{10}|^2 \sum\sum \exp\{-\lambda m\}\,|\Gamma_{mn}(\alpha^\circ)|^2 \exp\{-i(m - n)\omega^{\circ\circ}t\}$$
$$\times \exp\{i(\omega^\circ - 2\alpha^{\circ 2}\omega^{\circ\circ})t\} \exp\{-\gamma^{\{1,\,0\}}t\} \tag{371}$$

Now consider the Fourier transform of this last equation:

$$\{I_{\text{Rat}}(\omega)\} = [1/\sqrt{2\pi}]\int \{G_{\text{Rat}}(t)\} \exp\{-i\omega t\}\,dt \tag{372}$$

Performing the integration, one obtains:

$$\{I_{\text{Rat}}(\omega)\} = \sum\sum \{I_{mn}(\omega)\} \tag{373}$$
$$\{I_{mn}(\omega)\} = \varepsilon|\mu_{10}|^2[2/\sqrt{2\pi}]\exp\{-\lambda m\}\,|\Gamma_{mn}(\alpha^\circ)|^2$$
$$\times \{\gamma^{\{1,\,0\}}/[(\omega - \omega_{mn})^2 + \{\gamma^{\{1,\,0\}}\}^2]\} \tag{374}$$

3.   *The Limit Situation Where the Direct Damping Is Missing*

When the direct damping is missing (i.e., when the coupling of the dipole moment with the electric field, $\gamma^{\{1,\,0\}}$ and $\Delta^{\{1,\,0\}}$, become zero), the autocorrelation function (370) reduces, as required, to that (188) in the absence of damping:

$$\{G_{\text{Rat}}(t)\} \to \{G^0(t)\}$$

and the spectral density (372) reduces, as required, so that (194) in the absence of damping:

$$\{I_{Rat}(\omega)\} = \{I^0(\omega)\}$$

4. Numerical Simulation of Spectral Densities within the Quantum Direct Relaxation

We may now give some numerical simulations of spectral densities obtained with the aid of Eq. (370) for an angular frequency ω° of the fast mode equal to 3000 cm^{-1}, an angular frequency $\omega^{\circ\circ}$ of the slow mode equal to 150 cm^{-1}. Four series of calculations have been performed by changing progressively in each the value of the direct damping parameter γ^0. These series correspond respectively to $T = 30$ K and $T = 300$ K for $\alpha^\circ = \sqrt{2}$, (H-bonded species), and $T = 30$ K and $T = 300$ K for $\alpha^\circ = 1$ (D-bonded isotope for which, of course ω° becomes equal to 2120 cm^{-1} whereas $\omega^{\circ\circ}$ remains equal to 150 cm^{-1}). The results are given in Figs. 1.7a, 1.7b, 1.7c and 1.7d, without giving the intensity unities.

Inspection of these figures leads to the following remarks:

1. There is a shift in the angular frequency of the center of gravity of the spectrum towards low frequencies;
2. There is a decrease of the angular frequency shift with deuterium substitution;
3. There is a decrease in the half-width of the spectrum with deuterium substitution;
4. There is an increase of the half-width of the spectrum with temperature;
5. There is an increase of the asymmetry of the global spectrum with a lowering of the temperature;
6. There is an increase of the asymmetry with a decrease of the anharmonic constant;
7. There is a relatively weak influence of the damping parameter on the width of the sub-bands.

As it appears, the behavior is near what we have found for the quantum approach of the indirect damping. A deep examination would show that for the same physical parameters, and when there is coalescence, the asymmetry of the band shape is weakly increased when we pass from direct damping to indirect. Besides when the damping is moderately important, the shape of the spectral density narrows those obtained in the quantum and semiclassical approaches of the indirect relaxation within the fast-modulation limit but with an asymmetry that is intermediate between those

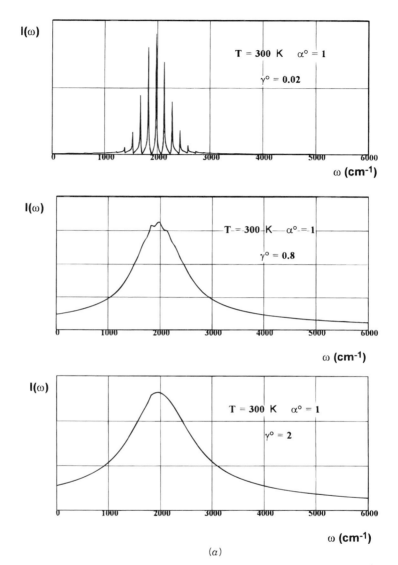

(a)

Figure 1.7. Examples of spectral densities obtained with the use of Eq. (370) for an angular frequency of the slow mode equal to 150 cm^{-1}. Two series of calculations have been considered. A first one in which the nondimensional anharmonic constant α° takes some suitable value, i.e. $\sqrt{2}$ and the angular frequency of the fast mode $\omega^\circ = 3000$ cm^{-1} and another one corresponding to isotopic substitution of the X–H bond by deuterium, leading for α° to the value 1 and for ω° to the value 2120 cm^{-1}. For each series, two values of temperature have been chosen, one corresponding to low temperatures (30 K) and the other one around room temperature (300 K). For each temperature, three different values of the damping parameter γ have been selected, all being compatible with the underdamped situation. The results are given in (a), (b), (c), and (d) without defining the intensity units.

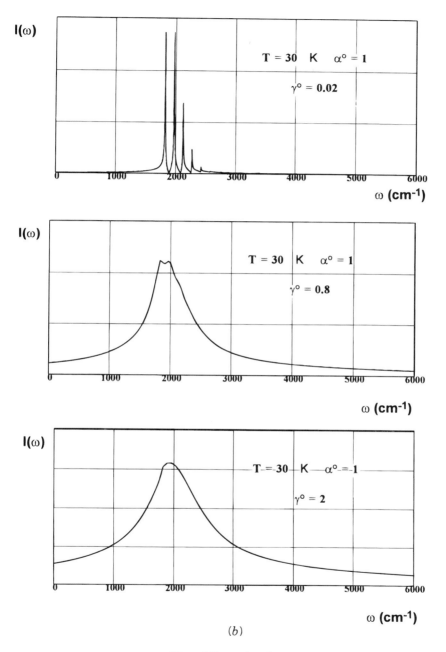

(b)

Figure 1.7—*continued*

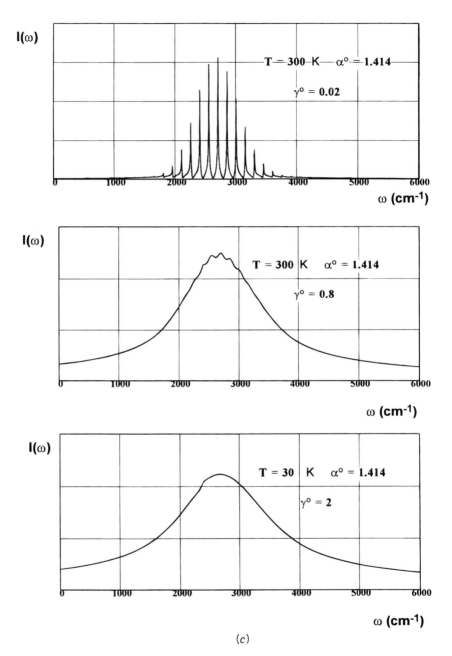

(c)

Figure 1.7—continued

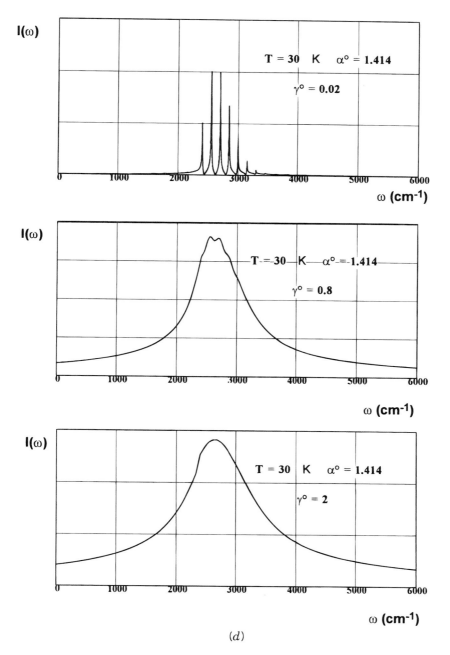

(d)

Figure 1.7—*continued*

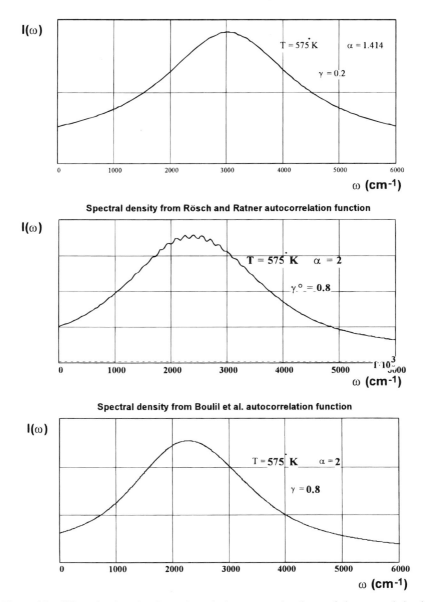

Figure 1.8. When the damping is moderately important, the shape of the spectral density narrows those obtained in the quantum and semiclassical approaches of the indirect relaxation within the fast modulation limit but with an asymmetry that is intermediate between those of the semiclassical and of the quantum approaches of the indirect relaxation. That is illustrated for a high temperature of 575 K, and for other physical parameters that are the same for both direct and indirect quantum theories and slightly different for the semiclassical theory, the anharmonic constant being taken as important in all the situations and the intensities unspecified.

of the semiclassical and of the quantum approaches of the indirect relaxation. That is illustrated in Fig. 1.8 for a high temperature of 575 K, and for other physical parameters that are the same for both direct and indirect quantum theories and slowly different for the semiclassical theory, the anharmonic constant being taken as important in all the situations and the intensities unspecified.

IX. QUANTUM APPROACH OF HYDROGEN-BONDED SOLIDS

After considering the quantum theory of the band shape of hydrogen-bonded species in the liquid phase, it may be of interest to look at the quantum theory of the stretching vibrational mode profile of hydrogen-bonded solids as proposed by Ratajczak and Yaremko (1995).

A. Recall on the Molecular Crystal Excitation Theory of Davydov

The Ratajczak and Yaremko approach is a transposition of the Davydov theory (1962) dealing with molecular exciton. Thus let us glance at this theory.

Consider a crystal in which to simplify we assume that there is only a single molecule in the unit cell. The total Hamiltonian describing the excitation of a crystal can be written:

$$\{H_{\mathrm{II}}^{\mathrm{tot}}\} = \{H_{\mathrm{II}}^{\mathrm{ex}}(R_n^f)\} + \{H_{\mathrm{II}}^{\mathrm{ph}}\} \tag{375}$$

with, respectively:

$$\{H_{\mathrm{II}}^{\mathrm{ex}}(R_n^f)\} = \sum \{E_n^f\}\{B_n^{f\dagger}B_n^f\} + \sum\sum \{D_{nm}^{fg}(R_n^f)\}\{B_n^{f\dagger}B_n^g\}$$
$$+ \sum\sum \{M_{nm}^{fg}(R_n^f)\}\{B_n^{f\dagger}B_m^g\} \tag{376}$$

$$\{H_{\mathrm{II}}^{\mathrm{ph}}\} = \sum (a_{sq}^\dagger a_{sq})\hbar\omega_{sq}^{\circ\circ} \tag{377}$$

The first term of the total Hamiltonian describes the change in the crystal energy due to the high-frequency intramolecular vibrational excitation; the second one describes the lattice phonon. In these equations, n and m are the number of unit cells in the crystal, that is, since there are single molecules in a unit cell, the number of molecules in the crystal. $\{E_n^f\}$ is the excitation energy of the nth molecule into the f state; f, g characterize the quasi-degenerate states of the molecules. $\{D_n^{fg}(R_n^f)\}$ determines the difference in the interaction energy between the excited and ground state of the nth molecule with all remaining molecules of the crystal. $\{M_n^{fg}(R_n^f)\}$ determines the exchange of excitation energy between the nth and mth molecules. $\{B_n^{f\dagger}\}$ and $\{B_n^f\}$ are the Fermion operators for the creation and annihilation of the excitation on molecule n. R is a short notation for the coordinates involving displacement and rotation of the nth molecule in the crystal. $a_{sq}^\dagger$ and a_{sq} are the Bose operators for the creation and annihilation of the lattice vibrations $\omega_{sq}^{\circ\circ}$, where s and q refer, respectively, to the phonon branches and wave vectors.

The realization condition of the strong coupling means that:

$$\{D_n^{fg}(R_n^f)\} > \{M_n^{fg}(R_n^f)\} \tag{378}$$

Therefore, neglecting the third term in Eq. (376), expanding the second one, $\{D_n^{fg}(R_n^f)\}$ as a power series of the coordinates R_n^f of the nth molecule, and passing for the linear terms in R_n^f to the corresponding boson operators a_{sq}, the full Hamiltonian becomes, following Davydov:

$$\begin{aligned}
\{H_{II}^{tot}\} &= \{E_n^f(0)\}\{B_n^\dagger B_n\} + \{B_n^\dagger B_n\} \\
&\quad \times [\sum \{a_{sq} + a_{s,-q}^\dagger\}\alpha_{sq}^\circ \hbar\omega_{sq}^{\circ\circ} + \sum a_{sq}^\dagger a_{sq} \hbar\omega_{sq}^{\circ\circ}] \\
&\quad + [1 - \{B_n^\dagger B_n\}] \sum a_{sq}^\dagger a_{sq} \hbar\omega_{sq}^{\circ\circ}
\end{aligned} \tag{379}$$

where:

$$\{E_n^f(0)\} = \{E_n^f\} + \{D_n^f(0)\} \tag{380}$$

α_{sq}° are dimensionless anharmonic couplings.

B.   Connection of the Davydov Theory with the Witkowski Model

If one applies the Davydov model to the situation of solids containing hydrogen bonds, $\{E_n^f(0)\}$ becomes the excitation energy of the high-frequency $v(X-H)$ mode, whereas α_{sq}° are the dimensionless anharmonic coupling constant between this fast mode and the low-frequency phonons. As a consequence, when there is only one single slow oscillator in the unit cell, the Hamiltonian reduces to:

$$\{H_{II}^{tot}\} = \{B_n^\dagger B_n\}[\{E_n^f(0)\} + \alpha^\circ(a^\dagger + a)\hbar\omega^{\circ\circ} + a^\dagger a\hbar\omega^{\circ\circ}]a^\dagger a\hbar\omega^{\circ\circ} \tag{381}$$

Again, the excitation energy of the fast mode is:

$$\{E_n^f(0)\} = \hbar\omega^\circ - \alpha^{\circ 2}\hbar\omega^{\circ\circ} \tag{382}$$

Besides, in the spirit of the analogy, the $B_n^\dagger$ and B_n fermion operators for the creation and annihilation of the excitation of the fast mode must be put in correspondence with the eigenkets of the harmonic oscillator of the fast mode involving bosons, and appearing in the Witkowski model.

$$B_n = |\{0\}\rangle\langle\{1\}| \quad\text{and}\quad B_n^\dagger = |\{1\}\rangle\langle\{0\}|$$

so that:

$$B_n^\dagger B_n = |\{1\}\rangle\langle\{0\}|\{0\}\rangle\langle\{1\}| = |\{1\}\rangle\langle\{1\}|$$

Moreover, owing to the fact that at the present level one considers the high-frequency mode as a two-level system, we have of course:

$$1 = |\{1\}\rangle\langle\{1\}| + |\{0\}\rangle\langle\{0\}| \tag{383}$$

The total Hamiltonian becomes, therefore:

$$\{H_{II}^{tot}\} = |\{1\}\rangle[\{E_n^f(0)\} + \alpha^\circ(a^\dagger + a)\hbar\omega^{\circ\circ} + a^\dagger a\hbar\omega^{\circ\circ}]\langle\{1\}| \\ + |\{0\}\rangle a^\dagger a\hbar\omega^{\circ\circ}\langle\{0\}| \tag{384}$$

Moreover, it may also be written:

$$\{H_{II}^{tot}\} = |\{0\}\rangle\{H_{II}^{0\{0\}}\}\langle\{0\}| + |\{1\}\rangle\{H_{II}^{0\{1\}}\}\langle\{1\}| \tag{385}$$

with, respectively, according to the fact that the fast mode is in its ground state or in its first excited state:

$$\{H_{II}^{0\{0\}}\} = a^\dagger a\hbar\omega^{\circ\circ}$$

$$\{H_{II}^{0\{1\}}\} = (a^\dagger a)\hbar\omega^{\circ\circ} + \{\alpha^\circ[a^\dagger + a] - \alpha^{\circ 2}\}\hbar\omega^{\circ\circ} + \hbar\omega^\circ$$

As it appears this Hamiltonian is nothing but the effective Hamiltonians (46) and (47), appearing in the Witkowski model within representation II.

C. From Representation {II} to Representation {III}

Now return to the general situation where there are many phonons in the unit cells. One may write the full Hamiltonian:

$$\{H_{II}^{tot}\} = |\{0\}\rangle\{H_{II}^{sol\{0\}}\}\langle\{0\}| + |\{1\}\rangle\{H_{II}^{sol\{1\}}\}\langle\{1\}| \tag{386}$$

with, respectively:

$$\{H_{II}^{sol\{0\}}\} = \sum \{a_{sq}^\dagger a_{sq}\}\hbar\omega_{sq}^{\circ\circ} \tag{387}$$

$$\{H_{II}^{sol\{1\}}\} = \{E_n^f(0)\} + \sum \{a_{sq}^\dagger a_{sq}\}\hbar\omega_{sq}^{\circ\circ} + \sum \{a_{sq} + a_{s,-q}^\dagger\}\alpha_{sq}^\circ \hbar\omega_{sq}^{\circ\circ} \tag{388}$$

Now consider the following unitary operator.

$$S = |\{1\}\rangle \prod_{sq} \{A^{\{1\}}(\alpha_{sq}^{\circ})\}\langle\{1\}| \tag{389}$$

where:

$$\{A^{\{1\}}(\alpha_{sq}^{\circ})\} = \exp\{(\alpha_{sq}^{\circ} a_{sq}^{\dagger} - \alpha_{sq}^{\circ*} a_{sq})\} = \{A_n^{\{1\}}(-\alpha_{sq}^{\circ})\}^{\dagger} \tag{390}$$

Now perform, following Davydov, the phase transformation on the total Hamiltonian (386) keeping in mind Eq. (A44):

$$\{H_{III}^{tot}\} = S\{H_{II}^{tot}\}S^{\dagger} \tag{391}$$

The result lying in representation III is a decoupling of the new Hamiltonian into two independent parts:

$$\{H_{III}^{tot}\} = |\{0\}\rangle\{H_{III}^{sol\{0\}}\}\langle\{0\}| + |\{1\}\rangle\{H_{III}^{sol\{1\}}\}\langle\{1\}| \tag{392}$$

with, respectively:

$$\{H_{III}^{sol\{1\}}\} = \{E_n\} + \sum (a_{sq}^{\dagger} a_{sq})\hbar\omega_{sq}^{\circ\circ} \tag{393}$$

$$\{H_{III}^{sol\{0\}}\} = \sum (a_{sq}^{\dagger} a_{sq})\hbar\omega_{sq}^{\circ\circ} \tag{394}$$

and where:

$$\{E_n\} = \{E_n^f(0)\} - \sum \alpha_{sq}^{\circ 2} \hbar\omega_{sq}^{\circ\circ} \tag{395}$$

D.   The Autocorrelation Function in Representation {III}

In this section we shall look at the spectral density of the fast mode interacting with the phonons within the unit cell. One possibility would be to consider the Green function of the system as obtained by Davydov that is directly related to the spectral density. That was the approach of Ratajczak and Yaremko (1995). Another possibility is to look at the autocorrelation function of the dipole moment operator of the high-frequency mode, which is in the spirit of our general approach of the spectral density in terms of Fourier transform of the autocorrelation function.

Thus let us look at the spectral density of the fast mode coupled with the phonons. It may be written:

$$\{G_{solid}(t)\} = tr'\{\{\rho_{solid}\}\{\mu_{III}^{\{0, 1\}}(0)\}\{\mu_{III}^{\{1, 0\}}(t)\}\} \tag{396}$$

where the trace must be performed on the basis of the slow and fast modes. The first term on the right-hand side is the Boltzmann density operator of the phonons given by:

$$\{\rho_{\text{solid}}\} = \prod \{\rho_{\text{III}\,sq}^{\{0\}}(0)\} \tag{397}$$

$$\{\rho_{\text{III}\,sq}^{\{0\}}(0)\} = \varepsilon_{sq} \exp\{-\lambda_{sq}\, a^{\dagger}a\} \tag{398}$$

In this equation, we have:

$$\lambda_{sq} = \hbar\omega_{sq}^{\circ\circ}/(k_{\text{B}}\,T) \qquad \text{and} \qquad \varepsilon_{sq} = [1 - \exp\{-\lambda_{sq}\}]$$

The dipole moment operators in representation $\{\text{III}\}$ appearing in the autocorrelation function (396) may be obtained by:

$$\{\mu_{\text{III}}^{\{0,\,1\}}(0)\} = \mu_{01}\,|\{0\}\rangle\langle\{1\}|\,S^{\dagger}$$

$$\{\mu_{\text{III}}^{\{1,\,0\}}(t)\} = \mu_{10}\,\exp\{i\{H_{\text{III}}^{\text{tot}}\}t/\hbar\}S\,|\{1\}\rangle\langle\{0\}|\,\exp\{-i\{H_{\text{III}}^{\text{tot}}\}t/\hbar\}$$

The autocorrelation function becomes therefore after simplification:

$$\{G_{\text{solid}}(t)\} = |\mu_{10}|^2\,\text{tr}'\{\prod\,\{\rho_{\text{III}\,sq}^{\{0\}}(0)\}\,|\{0\}\rangle\langle\{1\}|\,\{A^{\{1\}}(\alpha_{sq}^{\circ})\}^{\dagger}$$
$$\times\,\exp\{i\{H_{\text{III}}^{\text{tot}}\}t/\hbar\}\{A^{\{1\}}(\alpha_{sq}^{\circ})\}\,|\{1\}\rangle\langle\{0\}|\,\exp\{-i\{H_{\text{III}}^{\text{tot}}\}t/\hbar\}\}$$

$$(399)$$

Here the trace must be performed on both the bases $\{|\{k\}\rangle\}$ and $\{|m\rangle\}$. Again, owing to Eq. (392), one obtains:

$$\{G_{\text{solid}}(t)\} = |\mu_{10}|^2\,\text{tr}'\{\prod\,\{\rho_{\text{III}\,sq}^{\{0\}}(0)\}\,|\{0\}\rangle\langle\{1\}|\,\{A^{\{1\}}(\alpha_{sq}^{\circ})\}^{\dagger}$$
$$\times\,\exp\{i\{H_{\text{III}}^{\text{sol}\{1\}}\}t/\hbar\}\{A^{\{1\}}(\alpha_{sq}^{\circ})\}\,|\{1\}\rangle\langle\{0\}|$$
$$\times\,\exp\{-i\{H_{\text{III}}^{\text{sol}\{0\}}\}t/\hbar\}\} \tag{400}$$

Owing to the fact that the translation operators do not act within the space of high-frequency modes and according to Eqs. (393) and (394), this reduces, after performing the trace on $|\{0\}\rangle$ and $|\{1\}\rangle$, to:

$$\{G_{\text{solid}}(t)\} = |\mu_{10}|^2\,\text{tr}\{\prod\,\{\rho_{\text{III}\,sq}^{\{0\}}(0)\}\{A^{\{1\}}(\alpha_{sq}^{\circ})\}^{\dagger}$$
$$\times\,\exp\{[iE_n\,t/\hbar] + i\sum a_{sq}^{\dagger}a_{sq}\,\omega_{sq}^{\circ\circ}\,t\}\{A^{\{1\}}(\alpha_{sq}^{\circ})\}$$
$$\times\,\exp\{i\sum a_{sq}^{\dagger}a_{sq}\,\omega_{sq}^{\circ\circ}\,t\}\} \tag{401}$$

At last, using Eq. (398), giving the density operator, after rearrangements inside the exponential arguments and inside the space characterizing each phonon, one may write the autocorrelation function as a product of independent autocorrelation functions characterizing each phonon:

$$\{G_{\text{Solid}}(t)\} = |\mu_{10}|^2 \exp\{i\omega^\circ t\} \prod \{G^{\circ\circ}(\alpha_{sq}^\circ, t)\} \tag{402}$$

with, in view of Eq. (395):

$$\{G^{\circ\circ}(\alpha_{sq}^\circ, t) = \varepsilon_{sq} \, \text{tr}\{\exp\{-\lambda_{sq} a_{sq}^\dagger a_{sq}\}\{A^{\{1\}}(\alpha_{sq}^\circ)\}^\dagger \exp\{i a_{sq}^\dagger a_{sq} \omega_{sq}^{\circ\circ} t\}$$
$$\times \{A^{\{1\}}(\alpha_{sq}^\circ)\} \exp\{-i a_{sq}^\dagger a_{sq} \omega_{sq}^{\circ\circ} t\} \exp\{-i \alpha_{sq}^{\circ 2} \omega_{sq}^{\circ\circ} t\} \tag{403}$$

and $\omega^\circ = E_n(0)/\hbar$. Besides, according to the theorem (A21) dealing with bosons, one obtains:

$$\{G^{\circ\circ}(\alpha_{sq}^\circ, t)\} = \varepsilon_{sq} \, \text{tr}\{\exp\{-\lambda_{sq} a_{sq}^\dagger a_{sq}\}\{A^{\{1\}}(\alpha_{sq}^\circ(t))\}^\dagger$$
$$\times \{A^{\{1\}}(\Phi_{\text{III} \, sq}^0(t)^*)\}\} \exp\{-i \alpha_{sq}^{\circ 2} \omega_{sq}^{\circ\circ} t\} \tag{404}$$

with, respectively:

$$\{A^{\{1\}}(\Phi_{\text{III} \, sq}^0(t)) = \exp\{\{\Phi_{\text{III} \, sq}^0(t)\} a_{sq}^\dagger - \{\Phi_{\text{III} \, sq}^{0*}(t)\} a_{sq}\} \tag{405}$$

$$\{\Phi_{\text{III} \, sq}^0(t)\} = \alpha_{sq}^\circ \exp\{-i\omega_{sq}^{\circ\circ} t\} \tag{406}$$

At last, owing to theorem (A45) dealing with product of translation operators, one may write the autocorrelation function:

$$\{G^{\circ\circ}(\alpha_{sq}^\circ, t)\} = \varepsilon_{sq} \, \text{tr}\{\exp\{-\lambda_{sq} a_{sq}^\dagger a_{sq}\}\{A^{\{1\}}(\Phi_{\text{II} \, sq}^0(t)^*)\}\}$$
$$\times \exp\{i \alpha_{sq}^{\circ 2} \sin(\omega_{sq}^{\circ\circ} t)\} \exp\{-i |\alpha_{sq}^\circ|^2 \omega_{sq}^{\circ\circ} t\} \tag{407}$$

with:

$$\{\Phi_{\text{II} \, sq}^0(t)\} = \alpha_{sq}^\circ[\exp\{-i\omega_{sq}^{\circ\circ} t\} - 1] \tag{408}$$

Performing the thermal average within each phonon subspace, one obtains:

$$\{G^{\circ\circ}(\alpha_{sq}^\circ, t)\} = \exp\{-i \alpha_{sq}^{\circ 2} \omega_{sq}^{\circ\circ} t\} \exp\{i \alpha_{sq}^{\circ 2} \sin(\omega_{sq}^{\circ\circ} t)\}$$
$$\times \exp\{\alpha_{sq}^{\circ 2}(1 + 2\langle n_{sq} \rangle)[\cos(\omega_{sq}^{\circ\circ} t) - 1]\} \tag{409}$$

In this last equation appears the thermal average of the number occupation of each phonon:

$$\langle n_{sq} \rangle = [\exp\{\lambda_{sq}\} - 1]^{-1} \tag{410}$$

Of course, the spectral density is the Fourier transform of the autocorrelation function:

$$\{I_{\text{solid}}(\omega)\} = [1/\sqrt{2\pi}] \int_{-\infty}^{\infty} \{G_{\text{solid}}(t)\} \exp\{-i\omega t\} \, dt \tag{411}$$

$$\{I_{\text{solid}}(\omega)\} = [1/\sqrt{2\pi}] |\mu_{10}|^2 \int_{0}^{\infty} \exp\{i\omega^{\circ}t\} \prod \{G^{\circ\circ}(\alpha_{sq}^{\circ}, t)\} \exp\{-i\omega t) \, dt$$

$$\tag{412}$$

Note that when the fast mode is only coupled with a single slow mode, there is only one dimensionless anharmonic coupling parameter, which is the same as that α°, that is, $\alpha_{sq}^{\circ} = \alpha^{\circ}$. In a similar way, there is only one single slow-mode angular frequency given by $\omega_{sq}^{\circ\circ} = \omega^{\circ\circ}$. Besides, in the same situation, there is only one thermal average, which is that of the mean number occupation, $\langle n_{sq} \rangle = \langle n \rangle$.

Therefore, the autocorrelation function (409) reduces to:

$$\{G^{\circ\circ}(t)\} = |\mu_{10}|^2 \exp\{i\alpha^{\circ 2} \sin(\omega^{\circ\circ}t)\}$$
$$\times \exp\{\alpha^{\circ 2}(1 + 2\langle n \rangle)[\cos(\omega^{\circ\circ}t) - 1]\} \tag{413}$$

As it appears, this equation is that obtained for the quantum autocorrelation function of the hydrogen bond, in the absence of damping and given by Eq. (190). (See also Eq. 189).

X.   THE RECENT WITKOWSKI THEORY

In this section we shall make some comments on the pioneering work of Witkowski (1983, 1990) concerning the coupling between the slow movements of nuclei and the fast movements of electrons that may be applied to the question of hydrogen bonds and may illuminate what is at the basis of the starting equation of the strong anharmonic coupling that Witkowski has proposed 25 years ago, following the Stepanov (1945, 1946) and Sheppard (1959) ideas.

Here we start from the recent Löwdin statement (1993) concerning important unsolved problems in theoretical physical chemistry: "a fundamental problem is to treat the nuclear motion in the same level as the electronic motion, i.e., to go beyond the Born–Oppenheimer approximation—and to treat this problem one needs undoubtedly some

new fundamental ideas. Of particular importance is the lightest nucleus—
the proton—and its motion in the hydrogen bond."

We believe that the best response to this Löwdin question is the recent
theory of Witkowski. This theory concerns the separation of electronic and
molecular motion and accounts for the finite velocity of electrons and their
retardation in following the nuclear motions. Because of the importance of
this recent theory, we give here a short physical introduction to it.

A. The Basis Physical Ideas

As we have seen above, the strong coupling theory is the general frame of
the great majority of the theoretical approaches of the spectral density, and
we remember that at the basis of these theories there is the idea of a strong
anharmonic coupling between the high-frequency stretching mode and the
hydrogen bond bridge. Witkowski (1983, 1990) has reconsidered the question of the origin of the anharmonicity of the high-frequency stretching
mode from a very original dynamical viewpoint. His starting idea is that the
usual separation of the electronic and molecular motion is not so evident in
hydrogen bonds as in other molecules for the following reasons: While the
proton of the hydrogen bond must be moving faster than other nuclei since
it is the lightest nucleus, on the other hand, the electronic pairs of the
donor–acceptor of the hydrogen bond must be moving more slowly than
bonded electrons because of their nonbonding properties. As a consequence, the hydrogen bond appears to be the system in which is less easy
the usual separation between the electrons and nuclei motions that is at the
basis of the Born–Oppenheimer approximation. Recall that, in the Born–
Oppenheimer approximation, the electronic properties are determined for
fixed positions of the nuclei, and the nuclear dynamics determined by the
average time-independent distribution of electrons adjusting with no delay
to the nuclear position.

Starting from this Born–Oppenheimer physical picture, Witkowski was
going beyond the approximation of an infinitely quick electron, by including the change in the nuclear coordinate as a function of the ratio of the
nuclear and electronic velocities. That leads to a delay in adjustment of the
electrons to the nuclear position. The most important correction occurs
when the ratio is not too small, that is, for molecules involving simultaneously high harmonic frequencies and slow electronic velocities. That is
indeed the case of the hydrogen bond.

As a consequence of the Witkowski accounts for the finite velocity of the
electron, and its retardation following nuclear motion, the usual Hamiltonian of the simple quantum harmonic oscillator appears to be corrected by
a quadratic time-dependent term, the strength of which depends on the
velocity ratio. As a consequence, this time-dependent Hamiltonian govern-

ing the vibration of the fast mode, may be viewed, in a certain sense, as anharmonic.

More quantitatively, because of the time dependence of the Hamiltonian oscillator, some corrections will appear to the usual time-dependent expression of the dipole moment transition operator of a simple harmonic oscillator, involving the Airy function. Then, by series Fourier transform of the perturbed time-dependent dipole moment operator, one obtains a spectral density involving long tail on the high-frequency side of the spectrum and, thus, an asymmetric broadening.

Connected to this approach must be evoked a work of Y. Maréchal (1985) dealing with the infrared paradox and related to the above ideas of Witkowski (1983).

B.   Mathematical Considerations

1.   Starting Equations

In the spirit of the Born–Oppenheimer approximation, the electronic properties are determined for fixed positions of the nuclei and the nuclear dynamics determined by the average time-dependent distribution of electrons adjusting with no delay to the nuclear position. In the standard Born–Oppenheimer approximation, the potential energy of the harmonic oscillator is proportional to the square of the nuclear coordinate Q, which results from immediate electronic adjustment to any nuclear position. The Born–Oppenheimer Hamiltonian of the oscillator in the harmonic approximation is:

$$\{H^0_{\text{Wit}}\} = \wp^2/2m + \tfrac{1}{2}m\omega^{\circ 2}q^2 \tag{414}$$

with the usual commutation rule. Next passing to the dimensionless operators p and q with the aid of the transformations:

$$q = q[h/(m\omega^\circ)]^{1/2}, \qquad \wp = p[mh\omega^\circ]^{1/2}$$

with $[p, q] = -i$, we get the Born–Oppenheimer harmonic Hamiltonian

$$\{H^0_{\text{Wit}}\} = \tfrac{1}{2}(p^2 + q^2)\hbar\omega^\circ \tag{415}$$

Starting from the Born–Oppenheimer physical picture, the most straightforward way to go beyond the approximation of the infinitely quick electron is to include the change of the nuclear coordinate q as a function of the ratio of the nuclear electronic velocity. The real delay in adjustment compared with the dimensionless q coordinate appearing in the Born–Oppenheimer Hamiltonian is given, in the Witkowski approach, by the following Galilei transformation in the nuclear coordinate:

$$q(t) = q + \kappa pt \tag{416}$$

κ is a parameter characterizing the delay in adjustment between the electronic and nuclear motions given by:

$$\kappa = \omega^\circ / \langle v \rangle \tag{417}$$

$\langle v \rangle$ is the mean electronic velocity in the suitable dimensional units. All that implies that the electrons are seeing the nuclear position but not the nuclear velocity. Note that the retarded nuclear coordinate (416) presents the property to not commute with itself for different times:

$$[q(t_1), q(t_2)] = i\kappa(t_2 - t_1) \tag{418}$$

As may be observed by inspection of the Galilei transform (416), the most important correction to the Born–Oppenheimer approximation occurs, when the ratio k is not too small, that is, in view of Eq. (417), for molecules involving simultaneously high harmonic frequencies ω° and slow electronic velocities $\langle v \rangle$. That is, indeed, the case of hydrogen bonds where are met at the same time the lightest atom H and electronic pairs involving, because of their nonbonded character, slow average electronic velocities.

By aid of the Galilei transform, Witkowski passed from the usual dimensionless Hamiltonian (415) to the new one depending, by taking $q(t)$ in place of q:

$$\{H^0_{\text{Wit}}(t)\} = \tfrac{1}{2}[p^2 + q^2(t)]\hbar\omega^\circ \tag{419}$$

that is, in view of Eq. (416):

$$\{H^0_{\text{Wit}}(t)\} = \tfrac{1}{2}\{p^2 + (q + \kappa tp)^2\}\hbar\omega^\circ \tag{420}$$

This equation may also be written, owing to the noncommutative properties of the p and q operators:

$$\{H^0_{\text{Wit}}(t)\} = \tfrac{1}{2}\{(p^2 + q^2) + \kappa t(pq + qp) + (\kappa t)^2 p^2\}\hbar\omega^\circ \tag{421}$$

In order to return to the conventional meaning of coordinates, some rescaling is required in space–time with units that change from point to point. A very deep examination of this equation leads Witkowski to consider this transformed Hamiltonian, which depends on the p and q coordinates (involving the unusual commutation rules), in another new form that is a linear combination of three operators τ_k forming a closed Lie algebra:

$$\{H^0_{\text{Wit}}(t)\} = \{[1 + \tfrac{1}{2}(\kappa t)^2]\tau_3 + (\kappa t)\tau_2 - \tfrac{1}{2}(\kappa t)^2\tau_1\}\hbar\omega^\circ \qquad (422)$$

Here the τ_i, which are, respectively, given by:

$$\tau_1 = \tfrac{1}{2}(q^2 - p^2), \qquad \tau_2 = \tfrac{1}{2}(pq + qp), \qquad \tau_3 = \tfrac{1}{2}(p^2 + q^2)$$

obey the commutation rules of the following closed algebra:

$$[\tau_1, \tau_2] = 2i\tau_3, \qquad [\tau_2, \tau_3] = -2i\tau_1, \qquad [\tau_3, \tau_1] = -2i\tau_2$$

That was recognized by Witkowski to be related to the dynamical Schrödinger group $SO\,(2,1)$.

The time-dependent Hamiltonian may be diagonalized inside the closed algebra with successive approximations. This leads to the result:

$$\{H^0_{\text{Wit}}(t)\} = [1 + f(\kappa, t)]\tau_3 \qquad (423)$$

$f(\kappa, t)$ is a time-dependent function, whereas $\tau_3(0)$ is the expression of the operator at initial time $t = 0$. Up to second order in time, the diagonalized Galilei Hamiltonian is found by Witkowski to be:

$$\{H^0_{\text{Wit}}(t)\} = \tfrac{1}{2}\{[1 + (\kappa\omega^\circ t)^2][p^2 + q^2]\}\hbar\omega^\circ \qquad (424)$$

In view of Eq. (415), the last result allows one to write the Galilei Hamiltonian in terms of the Born–Oppenheimer one, according to:

$$\{H^0_{\text{Wit}}(t)\} = [1 + (\kappa\omega^\circ t)^2]\{H^0_{\text{Wit}}\} \qquad (425)$$

As a consequence of the Witkowski accounts of the finite velocity of the electron and its retardation following nuclear motion, the Born–Oppenheimer Hamiltonian $\{H^0_{\text{Wit}}\}$ appears to be corrected by the time-dependent term $(\kappa\omega t)^2$ so that it may be viewed in a certain sense as anharmonic. Thus, if we consider the hydrogen bond, where there are both the fast-mode X–H and slow X–H$\cdots$Y, the electronic retardation with respect to both nuclear motions of these modes must produce a coupling between these modes. That explains the anharmonic form of the Hamiltonian assumed by Witkowski and Maréchal in their old approach of the hydrogen bond. That may also be the basis of the unexplained anharmonic coupling claimed to be at the basis of the hydrogen bond features: Because of the time dependence of the Hamiltonian oscillator (425), some correction will appear in the usual time-dependent expression of the dipole moment transition. By series Fourier transforms, this must lead to asymmetry in the spectral density involving a long tail on the high-frequency side of the spectrum. That may be in correspondence with the Poisson distribution of the spectral density of the hydrogen bond at zero temperature in the absence of damping.

2.   The Spectral Density

It is now possible to look at the influence on the spectral density of the retarded motion of the electrons with respect to the nuclei. For this purpose, consider the Schrödinger equation:

$$i\hbar \; \partial\{U^0_{\text{Wit}}(t)\}/\partial t = \{H^0_{\text{Wit}}(t)\}\{U^0_{\text{Wit}}(t)\}$$

Owing to Eq. (424), it is:

$$i\hbar \; \partial\{U^0_{\text{Wit}}(t)\}/\partial t = \tfrac{1}{2}\{[1 + (\kappa\omega^\circ t)^2][p^2 + q^2]\}\hbar\omega^\circ\{U^0_{\text{Wit}}(t)\}$$

By integration over a dimensionless time this leads, owing to the initial situation $U^0_{\text{Wit}}(0) = 1$, to:

$$\{U^0_{\text{Wit}}(t)\} = \exp\{-\tfrac{1}{2}i\omega^\circ[p^2 + q^2]\int_0^t [1 + (\kappa\omega^\circ t')^2]\,dt'$$

Again, taking the matrix elements of this operator in the basis $\{|n\rangle\}$, where $\tfrac{1}{2}[p^2 + q^2]$ is diagonal, we find, neglecting the zero-point energy:

$$\langle n | \{U^0_{\text{Wit}}(t)\} | n\rangle = \{F_n(t)\} = \exp\{-in\omega^\circ t\}\{f_n(t)\} \tag{426}$$

with:

$$\{f_n(t)\} = \exp\left\{-in\omega^\circ \int_0^t (\kappa\omega^\circ t')^2 \, dt'\right\} \tag{427}$$

The Fourier transform of this matrix element is:

$$\{F_n(\omega)\} = [1/\sqrt{2\pi}] \int \{F_n(t)\} \exp\{-i\omega t\} \, dt \tag{428}$$

In the special situation where the coupling parameter κ is zero, the spectral density reduces to the usual Dirac delta peaks.

$$\{F_n(\omega)\} = \delta(n\omega^\circ - \omega), \qquad \text{if } \kappa = 0 \tag{429}$$

On the other hand, in the general situation, where the coupling parameter cannot be neglected, the integration involved in Eq. (428) gives:

$$\{I_n(\omega)\} = |\{F_n(\omega)\}|^2 = \pi^2 |(3\chi)^{-1/3}\{A_i((3\chi)^{-1/3}\Omega)\}|^2, \qquad \text{if } \kappa \neq 0 \quad (430)$$

χ and Ω are given, respectively, by:

$$\chi = \tfrac{1}{6}\kappa^2 n\omega^3, \qquad \Omega = n\omega^\circ - \omega$$

whereas A_i is the Airy function.

3. Numerical Simulation

In order to visualize the influence of the finite velocity of electron and its retardation following nuclear motion, it may be suitable to build up a damped function from the matrix element of the time evolution operator given by Eqs. (426) and (427) and involving the first excited state. For this purpose, we may define:

$$\{G_{\text{Wit}}(t)\} = \langle 1|\{U^0_{\text{Wit}}(t)\}|1\rangle \exp\{-\gamma^\circ t\}$$

that is:

$$\{G_{\text{Wit}}(t)\} = \exp\{-i\omega^\circ t\} \exp\{-i(\tfrac{1}{3})(\kappa\omega^\circ)^2 t^3\} \exp\{-\gamma^\circ t\} \qquad (431)$$

The Fourier transform of this function is:

$$\{I_{\text{Wit}}(\omega)\} = [1/\sqrt{2\pi}] \int \{G_{\text{Wit}}(t)\} \exp\{-i\omega t\}\, dt$$

Numerical Fourier transforms are given in Fig. 1.9, the angular frequency being taken to be 3000 cm^{-1}, the phenomenological damping parameter γ° being, respectively, equal to 0.02, 0.05, and 0.08, the intensity units unspecified. Of course, such numerical simulation holds for zero temperature.

Note that the line shape is asymmetric. Note also, for the smallest damping parameter, the high-frequency tail, which is deeply connected to the Airy function appearing in the analytical Fourier transform (430). It may also be of interest to observe that the asymmetry shape for the intermediate relaxation may be put in correspondence with that obtained within the quantum indirect relaxation approach for low temperatures (see Fig. 6b for a reduced damping parameter equal to 0.8 in $\omega^{\circ\circ}$ units, and equal to 0.05 in ω° units). Finally, note the analogy between the asymmetric shape involving the strongest damping parameter with that obtained in the

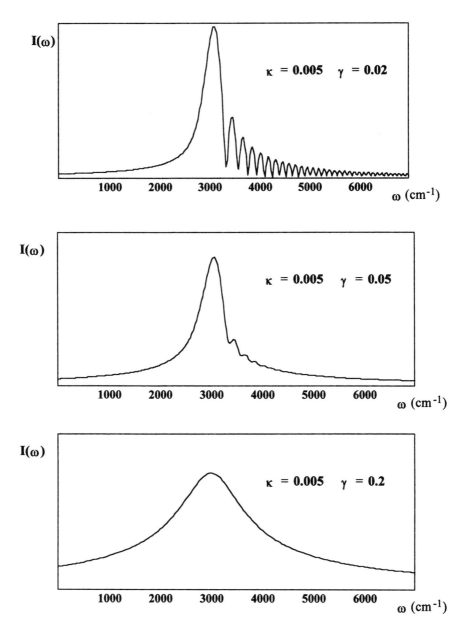

Figure 1.9. Numerical Fourier transforms (431) are given for several values of the retardation parameter κ, the angular frequency always taken to be 3000 cm^{-1}, the phenomenological damping parameter γ, respectively, equal to 0.02, 0.05, and 0.08, the intensity units being unspecified. Of course, such a numerical simulation holds for zero temperature.

quantum direct relaxation approach (given in Fig. 7b for a relative damping parameter equal to 2 in units of $\omega^{\circ\circ}$, and equal to 0.1 in units of ω°).

4. Recent Application of the Witkowski Theory

The present Witkowski theory has been used recently by Witkowski et al. (1997) to explain the inhomogeneous broadening of the infrared line shapes of high harmonics of hydrogen-containing modes observed by the opto-acoustic method and thermal lensing spectroscopy in the liquid phase of benzene and other organic transparent liquids. Attention is focused on the experimental facts that both the linewidth and asymmetry factor of the harmonic series increase linearly with the harmonic number. The basic idea is to account for the finite velocity of the π electronic cloud of vicinal molecules involving π hydrogen bonds and its retardation in following the protonic motion.

XI. PERSPECTIVES DEALING WITH FERMI RESONANCE, DAVYDOV SPLITTING, AND THE TUNNELING EFFECT

Up to now, within either the direct or the indirect damping, we have considered the spectral density of the hydrogen bond in the special situation where, in the absence of damping, the fundamental mechanism is the strong coupling between the high- and low-frequency stretching mode. We have ignored the mechanisms that must complicate the spectral density of the hydrogen bond in the presence of damping, such as the Fermi resonances, the Davydov splitting, and the tunneling effect. This area, which is very puzzling, is susceptible to be considered within the linear response theory. However, the present approaches within this theory, when they are quantum mechanical, do not present the same straightforward qualities as those we have considered in the above sections. Indeed, the most important references in the literature remain within the static quantum harmonic method, and the subject is therefore a perspective to develop in this area. For all these reasons, we leave the questions of Fermi resonances, Davydov splitting, and the tunnel effect for a further review. We shall only glance at this subject.

A. Fermi Resonances

As seen above, the influence of Fermi resonance on the spectral density of the hydrogen bond is a consequence of anharmonicity and, more precisely, for simple $X-H\cdots Y$ hydrogen bonds, of the anharmonic potential coupling the fast stretching mode and the in-plane bending mode. It must be observed that if the high-frequency and low-frequency stretching vibrations may be satisfactorily viewed as normal modes, the bending motion is not usually a normal mode, as observed by Fukushima (1969). As a consequence many vibrational modes must participate in Fermi resonance.

The semiclassical approaches of Fermi resonances by Bratos (1975) and Bratos and Ratajczak (1982) are applied to weak, intermediate, and strong hydrogen bonds in the liquid phase. They treat the Fermi resonances in connection with the Evans windows (1960), within the linear response theory and by aid of the Bratos formalism quoted above.

Witkowski and Wojcik (1973b) and Wojcik (1978) have considered quantum mechanically the question of Fermi resonances in situations where the relaxation phenomena are neglected. If the first excited state of the fast q mode is quasidegenerate with the second harmonic of a bending mode as is generally supposed, then, because of the anharmonic coupling between these two modes, there is the possibility for the excitation to come back and forth from the first excited state of the fast mode to the second excited state of the bending mode. Now, when the excitation is on the stretching fast mode, the effective Hamiltonian of the slow mode is given by Eq. (35). On the other hand, when the excitation is on the second harmonic of the bending mode, the stretching mode is in its ground state, and thus the effective Hamiltonian of the slow mode is described by Eq. (34). As a consequence, the effective Hamiltonian characterizing the slow mode in the presence of Fermi resonances between the fast stretching mode and the bending mode takes the form of a matrix of operators where the nondiagonal matrix operator is the anharmonic potential coupling the first excited state of the fast mode to the second excited state of the bending mode. The theory leads to delta peaks, since it neglects all relaxation phenomena. The direct damping may be introduced, within the linear response theory, into the Witkowski and Wojcik model. That has been performed by Chamma et al. (1998). See also Henri-Rousseau et al. (1998).

Dealing with Fermi resonance but from a very different perspective, Maréchal (1983) has proposed a model that enables one to eliminate in the experimental spectra of a hydrogen bond compound the irrelevant features that are due to Fermi resonance and are masking more subtle features that may give information on the dynamics of the hydrogen bonds. Maréchal has named this method the *peeling-off* procedure. He establishes an equation relating the line shape of the experimental spectrum to the ideal line shape that the fast stretching band is supposed to exhibit after a complete elimination of these Fermi resonances. In practice, Maréchal has reversed the problem by assuming that the ideal line shape is known, and then calculating the line shape involving Fermi resonances, which will be compared with the experimental spectrum. Practically, in order to perform such calculations, the empirical parameters characterizing Fermi resonances must be introduced. The test of the thoery finally consists in the ability to reproduce the spectrum at some definite temperature and also to predict the change in the spectra with temperature without introducing any further empirical parameters. Maréchal considered that the hydrogen bond is strongly modulated by the low-frequency vibration of the hydrogen bond and also that the hydrogen bond interacts at the same time with some binary combinations of bending modes (Fermi resonances). In this study, Maréchal has shown that his approach leads to solutions that have a good qualitative behavior: In the absence of Fermi resonance, the spectral density he obtains reduces to the usual Franck–Condon transitions of the strong coupling approach. When the low-frequency vibrations are considered as forced in their classical limit, the Maréchal spectral density in solution reduces explicitly to that found by Bratos (1975) for the case of a hydrogen bond diluted in an inert solvent and also to that obtained by Weidemann and Hayd (1977) in their study of a single Fermi resonance. For applications, see Maréchal (1983b, 1985, 1987b).

B. Davydov Coupling

Consider the situation where there is a symmetric double hydrogen bond as, for instance, in acetic acid. Then, as observed by Maréchal and Witkowski (1968), there is the possibility of Davydov coupling in the first excited state of the fast vibrational mode; this is because the excitation may pass from the first excited state of the fast mode of one hydrogen bond to the corresponding degenerate first excited state of the second hydrogen bond. The study of Maréchal and Witkowski, is a consequence of an initial work of Witkowski (1967), related to the pioneering work of Witkowski and Moffitt (1960).

Consider the effective Hamiltonian of the first excited state in the presence of Davydov coupling. There are two possibilities because of the degeneracy: Either the first oscillator is in its excited state with an effective Hamiltonian given by Eq. (35) and the second oscillator in its ground state with effective Hamiltonian given by an equation similar to Eq. (34), or vice versa. Now let us consider the situation where the dimer is embedded in a thermal bath. One possibility is then to simulate the influence of damping in the spectral lines given by Eq. (584) by taking each line as a Gaussian, as has been performed by Flakus (1978). Another possibility is to take explicitly into account the influence of the thermal bath, as has been performed by Boulil et al. (1994b).

C. Tunneling Effect

In weak hydrogen bond complexes an asymmetrical structure is generally observed and the potential energy curve of the proton stretching vibration is considered as involving either a single or a double asymmetrical minimum. In moderately strong symmetrical hydrogen bonds, it is believed that a double-minimum potential does exist. For increasing the strength of the hydrogen bonds, there is the possibility that the hump involved in the double minimum potential is lowered enough to allow a tunneling through the barrier. In this case, the excited state of the high-freuqency mode and, at a least degree, that of the corresponding ground state, may be split. This will lead to a superposition of bands.

Thinking one dimensionally in terms of the effective coordinate along the hydrogen bond, the potential energy profile of the proton transfer may then possess a double minimum symmetric or asymmetric in dependence of the interplay of diverse effects. Such a double minimum curve may lead in the infrared to a doublet. It may be observed that the experimental argument according to which an observed doublet peak requires a double minimum is clearly fallacious, since an anharmonic single minimum leading to Fermi resonance can easily give a doublet infrared spectrum. The one-dimensional picture, which is rather crude and may be improved by considering multidimensional hypersurface, may constitute the widespread mode of thinking about hydrogen tunneling in hydrogen bonds.

Rösch (1973) has studied the damped motion in a two-dimensional double-well suitable to describe tunneling proton transfer between two sites of a hydrogen bond forming part of a larger molecular complex in a thermal environment.

Robertson and Lawrence (1981a,b, 1988) analyzed the infrared absorption spectra of chromous acid by assuming proton tunneling, which they modeled with the aid of a one-dimensional Morse potential. The major source of the observed bandwidth being the thermal motion of the oxygen atoms to which the protons are anharmonically coupled, they assumed that the relative motion of nearest-neighbor oxygen atoms can be treated as a Gaussian stochastic process, and that the modulation of the proton energy levels by the random fluctuations in the double Morse potential is quasistatic.

Sokolov and Vener (1992) have recently studied the proton tunneling assisted by the intermolecular vibration. They used a two-dimensional energy surface to treat the proton transfer dynamics. This energy surface describes semiquantitatively the main experimental regularities for a strong hydrogen bond. The strong coupling between the slow and fast modes and the dynamic asymmetry of the potential energy surface are both taken into account. Quantum jumps between vibrational levels of both subsystems under the random force action of the environment are assumed to achieve the proton transfer from one well into another. Kryachko (1994) has also been dealing recently with the tunneling effect within a two-level state model.

Abramczyk (1990) has proposed a model in which the proton involves motions in the Born–Oppenheimer potential energy, which is accompanied by the anharmonic coupling to the low-frequency hydrogen bridge stretching motion. Both the potential energy well and the anharmonic coupling are randomly modulated with time. For the fluctuations of the potential energy arising from the direct coupling to the bath and obeying the stochastic Liouville equation, Abramczyk used the statistical treatment of Lami and Villani (1987) and, for the random modulation of the slow mode, the results of Boulil et al. (1988).

There are also some recent important papers that indicate progress in the area of the tunneling effect: that of Staib et al. (1995), dealing with proton transfer in hydrogen-bonded acid–base complexes in polar solvents, and that of Scheurer and Saalfrank (1996), concerning hydrogen transfer in vibrational relaxing acid dimers.

XII. CONCLUSION

Recapitulating the present study, the spectral density $\{I(\omega)\}$ of the high-frequency stretching mode is within the linear response theory given by the Fourier transform of the autocorrelation function $\{G(t)\}$ of the dipole moment operator.

$$\{I(\omega)\} = \text{const} \int_{-\infty}^{\infty} \{G(t)\} \exp\{-i\omega t\} \, dt$$

with

$$\{G(t)\} = \langle \boldsymbol{\mu}^{\dagger}(0) \cdot \boldsymbol{\mu}(t) \rangle_0$$

and where $\boldsymbol{\mu}$ is the dipole moment operator.

For the calculation of the autocorrelation function, it is assumed that the influence of Davydov splitting, Fermi resonance, and tunneling effect on the spectral density can be ignored. In such a situation, which is a special one from the experimental viewpoint, the fundamental mechanism is the anharmonic coupling between the high- and low-frequency stretching modes of the hydrogen bond.

A. The Anharmonic Coupling Theory and the Indirect Relaxation Mechanism

Within the anharmonic theory, the high-frequency $v_s(X\text{–}H)$ mode is anharmonically coupled to the low-frequency $v_s(X\text{–}H\cdots Y)$ stretching mode. That leads to the Hamiltonian H of both the slow and fast modes as:

$$H = [p^2/(2m) + \tfrac{1}{2}m\omega^{\circ 2}q^2] + [P^2/(2M) + \tfrac{1}{2}M\omega^{\circ\circ 2}Q^2] + [K_{112}Qq^2]$$

with q being the position coordinate of the fast mode; p, the conjugate momentum, with $[p, q] = -i\hbar$; Q, the position coordinate of the slow mode; P, the conjugate momentum, with $[P, Q] = -i\hbar$ or $[P, Q] = 0$; m, the reduced mass of the fast mode; M, the reduced mass of the slow mode; K_{112}, the anharmonic constant; ω°, the angular frequency of the fast mode; and $\omega^{\circ\circ}$, the angular frequency of the slow mode. It is suitable to take a nondimensional anharmonic constant defined according to:

$$\alpha^{\circ} = K_{112}[Q^{\circ\circ}/(m\omega^{\circ}\omega^{\circ\circ})]$$
$$Q^{\circ\circ} = [\hbar/(2M\omega^{\circ\circ})]^{1/2}$$

The anharmonicity may be treated either semiclassically or quantum mechanically. In the semiclassical approach, the angular frequency, which is involved in the Hamiltonian of the fast mode, is expanded up to first order with respect to the position coordinate of the slow mode, considered a classical variable. On the other hand, in the quantum-mechanical approach, the expansion of the angular frequency of the fast mode is expanded with respect to the position coordinate of the slow mode viewed as a quantum operator. This difference, as compared to the semiclassical approach, involves the existence of different effective Hamiltonians for the slow mode, according to the fact that the fast mode is in its ground state or in excited states. These effective Hamiltonians may be viewed as a consequence of different effective potentials which are each other displaced. Changing the quantum representation, such a difference leads to consider the slow mode as a driven oscillator.

Within the indirect mechanism of relaxation, the fast mode relaxes through the slow mode, which is strongly damped because of its anharmonic coupling with it. Thus the semiclassical approach of the spectral density leads one to take into account the behavior of a classical Brownian oscillator, whereas the quantum approach has to consider the slow mode as a driven damped quantum harmonic oscillator.

Note that within the indirect relaxation mechanism, the physical parameters that may be fitted with experiment are the same in both quantum and classical approaches and are: ω°, the angular frequency of the fast mode; $\omega^{\circ\circ}$, the angular frequency of the slow mode; α°, the nondimensional anharmonic coupling; γ, the parameter characterizing the damping of the slow mode; and T, the absolute temperature.

B. The Classical Approaches

From the semiclassical viewpoint of the theory of the spectral density of the hydrogen bond, there are the commutators $[p, q] = -i\hbar$ and $[P, Q] = 0$.

1.   The Autocorrelation Function in the Underdamped Situation

Within the indirect damping mechanism, the autocorrelation function of the dipole moment operator becomes, according to Robertson and Yarwood, and writing $g = |\mu_{10}|^2$.

$$\{G_{\text{Rob}}(t)\} = g \exp\{i\omega^\circ t\}\left\langle \exp\left\{i\omega^{\circ\circ}\alpha^\circ \int_0^t [Q(t')/Q^{\circ\circ}] \, dt'\right\}\right\rangle_{Q(t)}$$

The thermal average involved in this question may be obtained with a cumulant expansion. In the underdamped situation and within the Uhlenbeck–Ornstein approach of Brownian motion, the autocorrelation function is:

$$\{G_{\text{Rob}}(t)\} = g \exp\{i\omega^\circ t\} \exp\{-\alpha^{\circ 2}[2k_B T/(\hbar\omega^{\circ\circ})]$$
$$\times \{\gamma t + [(\gamma^2 - \omega^{\circ\circ 2})/(\omega^{\circ\circ 2})][\exp\{-\gamma t/2\} \cos(\omega^{\circ\circ}_{\text{und}} t) - 1]$$
$$+ [\gamma/(2\omega^{\circ\circ}_{\text{und}})][(\gamma^2 - 3\omega^{\circ\circ 2})/\omega^{\circ\circ 2}] \exp\{-\gamma t/2\} \sin(\omega^{\circ\circ}_{\text{und}} t)\}\}$$
$$\omega^{\circ\circ 2}_{\text{und}} = [\omega^{\circ\circ 2} - \gamma^2/4]$$

where γ is a parameter characterizing the damping of the slow mode.

In the spirit of the classical approach of Robertson, where the slow mode is considered as classical, the half-width of the spectrum must be given by:

$$\Delta\omega_{\text{cl}} = \alpha^\circ \omega^{\circ\circ}[2k_B T/(\hbar\omega^{\circ\circ})]^{1/2}$$

2.   Special Situations

Within the Robertson model, and according to Kubo, it appears that there are two limit situations:

a. Slow-Modulation Limit for Which the Shape is Gaussian. In the slow-modulation limit, the relation between the basic parameters is:

$$\alpha^\circ[2k_B T/(\hbar\omega^{\circ\circ})]^{1/2} > [\omega^{\circ\circ}/\gamma]$$

The Robertson autocorrelation function reduces in this situation to:

$$\{G_{\text{Rob}}(t)\} = g \exp\{i\omega^\circ t\} \exp\{-\tfrac{1}{2}[\alpha^\circ \omega^{\circ\circ}]^2[2k_B T/(\hbar\omega^{\circ\circ})]t^2\}$$

the Fourier transform of which is the Bratos spectral density:

$$\{I_{Brat}(\omega)\} = g[[\hbar\omega^{\circ\circ}/(k_B T)]^{1/2}/(2\alpha^{\circ}\omega^{\circ\circ})]$$
$$\times \exp\{-[\hbar\omega^{\circ\circ}/(2k_B T\alpha^{\circ 2})][(\omega - \omega^{\circ})/\omega^{\circ\circ}]^2\}$$

b. *Fast-Modulation Limit for Which the Shape is Lorentzian.* In the fast-modulation limit the relation between the basic parameters is:

$$\alpha^{\circ}[2k_B T/(\hbar\omega^{\circ\circ})]^{1/2} < [\omega^{\circ\circ}/\gamma]$$

In this situation, the Robertson autocorrelation function becomes:

$$\{G_{Rob}(t)\} = g \exp\{i\omega^{\circ}t\} \exp\{-2\gamma[\alpha^{\circ 2}[k_B T/(\hbar\omega^{\circ\circ})]]t\}$$

The Fourier transform of this autocorrelation function is a Lorentzian of the form:

$$\{I_{Rob}(\omega)\} = g(2/\pi)2\gamma\alpha^{\circ 2}[k_B T/(\hbar\omega^{\circ\circ})]/[(\omega - \omega^{\circ})^2$$
$$+ (2\gamma\alpha^{\circ 2}[k_B T/(\hbar\omega^{\circ\circ})])^2]$$

As it appears, the factors that favor the passage from the Gaussian shape to Lorentzian are a decrease of the anharmonic coupling, a lowering of the temperature, and an underdamped situation.

c. *Unphysical Limit Situation in Which the Damping is Missing.* In the special situation of no damping, where the theory fails, the Robertson auto-correlation function reduces to:

$$\{G_{Rob}^{0}(t)\} = g \exp\{i\omega^{\circ}t\} \exp\{\alpha^{\circ 2}[2k_B T/(\hbar\omega^{\circ\circ})][\cos(\omega^{\circ\circ}t) - 1]\}$$

3. Limit of Validity of the Semiclassical Approach

The semiclassical theory of Robertson understands that in solution of the profile of the infrared spectra of the high-frequency mode often diverges from a Gaussian shape. It gives the possibility to reproduce some features of the band shape. However, it should be noted that the Robertson model is not satisfactory for low temperatures or when the system becomes very underdamped, as might be the case in the gas phase since the basic assumption that the character of the slow mode is determined largely by its inter-action with the environment is no longer valid. In the very underdamped situation, each excited state of the high-frequency vibrational mode is prob-ably sufficiently long-lived to define for the low-frequency mode an effective potential, in which a number of complete oscillations can take place, so that vibrational Franck–Condon transitions need to be considered. That is the spirit of the quantum approach. More generally, the asymmetry of the band shape everywhere it appears cannot be taken into account from the semi-classical model, which neglects the possibility of an effective potential. For all these situations a full quantum-mechanical theory is required.

C. The Quantum Approach

In the quantum approach, there are the commutators $[p, q] = -i\hbar$ and $[P, Q] = -i\hbar$. The basic idea of this approach is that each excited state of the high-frequency vibrational mode is sufficiently long-lived to define for the low-frequency mode an effective potential in which a number of complete oscillations can take place, so that vibrational Franck–Condon transitions need to be considered.

1. The Underdamped Autocorrelation Function within the Indirect Mechanism of Relaxation

The autocorrelation function of the dipole moment operator of the fast mode damped by the medium through the slow mode, both coupled with the fast mode and with the medium, is given, within the quantum representation {II}, and with the same g as above by:

$$\{G_{II}(t)\} = \varepsilon g \ \text{tr}\{\exp\{-\lambda a^\dagger a\}\{U_{II}^{IP}(t)\}^\dagger\}$$

$$\lambda = \hbar\omega^{\circ\circ}/(k_B T)$$

$$\varepsilon = 1 - \exp\{-\lambda\}$$

where T is the temperature, and with the commutation rule $[a, a^\dagger] = 1$. In the expression of the autocorrelation function appears the time evolution operator in the interaction picture of a driven damped quantum harmonic oscillator (related to the existence of an effective potential). It is given according to the following equations:

$$\{U_{II}^{IP}(t)\} = \exp\{i(\omega^\circ - \alpha^{\circ 2}\omega^{\circ\circ})t\} \ \exp\{-i(|\beta|^2\omega^{\circ\circ})t\}$$

$$\times \exp\{i|\beta|^2 \exp\{-\gamma t/2\} \ \sin(\omega^{\circ\circ}t)\}\{A^{\{1\}}(\Phi_{II}(t)^*)\}$$

$$\{A^{\{1\}}(\Phi_{II}(t))\} = \exp\{\Phi_{II}(t)a^\dagger - \Phi_{II}^*(t)a\}$$

$$\{\Phi_{II}(t)\} = \beta[\exp\{-\gamma t/2\} \ \exp\{-i\omega^{\circ\circ}t\} - 1]$$

$$|\beta| = \alpha^\circ[4\omega^{\circ\circ 4} + \gamma^2\omega^{\circ\circ 2}]^{1/2}/[2(\omega^{\circ\circ 2} + \gamma^2/4)]$$

where γ is the parameter characterizing the damping of the slow mode. Owing to the above equations, the autocorrelation function takes the final form:

$$
\begin{aligned}
\{G_{\mathrm{II}}(t)\} = {}& g \, \exp\{-\tfrac{1}{2}|\beta|^2(1 + 2[\exp\{\hbar\omega^{\circ\circ}/(k_B T)\} - 1]^{-1})\} \\
& \times \exp\{\tfrac{1}{2}|\beta|^2(1 + 2[\exp\{\hbar\omega^{\circ\circ}/(k_B T)\} - 1]^{-1}) \\
& \times [2 \exp\{-\gamma t/2\} \cos(\omega^{\circ\circ}t) - \exp\{-\gamma t\}]\} \\
& \times \exp\{i|\beta|^2 \exp\{-\gamma t/2\} \sin(\omega^{\circ\circ}t)\} \\
& \times \exp\{-i|\beta|^2\omega^{\circ\circ}t\} \exp\{i(\omega^\circ - \alpha^{\circ 2}\omega^{\circ\circ})t\}
\end{aligned}
$$

It must be emphasized that in view of the remarks of Section IV.D.2, the asymmetry of the fine structure of the spectral density obtained by Fourier transformation is mainly caused by the factor:

$$
\exp\{i|\beta|^2 \exp\{-\gamma t/2\} \sin(\omega^{\circ\circ}t)\}
$$

The dependence on the temperature of the half-width is given by:

$$
[\Delta\omega(T_1)/\Delta\omega(T_2)] = \coth^{1/2}[\hbar\omega^{\circ\circ}/(2k_B T_1)]/\coth^{1/2}[\hbar\omega^{\circ\circ}/(2k_B T_2)]
$$

This result is in agreement with the experimental results of Asselin and Sandorfi (1971, 1972). For high temperatures, this temperature dependence reduces to the behavior of the semiclassical theory:

$$
[\Delta\omega(T_1)/\Delta\omega(T_2)] = [T_1/T_2]^{1/2}
$$

2. The Maréchal and Witkowski Approach as Special Situation

Observe that when the indirect damping is missing, the above autocorrelation function reduces to:

$$
\begin{aligned}
\{G^0(t)\} = {}& g \, \exp\{i\alpha^{\circ 2} \sin(\omega^{\circ\circ}t)\} \exp\{i(\omega^\circ - 2\alpha^{\circ 2}\omega^{\circ\circ})t\} \\
& \times \exp\{\alpha^{\circ 2}(1 + 2[\exp\{\hbar\omega^{\circ\circ}/(k_B T)\} - 1]^{-1})[\cos(\omega^{\circ\circ}t) - 1]\}
\end{aligned}
$$

The spectral density, that is, the Fourier transform of this undamped autocorrelation function, is the Franck–Condon progression obtained by Maréchal and Witkowski in their time-independent approach:

$$
\begin{aligned}
I^0(\omega) = {}& \varepsilon g \sum \exp\{-m\hbar\omega^{\circ\circ}/(k_B T)\} \\
& \times \sum |\Gamma_{mn}(\alpha^\circ)|^2 \, \delta(\omega - [(m - n)\omega^{\circ\circ} + (\omega^\circ - 2\alpha^{\circ 2}\omega^{\circ\circ})])
\end{aligned}
$$

with the Franck–Condon factors:

$$|\Gamma_{mn}(\alpha^{\circ})|^2 = \exp\{-\alpha^{\circ 2}\}\alpha^{\circ 2(m-n)}(m!\,n!)$$

$$\times \left(\sum_{k=0}^{m}\{(-1)^k\alpha^{\circ 2k}/[(n-k)!\,k!\,(m-n+k)!]\}\right)^2, \qquad \text{for } m \geq n$$

The half-width of the spectrum is:

$$\Delta\omega_{\text{quant}} = \alpha^{\circ}\omega^{\circ\circ}\,\text{coth}^{1/2}[\hbar\omega^{\circ\circ}/(2k_B\,T)]$$

which reduces, for high temperature, to:

$$\Delta\omega_{\text{quant}} = \Delta\omega_{\text{class}} = \alpha^{\circ}\omega^{\circ\circ}[2k_B\,T/(\hbar\omega^{\circ\circ})]^{1/2}$$

The quantum expression giving the half-width, as the classical one, also takes into account the isotope effect, that is, the substitution of deuterium for hydrogen in X–H, since the half-width must decrease by roughly $1/\sqrt{2}$, through α°.

3. The Direct Mechanism of Relaxation

In the presence of direct damping, according to Rösch and Ratner, the autocorrelation function of the high-frequency stretching mode may be expressed simply in terms of the undamped autocorrelation function.

$$\{G_{\text{Rat}}(t)\} = \{G^0(t)\}\,\exp\{\gamma^0 t\}$$

Then the spectral density is given by a double sum of Lorentzians, the intensities of which are the Boltzmann probabilities times the Franck–Condon factors between the displaced and undisplaced wave functions of the Hamiltonian of the slow mode:

$$\{I_{\text{Rat}}(\omega)\} = \sum\sum\{I_{kl}(\omega)\}$$

$$\{I_{kl}(\omega)\} = \varepsilon g[2/\sqrt{\pi}]\,\exp\{-k\hbar\omega^{\circ\circ}/(k_B\,T)\}\,|\Gamma_{kl}(\alpha^{\circ})|^2$$

$$\times \{\gamma^0/[(\omega-\omega_{kl})^2 + (\tfrac{1}{2}\gamma^0)^2]\}$$

It must be emphasized that there are some approximations in the quantum treatment of the direct relaxation.

4. Properties of the Spectral Density within both the Direct and Indirect Mechanisms

Within the direct and indirect relaxation mechanisms, the spectral density obtained by numerical Fourier transform of the quantum autocorrelation function exhibits the following features:

1. A shift in the angular frequency of the center of gravity of the spectrum towards low frequencies;
2. A decrease of the angular frequency shift with deuterium substitution;
3. A decrease in the half-width of the spectrum with deuterium substitution;
4. An increase of the half-width with temperature;
5. An increase of the asymmetry of the global spectrum with a lowering of the temperature;
6. An increase of the asymmetry with a decrease of the anharmonic constant.

Among these properties, only 3 and 4 are obtained in the semiclassical approach. Note also that the influence of damping is less sensitive to the coalescence of the fine structure in the quantum approaches (of either the direct or indirect relaxation) than in the classical approach (of the indirect relaxation). An important difference between the direct and indirect underdamped relaxation mechanism is a tendency to an increase in the half-width of the spectrum with the damping in the direct mechanism and to a decrease of the half-width in the indirect mechanism.

ACKNOWLEDGMENTS

We thank Professor D. Hadzi of the University of Lublijana for helpful suggestions and Professor A. Witkowski of the University of Cracow for enlighting discussions.

REFERENCES

Abramczyk, H. (1985) *Chem. Phys.* **94**, 91.

Abramczyk, H. (1987) *Chem. Phys.* **116**, 249.

Abramczyk, H. (1990) *Chem. Phys.* **144**, 305; 319.

Asselin, M., Sandorfy, C. (1971) *Chem. Phys. Letters* **8**, 601.

Asselin, M., Sandorfy, C. (1972) *Can. J. Spectro.* **17**, 24.

Badger, R., Bauer, S. (1937) *J. Chem. Phys.* **5**, 839.

Barton, S., Thorson, W. (1979) *J. Chem. Phys.* **71**, 4263.

Beswick, J., Jortner, J. (1978) *J. Chem. Phys.* **68**, 2277; **69**, 512.

Blaise, P., Durand, Ph., Henri-Rousseau, O. (1994) *Physica A* **209**, 51.

Blaise, P., Giry, M., Henri-Rousseau, O. (1992) *Chem. Phys.* **159**, 169.

Borstnik, B. (1976) *Chem. Phys.* **15**, 391.

Boulil, B., Blaise, P., Henri-Rousseau, O. (1994) *J. Mol. Struc. Theochem.* **314**, 101.

Boulil, B., Déjardin, J-L., El-Ghandour, N., Henri-Rousseau, O. (1994) *J. Mol. Struct. Theochem.* **314**, 83.

Boulil, B., Henri-Rousseau, O., Blaise, P. (1988) *Chem. Phys.* **126**, 263.

Bournay, J., Robertson, G. (1978) *Nature* **275**, 47.

Bratos, S. (1975) *J. Chem. Phys.* **63**, 3499.

Bratos, S., Hadzi, D. (1957) *J. Chem. Phys.* **27**, 991.

Bratos, S., Ratajczak, H. (1982) *J. Chem. Phys.* **76**, 77.

Bratos, S., Rios, J., Guissani, Y. (1970). *J. Chem. Phys.* **52**, 439.

Carruthers, P., Nieto, M. (1965) *Am. J. Phys.* **33**, 537.

Chamma, D., Henri-Rousseau, O. (1998). *Chem. Phys.* (in press).

Coffey, W. (1985) *Adv. Chem. Phys.* **63**, 69.

Cohen-Tannoudji, C., Dupont-Roc, J., Grinberg, G. (1992) *Atom Photon Interactions, Basic Processes and Applications*, Wiley, New York.

Coulson, C., Robertson, G. (1974) *Proc. R. Soc. Lond.* **A337**, 167.

Coulson, C., Robertson, G. (1975) *Proc. R. Soc. Lond.* **A342**, 380.

Davydov, A. (1962) *Theory of Molecular Excitons*, McGraw-Hill, New York.

Davydov, A. (1991) *Quantum Mechanics*, 2nd ed. Pergamon Press, Oxford.

Evans, J. (1960) *Spectrochimica Acta* **16**, 994.

Ewing, G. (1978) *Chem. Phys.* **29**, 253.

Ewing, G. (1980) *J. Chem. Phys.* **72**, 2096.

Fischer, S., Hofacker, G., Ratner, M. (1970) *J. Chem. Phys.* **52**, 1934.

Flakus, H. (1978) *Acta Physica Pol.* **A53**, 287.

Fong, F. (1975) *Theory of Molecular Relaxation*, Wiley, New York.

Fukushima, K., Zvolinski, B. (1969) *J. Chem. Phys.* **50**, 737.

Gordon, R. (1968) *Adv. Magn. Res.* **3**, 1.

Hadzi, D., Bratos, S. (1976) in Schuster, P., Zundel, G., Sandorfi, C., *The Hydrogen Bond Theory*, North Holland, Amsterdam, p. 567.

Henri-Rousseau, O., Blaise, P., in Hadzi, D., *Theoretical Treatment of Hydrogen Bonding*, Wiley, New York. (1997) p. 165.

Henri-Rousseau, O., Chamma, A. (1998) *Chem. Phys.* (in press).

Hofacker, G., Maréchal, Y., Ratner, M. (1976) in Schuster, P., Zundel, G., Sandorfi, C., *The Hydrogen Bond Theory*, North Holland, Amsterdam, p. 297.

Janoschek, R., Weidemann, G., Zundel, G. (1973) *J.C.S. Faraday II* **69**, 505.

Kryachko, E. (1994) *J. Mol. Struc. Theochem.* **314**, 133.

Kubo, R. (1961) in D. Ter Haar (Ed.) *Fluctuations, Relaxation and Resonance in Magnetic Systems*, Scottish Universities' Summer School, Oliver and Boyd, Edinburgh, p. 23.

Lami, A., Villani, G. (1987) *Chem. Phys.* **115**, 391.

Lansberg, G., Baryshanskaya, F. (1946) *Izv. Akad. Nauk. SSSR Ser. Fiz.* **10**, 509.

Lascombe, J., Lassègues, J., Huong, P. (1973) *J. Phys. Chem.* **77**, 2779.

Lippincott, E., Schroeder, R. (1956) *J. Am. Chem. Soc.* **78**, 517.

Lippincott, E., Schroeder, R. (1957) *J. Chem. Phys.* **23**, 1099.

Lippincott, E., Schroeder, R. (1957) *J. Phys. Chem.* **61**, 921.

Louisell, H. (1973) *Quantum Statistical Properties of Radiations*, Wiley, New York.

Louisell, W., Walker, L. (1965) *Phys. Rev.* **137**, 204.

Löwdin, P. (1993) International Congress in Quantum Chemistry, June, Girona.

Maradudin, A. (1966) *Theoretical and Experimental Aspects of the Effects of Point Effects and Disorder on the Vibration of Crystals*. Academic Press, New York.

Maréchal, Y. (1968) Thesis, Grenoble.

Maréchal, Y. (1972) *Chem. Phys. Letters* **13**, 237.

Maréchal, Y. (1980) in H. Ratajczak, W. Orville-Thomas, Wiley, New York, p. 230.

Maréchal, Y. (1983a) *Chem. Phys.* **79**, 69.

Maréchal, Y. (1983b) *Chem. Phys.* **79**, 85.

Maréchal, Y. (1985) *J. Chem. Phys.* **83**, 247.

Maréchal, Y. (1987a) in J. Durig, *Vibrational Spectra and Structure*, Elsevier, Amsterdam, Vol. 16, p. 312.

Maréchal, Y. (1987b) *J. Chem. Phys.* **87**, 6344.

Maréchal, Y., Witkowski, A. (1968) *J. Chem. Phys.* **48**, 3697.

Mühle, S., Süsse, K., Welsch, D. (1978) *Phys. Lett.* **66-A**, 25.

Mühle, S., Süsse, K., Welsch, D. (1980) *Ann. der Phys.* **7**, 213.

Novak, A. (1974) *Structure and Bonding*, Ed. J. Dunitz, **18**, 177.

Pimentel, G., McClellan, A. (1960) *The Hydrogen Bond*, Freeman, San Francisco.

Ratajczak, H., Yaremko, A. (1995) *Chem. Phys. Lett.* **243**, 348.

Reid, C. J. (1959) *J. Chem. Phys.* **30**, 182.

Rice, S., Bergren, M., Belch, A., Nielson, G. (1983) *J. Phys. Chem.* **87**, 4295.

Rice, S., Sceats, M. (1981) *J. Phys. Chem.* **85**, 1108.

Robertson, G. (1976) *J. Chem. Soc. Faraday Trans.* **72**, 1153.

Robertson, G. (1977) *Phil. Trans. R. Soc. Lond.* **A286**, 25.

Robertson, G., Lawrence, M. (1981) *Chem. Phys.* **62**, 131.

Robertson, G., Lawrence, M. (1981) *Mol. Phys.* **43**, 193.

Robertson, G., Lawrence, M. (1988) *J. Chem. Phys.* **89**, 5352.

Robertson, G., Yarwood, J. (1978) *Chem. Phys.* **32**, 267.

Rösch, N. (1973) *Chem. Phys.* **1**, 220.

Rösch, N., Ratner, M. (1974) *J. Chem. Phys.* **61**, 3344.

Sakun, V. (1981) *Chem. Phys.* **55**, 27.

Sakun, V. (1985) *Chem. Phys.* **99**, 457.

Sandorfi, C. (1984) in *Topics in Current Chemistry*, **120**, 41.

Sandorfi, C., Schuster, P., Zundel, G., Sandorfi, C. (1976) ed., *The Hydrogen Bond Theory*, North Holland, Amsterdam, p. 616.

Scheurer, C., Saalfrank, P. (1996) *J. Chem. Phys.* **104**, 2869.

Sheppard, N. (1959) *Hydrogen Bonding*, Hadzi, D., ed., Pergamon, London, pp. 85–105.

Singh, J., Wood, J. (1968) *J. Chem. Phys.* **48**, 4567.

Sokolov, N., Savel'ev, V. (1977) *Chem. Phys.* **22**, 383.

Sokolov, N., Vener, M. (1992) *Chem. Phys.* **168**, 29.

Staib, A., Borgis, D., Hynes, J. (1995) *J. Chem. Phys.* **102**, 2487.

Stepanov, B. (1945) *Zh. Fiz. Khim.* **19**, 507.

Stepanov, B. (1946) *Nature* **157**, 808.

Uhlenbeck, G., Ornstein, L. (1930) *Phys. Rev.* **36**, 823.

Wang, M., Uhlenbeck, G. (1945) *Rev. Mod. Phys.* **17**, 323.

Weidemann, E., Hayd, A. (1977) *J. Chem. Phys.* **67**, 3713.

Witkowski, A., Moffitt, W. (1960) *J. Chem. Phys.* **33**, 872.

Witkowski, A. (1967) *J. Chem. Phys.* **47**, 3645.

Witkowski, A. (1983) *J. Chem. Phys.* **79**, 852.

Witkowski, A. (1990) *Phys. Rev.* **A41**, 3511.

Witkowski, A., Henri-Rousseau, O., Blaise, P. (1997) *Acta Phys. Pol., A* **91**, 495.

Witkowski, A., Wojcik, M. (1973) *Chem. Phys.* **1**, 9.

Wojcik, M. (1978) *Mol. Phys.* **36**, 1757.

Yarwood, J., Ackroyd, R., Robertson, G. (1978) *Chem. Phys.* **32**, 283.

Yarwood, J., Robertson, G. (1975) *Nature* **257**, 41.

APPENDIX A

Some Basic Theorems

1. *Some Important Phase Transformations*

H is a time-independent Hamiltonian, B a Hermitean operator, ρ the Hermitean density operator, A a unitary operator, and $U(t)$ the unitary time evolution operator.

a. The Schrödinger equation governing the time evolution operator and the corresponding time evolution operator

$$i\hbar\{\partial U/\partial t\} = \{H\}\{U(t)\} \tag{A1}$$

$$\{U(t)\} = \exp\{-iHt/\hbar\} \tag{A2}$$

b. Some unitary transformations of importance

• Unitary transformations involving translation of time.

$$\{\rho(t)\} = \{U(t)\}\{\rho(0)\}\{U(t)\}^{\dagger} \tag{A3} \text{ p. 79}$$

$$\{B(t)\} = \{U(t)\}^{\dagger}\{B(0)\}\{U(t)\} \tag{A4} \text{ p. 68}$$

• Other unitary phase transformation leading to a change in the representation.

$$\rho_{II} = A^{\dagger}\rho_{I}A \tag{A5}$$

$$B_{II} = A^{\dagger}B_{I}A \tag{A6}$$

c. Schrödinger picture

In the Schrödinger picture the density operator change with time as the kets, whereas the operators do not change.

$$\{\rho^{SP}(t)\} = \exp\{-iHt/\hbar\}\{\rho(0)\} \exp\{iHt/\hbar\} \tag{A7}$$

$$\{B^{SP}(t)\} = \{B(0)\} \tag{A8}$$

d. Heisenberg picture

In the Heisenberg picture the density operator does not change with time as the kets, whereas the operators are changing.

$$\{\rho^{HP}(t)\} = \{\rho(0)\} \tag{A9}$$

$$\{B^{HP}(t)\} = \exp\{iHt/\hbar\}\{B(0)\} \exp\{-iHt/\hbar\} \tag{A10} \text{ p. 55}$$

e. Interaction picture

● Hamiltonian partition:

$$H = H^0 + V \tag{A11}$$

● $\{U^0(t)\}$, $\{U^{IP}(t)\}$, $\{U(t)\}$ are, respectively, the unperturbed, IP, and full-time evolution operators and $\{V^{IP}(t)\}$ is the IP perturbation Hamiltonian.

The basic equations are:

$$\{U_0(t)\} = \exp\{-iH^0t/\hbar\} \tag{A12} \text{ p. 58}$$

$$\{U(t)\} = \exp\{-iH^0t/\hbar\}\{U^{IP}(t)\} \tag{A13} \text{ p. 60}$$

with

$$i\hbar\{\partial U^{IP}/\partial t\} = \{V^{IP}(t)\}\{U^{IP}(t)\} \tag{A14} \text{ p. 59}$$

$$\{V^{IP}(t)\} = \exp\{iH^0t/\hbar\}V \exp\{-iH^0t/\hbar\} \tag{A15} \text{ p. 58}$$

In the interaction picture the density operator and the other operators change with time according to:

$$\{\rho^{IP}(t)\} = \exp\{iH^0t/\hbar\}\{\rho^{SP}(t)\} \exp\{-iH^0t/\hbar\} \tag{A16} \text{ p. 125}$$

$$\{\rho^{IP}(t)\} = \exp\{-i\{V^{IP}(t)\}t/\hbar\}\{\rho^{SP}(0)\} \exp\{i\{V^{IP}(t)\}t/\hbar\}$$

$$\{B^{IP}(t)\} = \exp\{iH^0t/\hbar\}\{B(0)\} \exp\{-iH^0t/\hbar\} \tag{A17} \text{ p. 125}$$

The pages refer to the Louisell book (1973), except for Eqs. (A16) and (A17), which refer to the Davydov book (1991).

2. Theorems Dealing with Fermions

a. *Glauber–Weyl relations:* A and B are two operators obeying

$$\text{if } [A, [A, B]] = 0 \quad \text{and} \quad \text{if } [B, [A, B]] = 0$$

$$\exp\{A\} \exp\{B\} = \exp\{A + B\} \exp\{\tfrac{1}{2}[A, B]\} \qquad \text{(A18) p. 137}$$

b. *Some transformations on function* $\{f(a, a^\dagger)\}$ *of the boson operators*
$a^\dagger$ and a obeying: $[a, a^\dagger] = 1$ with x being a scalar.

$$\exp\{-xa^\dagger\}\{f(a, a^\dagger)\} \exp\{xa^\dagger\} = \{f(a + x, a^\dagger)\} \qquad \text{(A19) p. 151}$$

$$\exp\{xa\}\{f(a, a^\dagger)\} \exp\{-xa\} = \{f(a, a^\dagger + x)\} \qquad \text{(A20) p. 151}$$

$$\exp\{xa^\dagger a\}\{f(a, a^\dagger)\} \exp\{-xa^\dagger a\} = \{f(a \exp\{-x\}, a^\dagger \exp\{x\})\}$$

$$\text{(A21) p. 154}$$

c. *The limit of zero temperature* T *of the coherent state density operator*

$$\varepsilon \exp\{-\lambda(a^\dagger + \alpha)(a + \alpha)\} \to |\alpha\rangle\langle\alpha|, \qquad \text{if } T \to 0 \quad \text{(A22) p. 159}$$

For explanation of the symbols λ and ε, see Eqs. (H36) and (H37). α is a scalar.

d. *Theorem dealing with the thermal average of a function* $\{f(a, a^\dagger)\}$ *of the boson operators*
$a^\dagger$ and a obeying $[a, a^\dagger] = 1$, performed on the Boltzmann density operator.

$$\varepsilon \, \mathrm{tr}\{\exp\{-\lambda a^\dagger a\} f(a, a^\dagger)\} = \langle 0|\langle 0| f[(1 + \langle n\rangle)^{1/2} a + \langle n\rangle^{1/2} b^\dagger,$$

$$(1 + \langle n\rangle)^{1/2} a^\dagger + \langle n\rangle^{1/2} b]|0\rangle|0\rangle \quad \text{(A23) p. 161}$$

Application to a translation operator $A(\Phi(t))$ [see Eq. (D3) of Appendix D]:

$$\varepsilon \, \mathrm{tr}\{\exp\{-\lambda a^\dagger a\} A(\Phi(t))\} = \exp\{-\tfrac{1}{2}|\Phi(t)|^2[1 + 2\langle n\rangle]\} \qquad \text{(A24)}$$

$\langle n\rangle$ is the thermal mean number occupation of the quantum harmonic oscillator given by Eq. (H90).

e. *Normal ordering relations*
Let

$$f^{(n)}(\alpha, \alpha^*) = \sum \{f^{(n)}_{rs}\}\alpha^{*r}\alpha^s \qquad \text{(A25)}$$

and

$$f^{(n)}(a, a^\dagger) = \sum \{f^{(n)}_{rs}\}a^{\dagger r}a^s \tag{A26}$$

then the N operator is by definition:

$$f^{(n)}(a, a^\dagger) = N\{f^{(n)}(\alpha, \alpha^*)\} \tag{A27} \text{ p. 140}$$

$$f^{(n)}(\alpha, \alpha^*) = N^{-1}\{f^{(n)}(a, a^\dagger)\} \tag{A28} \text{ p. 140}$$

$$f^{(n)}(a, a^\dagger) = N\{f^{(n)}(\alpha + \partial/\partial\alpha^*, \alpha^*)\} \tag{A29} \text{ p. 142}$$

$$af(a, a^\dagger) = N\{(\alpha + \partial/\partial\alpha^*)f^{(n)}(\alpha, \alpha^*)\} \tag{A30} \text{ p. 154}$$

$$N\{\exp\{[e^{-x} - 1]\alpha^*\alpha\} = \exp\{-xa^\dagger a\} \tag{A31} \text{ p. 156}$$

The pages refer to the Louisell book (1973).

3. Some Properties of the Translation Operator $A(\alpha)$

a. Different equivalent expressions of the translation operator

$$\{A(\alpha)\} = \exp\{\alpha a^\dagger - \alpha^* a\} = \exp\{\alpha a^\dagger\}\exp\{-\alpha^* a\}\exp\{-\tfrac{1}{2}|\alpha|^2\} \tag{A32}$$

$$\{A(\alpha)\} = \exp\{-\alpha^* a + \alpha a^\dagger\} = \exp\{-\alpha^* a\}\exp\{\alpha a^\dagger\}\exp\{\tfrac{1}{2}|\alpha|^2\} \tag{A33}$$

$$\{A(\alpha)\}^\dagger = \exp\{\alpha^* a - \alpha a^\dagger\} = \exp\{\alpha^* a\}\exp\{-\alpha a^\dagger\}\exp\{\tfrac{1}{2}|\alpha|^2\} \tag{A34}$$

$$\{A(\alpha)\}^\dagger = \exp\{-\alpha a^\dagger + \alpha^* a\} = \exp\{-\alpha a^\dagger\}\exp\{\alpha^* a\}\exp\{-\tfrac{1}{2}|\alpha|^2\} \tag{A35}$$

$$\{A(\alpha)\}^\dagger = \{A(-\alpha)\}$$

b. The action of the translation operator $A(\alpha)$ on the eigenkets $|n_{III}\rangle$ of the number occupation operator and the Franck–Condon factors $\Gamma_{mn}(\alpha)$

$$A(\alpha)|n_{III}\rangle = |n_{II}\rangle \tag{A36}$$

$$\langle Q|n^{\{k\}}_{II}\rangle = \{\chi^{\{k\}}_m(Q)\}, \quad \langle Q|n^{\{k\}}_{III}\rangle = \{\chi^{\{k\}}_m(Q - 2k\alpha Q^{\circ\circ})\} \tag{A37}$$

$$\langle n_{II}|n_{III}\rangle = \Gamma_{mn}(\alpha) \tag{A38}$$

$$\Gamma_{mn}(\alpha) = \int \{\chi^{\{k\}}_m(Q)\}\{\chi_n(Q - 2k\alpha Q^{\circ\circ})\}\, dQ \tag{A39}$$

$$\langle m|A(\alpha)|n\rangle = \Gamma_{mn}(\alpha) = \Gamma_{nm}(-\alpha) \tag{A40}$$

$$\Gamma_{mn}(\alpha) = \exp\{-\alpha^2/2\}\alpha^{m-n}(m!\,n!)^{1/2}$$

$$\times \sum_{k=0}^{m} \{(-1)^k\alpha^{2k}/[(n-k)!\,k!\,(m-n+k)!\},\quad \text{for } m \geq n \quad \text{(A41)}$$

$$\Gamma_{mn}(\alpha) = \exp\{-\alpha^2/2\}\alpha^{n-m}(m!\,n!)^{1/2}$$

$$\times \sum_{k=0}^{n} \{(-1)^k\alpha^{2k}/[(m-k)!\,k!\,(n-m+k)!\]},\quad \text{for } n \geq m \quad \text{(A42)}$$

$$\Gamma_{0n}(\alpha) = \exp\{-\alpha^2/2\}\alpha^n/\sqrt{n!} \quad \text{(A43)}$$

c. Some properties of the translation operator

$$\{A(\alpha)\}^\dagger f(a^\dagger, a)\{A(\alpha)\} = \{f(a^\dagger + a^*, a + \alpha)\} \quad \text{(A44)}$$

$$\{A(\alpha)\}\{A(\beta)\} = \{A(\alpha + \beta)\} \exp\{\tfrac{1}{2}[\alpha\beta^* - \alpha^*\beta]\} \quad \text{(A45)}$$

The first expression is a simple consequence of theorems (A19) and (A20). The proof of the last expression is given in Appendix E.

d. The translation operator generates coherent state at zero temperature
Definition of the coherent state $|\alpha\rangle$

$$a|\alpha\rangle = \alpha|\alpha\rangle \quad \text{(A46)}$$

How to pass from the ground state $|0\rangle$ of the number occupation operator to the coherent state

$$A(\alpha)|0\rangle = |\alpha\rangle \quad \text{(A47)}$$

The scalar product of the coherent state on an eigenstate $|n\rangle$ of the number occupation operator

$$\langle n|\alpha\rangle = \exp\{-\tfrac{1}{2}|\alpha|^2\}\alpha^n/n! = \Gamma_{0n}(\alpha) \quad \text{(A48)}$$

e. The density operator of a coherent state at any temperature as a result of the phase transformation on the Boltzmann density operator involving the translation operator

$$A^\dagger(\alpha)\varepsilon \exp\{-\lambda a^\dagger a\}A(\alpha) = \varepsilon \exp\{-\lambda(a^\dagger + \alpha)(a + \alpha^*)\} \quad \text{(A49)}$$

ε and λ are given by expressions similar to Eqs. (H36) and (H37) in which T is the absolute temperature.

4.   Properties Dealing with Distributions and Diffraction Function

a.   General Relation

$$\{\delta(\omega)\} = 1/\sqrt{2\pi} \int_{-\infty}^{\infty} e^{i\omega t}\, dt \tag{A50}$$

b.   Properties of the Diffraction Function

$$\{\delta^T(E_k - E_f)\} = 1/\sqrt{2\pi} \int_{-T/2}^{T/2} \exp\{i(E_k - E_f)t/\hbar\}\, dt \tag{A51}$$

$$\{\delta^T(E_k - E_f)\} = (1/\pi) \sin\{(E_k - E_f)T/(2\hbar)\}/[(E_k - E_f)/\hbar]$$

$$\{\delta^T(E_k - E_f)\}^2 = T/\sqrt{2\pi\hbar}\,\{\delta^T(E_k - E_f)\} \tag{A52}$$

$$\{\delta^T(E_k - E_f)\} = \{\delta(E_k - E_f)\}, \qquad \text{if } T \to \infty \tag{A53}$$

c.   Theorems

$$\{\delta(f(x))\} = \sum \left[1/(df/dx)_{x=x_i}\right]\{\delta(x - x_i)\} \tag{A54}$$

where the x_i are the roots of equation $f(x) = 0$

$$(\tfrac{1}{2}\varepsilon) \exp\{-|z|/\varepsilon\} \to \{\delta(z)\} \tag{A55}$$

APPENDIX B

The Linear Response Theory

1.   The Time Evolution Operator of a Two-Level Molecule Excited by a Monochromatic Field in the Dipolar Electric Approximation

Consider a group of molecules that are illuminated by a monochromatic electromagnetic radiation of angular frequency ω. Let H^0 be the Hamiltonian of each molecule, the eigenvalue equation of which is:

$$H^0|k\rangle = E_k|k\rangle$$

Of course, the eigenvectors are orthonormalized so that we may write:

$$\langle j|k\rangle = \delta_{jk} \qquad \text{and} \qquad \sum |k\rangle\langle k| = 1$$

The monochromatic electromagnetic field $E(t, \omega)$ is of the form:

$$E(t, \omega) = E^0\varepsilon \cos(\omega t) = \tfrac{1}{2}E^0\varepsilon[e^{i\omega t} + e^{-i\omega t}]$$

Here ε is the unit vector along the electric field of the radiation. This field will induce a transition between the energy levels of the molecule. If the interaction between the radiation and matter is approximated by the electric dipole interaction, which is valid when the wavelength is large compared to the molecular dimensions, the interaction Hamiltonian describing the coupling between the field and the dipole moment μ of the molecule is given by:

$$V(t, \omega) = E(t, \omega) \cdot \mu \tag{B1}$$

The full Hamiltonian of the system formed by the molecule interacting with the electromagnetic field, which is classical and can be omitted, is:

$$\{H(t, \omega)\} = H^0 + \{V(t, \omega)\}$$

The time evolution operator governing the transition between the energy levels of each molecule is given by the Schrödinger equation:

$$i\hbar\, \partial U/\partial t = H(t, \omega)U(t)$$

with the boundary condition $\{U(0)\} = 1$. According to time-dependent perturbation theory, the time evolution operator is given by:

$$\{U(t)\} = 1 + [1/(i\hbar)] \int_0^t \{H(t')\}\, dt'$$

$$+ [1/(i\hbar)]^2 \int_0^t \int_0^{t'} \{H(t')\}\{H(t'')\}\, dt'\, dt'' + \cdots$$

2. The Linear Response of the Molecule to the Electrical Field

In the interaction picture, this time evolution operator may be written, according to Eq. (A13):

$$\{U(t)\} = \{U^0(t)\}\{U^{IP}(t)\}$$

$U^0(t)$ is the unperturbed time evolution operator given by:

$$\{U^0(t)\} = \exp\{-iH^0 t/\hbar\} \tag{B2}$$

$U^{IP}(t)$ is the interaction picture time evolution operator by:

$$\{U^{IP}(t)\} = 1 + [1/(i\hbar)] \int_0^t \{V^{IP}(t')\} \, dt'$$
$$+ [1/(i\hbar)]^2 \int_0^t \int_0^{t'} \{V^{IP}(t')\}\{V^{IP}(t'')\} \, dt' \, dt''$$

Here the perturbation operator in the interaction picture is, according to Eq. (A15) and (B2):

$$\{V^{IP}(t, \omega)\} = \{U^0(t)\}^\dagger \{V(t, \omega)\}\{U^0(t)\} \tag{B3}$$

If the perturbation due to the electromagnetic field is small, we may limit the perturbative expansion to first order in the perturbation:

$$\{U^{IP}(t)\} = 1 + [1/(i\hbar)] \int_0^t \{V^{IP}(t', \omega)\} \, dt' \tag{B4}$$

Then, in view of Eqs. (B1), (B3), and (B4), the time evolution operator up to first order becomes:

$$\{U(t)\} = \{U^0(t)\} + [1/(i\hbar)]\{U^0(t)\} \int_0^t \{U^0(t')\}^\dagger \{E(t', \omega)\mu\}\{U^0(t')\} \, dt'$$

As it appears, it is linear in the electrical field.

3.   The Rate of Energy Loss from the Radiation

For symmetry in time, we shall suppose that the electromagnetic field is switched on at initial time $-\tau/2$. Now let us look, at final time $+\tau/2$, at the probability that the molecule passes during the time interval τ, because of the action of the electromagnetic field, from the kth to the fth eigenstate of the molecule.

$$\{P_{f \leftarrow k}(\tau)\} = |C_{f \leftarrow k}(\tau)|^2 \tag{B5}$$

The amplitude probability of the transition is given by:

$$\{C_{f \leftarrow k}(\tau)\} = \langle f | \{U(\tau)\} | k \rangle \tag{B6}$$

Owing to the two above equations, one obtains, for the transition probability:

$$\{P_{f\leftarrow k}(\tau)\} = |\langle f | \{U^0(\tau)\}\{U^{IP}(\tau)\} | k\rangle|^2$$

Because of Eq. (B2) and of the first equation of this appendix, one has:

$$\{P_{f\leftarrow k}(\tau)\} = | \exp\{-iE_f t/\hbar\}\langle f | \{U^{IP}(\tau)\} | k\rangle|^2$$

And after simplification:

$$\{P_{f\leftarrow k}(\tau)\} - |\langle f | \{U^{IP}(\tau)\} | k\rangle|^2$$

The transition probability becomes, up to first order and according to Eq. (B4):

$$\{P_{f\leftarrow k}(\tau)\} = \left|\langle f | \left\{1 + [1/(i\hbar)] \int_{-\tau/2}^{\tau/2} \{V^{IP}(t, \omega)\} \, dt\right\} | k\rangle\right|^2$$

Then, in view of Eq. (B3), this becomes,

$$\{P_{f\leftarrow k}(\tau)\} = \left|\langle f | k\rangle + [1/(i\hbar)] \right.$$

$$\left. \times \int_{-\tau/2}^{\tau/2} \langle f | \exp\{iH^0 t/\hbar\} V(t, \omega) \exp\{-iH^0 t/\hbar\} | k\rangle \, dt\right|^2$$

Besides, owing to the first equation of this appendix, it becomes, after simplifications:

$$\{P_{f\leftarrow k}(\tau)\} = [1/\hbar]^2 \left| \int_{-\tau/2}^{\tau/2} \langle f | V(t, \omega) | k\rangle \exp\{i\omega_{fk} t\} \, dt\right|^2$$

with:

$$\omega_{fk} = (E_f - E_k)/\hbar$$

As a consequence, using Eq. (B1) and after passing from the cosine to the exponential terms, and inserting the constants such as $\hbar$, π, and numerical factors such as 1 or 2 in the term const, one obtains:

$$\{P_{f \leftarrow k}(\tau, \omega)\} = \text{const } E^{02} |\langle f | \varepsilon \cdot \mu | k \rangle|^2$$

$$\times \left| \int_{-\tau/2}^{\tau/2} [\exp\{i(\omega_{fk} + \omega)t\} + \exp\{i(\omega_{fk} - \omega)t\}] \, dt \right|^2$$

Now, neglecting as usually the antiresonant term, this result reduces to:

$$\{P_{f \leftarrow k}(\tau, \omega)\} = \text{const } E^{02} |\langle f | \varepsilon \cdot \mu | k \rangle|^2 \left| \int_{-\tau/2}^{\tau/2} \exp\{i(\omega_{fk} - \omega)t\}] \, dt \right|^2$$

According to Eq. (A51), the result of integration leads to the diffraction delta function.

$$\{P_{f \leftarrow k}(\tau, \omega)\} = \text{const}' E^{02} |\langle f | \varepsilon \cdot \mu | k \rangle|^2 \{\delta^\tau(\omega_{fk} - \omega)\}^2$$

Besides, owing to Eq. (A52), this leads to:

$$\{P_{f \leftarrow k}(\tau, \omega)\} = \text{const}'' E^{02} |\langle f | \varepsilon \cdot \mu | k \rangle|^2 \{\delta^\tau(\omega_{fk} - \omega)\} \tau$$

In the usual absorption, one measures the rate of energy loss from the radiation, that is, W, which is related to the above transition probability by:

$$W = - \sum \sum [\hbar\omega_{fk}] \rho_k \{P_{f \leftarrow k}(\tau, \omega)/\tau\}_{\tau \to \infty} \qquad (B7)$$

where ρ_k is the probability of finding the molecules in the kth state in the initial ensemble from which it came. We have, therefore:

$$W = -\text{const}'' \{\hbar\omega E^{02}\} \sum \sum \rho_k |\langle f | \varepsilon \cdot \mu | k \rangle|^2 \{\delta^\tau(\omega_{fk} - \omega)\}$$

Finally, if we consider very long times, we may use Eq. (A53), which leads to:

$$W = -\text{const}'' \{\hbar\omega E^{02}\} \sum \sum \rho_k |\langle f | \varepsilon \cdot \mu | k \rangle|^2 \{\delta(\omega_{fk} - \omega)\} \qquad (B8)$$

4.   The Absorption Shape in the Heisenberg and Schrödinger Pictures

Following Gordon, the spectral density is closely related to W. This spectral density is:

$$\{I(\omega)\} = 3 \sum \sum \rho_k |\langle f | \varepsilon \cdot \mu | k \rangle|^2 \{\delta(\omega_{fk} - \omega)\}$$

which may also be written:

$$\{I(\omega)\} = 3 \sum \sum \rho_k \langle k | \boldsymbol{\varepsilon} \cdot \boldsymbol{\mu}^\dagger | f \rangle \langle f | \boldsymbol{\varepsilon} \cdot \boldsymbol{\mu} | k \rangle \{\delta(\omega_{fk} - \omega)\} \qquad \text{(B9)}$$

This formula represents the Bohr–Schrödinger view of spectroscopy as a transition between Bohr stationary states, represented by the time-dependent Schrödinger states $| k \rangle$ and $| f \rangle$.

The Heisenberg form of quantum mechanics gives an equivalent expression that stresses the time dependence of the operators rather than the states. Now it is suitable to introduce the Fourier expansion of the Dirac δ function according to Eq. (A50). With this expansion, the absorption shape becomes:

$$\{I(\omega)\} = [3/\sqrt{2\pi}] \sum \sum \rho_k \langle k | \boldsymbol{\varepsilon} \cdot \boldsymbol{\mu}^\dagger | f \rangle \langle f | \boldsymbol{\varepsilon} \cdot \boldsymbol{\mu} | k \rangle$$
$$\times \int_{-\infty}^{\infty} \exp\{i[(\mathbf{E}_f - \mathbf{E}_k)/\hbar - \omega]t\} \, dt$$

which may be also written:

$$\{I(\omega)\} = [3/\sqrt{2\pi}] \sum \sum \rho_k \int_{-\infty}^{\infty} \langle k | \boldsymbol{\varepsilon} \cdot \boldsymbol{\mu}^\dagger | f \rangle \langle f | \exp\{iH^0 t/\hbar\}$$
$$\times [\boldsymbol{\varepsilon} \cdot \boldsymbol{\mu}] \exp\{-iH^0 t/\hbar\} | k \rangle \exp\{-i\omega t\} \, dt$$

Moreover, in the interaction picture, the dipole moment operator changes with time according to Eq. (A15):

$$\{\boldsymbol{\mu}^{IP}(t)\} = \exp\{iH^0 t/\hbar\} \{\boldsymbol{\mu}(0)\} \exp\{-iH^0 t/\hbar\}$$

As a consequence, the line shape becomes:

$$\{I(\omega)\} = [3/\sqrt{2\pi}] \sum \sum \rho_k$$
$$\times \int_{-\infty}^{\infty} \langle k | \boldsymbol{\varepsilon} \cdot \boldsymbol{\mu}^\dagger(0) | f \rangle \langle f | \boldsymbol{\varepsilon} \cdot \boldsymbol{\mu}^{IP}(t) | k \rangle \exp\{-i\omega t\} \, dt$$

Again, removing the closeness relation $\sum | f \rangle \langle f | = 1$, the line shape yields:

$$\{I(\omega)\} = [3/\sqrt{2\pi}] \int_{-\infty}^{\infty} \exp\{-i\omega t\} \sum \rho_k \langle k | \boldsymbol{\varepsilon} \cdot \boldsymbol{\mu}^\dagger(0) \boldsymbol{\varepsilon} \cdot \boldsymbol{\mu}^{IP}(t) | k \rangle \, dt$$

Now, we may observe that the sum over initial states i weighted by ρ_k is simply an equilibrium average, so that the above equation gives:

$$\{I(\omega)\} = [3/\sqrt{2\pi}] \int_{-\infty}^{\infty} \exp\{-i\omega t\}\langle \boldsymbol{\varepsilon} \cdot \boldsymbol{\mu}^\dagger(0)\boldsymbol{\varepsilon} \cdot \boldsymbol{\mu}^{IP}(t)\rangle_0 \, dt$$

Finally, for an isotropic absorbing system, such as a liquid or a gas, the same result is obtained for any direction of polarization ε, and we have:

$$\{I(\omega)\} = [1/\sqrt{2\pi}] \int_{-\infty}^{\infty} \exp\{-i\omega t\}\langle \boldsymbol{\mu}^\dagger(0) \cdot \boldsymbol{\mu}^{IP}(t)\rangle_0 \, dt$$

or:

$$\{I(\omega)\} = [2/\sqrt{2\pi}] \, \text{Re} \int_{0}^{\infty} \exp\{-i\omega t\}\langle \boldsymbol{\mu}^\dagger(0) \cdot \boldsymbol{\mu}^{IP}(t)\rangle_0 \, dt$$

As it appears, the line shape is the Fourier transform of the autocorrelation function of the dipole moment operator for the absorbing molecules. Recall that this expression holds if the response of the molecule to the dipolar electric excitation by the light is linear in the electrical field. Note that the dipole moment operator is in the interaction picture, that is, time dependent with respect to the Hamiltonian of the molecular system in the absence of electromagnetic field. As a consequence, in the following, we shall omit this specification, and we shall write:

$$\{I(\omega)\} = [1/\sqrt{2\pi}] \int_{-\infty}^{\infty} \exp\{-i\omega t\}\langle \boldsymbol{\mu}^\dagger(0) \cdot \boldsymbol{\mu}(t)\rangle_0 \, dt \qquad (B10)$$

APPENDIX C

The Time Evolution Operator of the Driven Quantum Harmonic Oscillator

1. Interaction Picture of the Effective Hamiltonian (47)

Consider the effective Hamiltonian (47) of the slow mode in representation {II}. We shall perform the following change in notation: $\beta = \alpha^\circ$ and $\omega = \omega^{\circ\circ}$.

$$\{H_{II}^{\circ\{1\}}\} = \hbar\omega[a^\dagger a + \beta(a^\dagger + a) - \beta^2] + \hbar\omega^\circ$$

Recall that $[a, a^\dagger] = 1$, and that β and ω are scalars.

Now let us look at the time evolution operator generated by this Hamiltonian. It is given by the Schrödinger equation:

$$i\hbar\ \partial\{U_{\mathrm{II}}^{0\{1\}}(t)\}/\partial t = \{H_{\mathrm{II}}^{0\{1\}}\}\{U_{\mathrm{II}}^{0\{1\}}(t)\}$$

In order to solve this equation, we may perform the following partition of the Hamiltonian.

$$\{H_{\mathrm{II}}^{0\{1\}}\} = H + H^{00}$$
$$H = [a^{\dagger}a + \beta(a^{\dagger} + a)]\hbar\omega$$
$$H^{00} = \hbar\omega^{\circ} - \beta^{2}\hbar\omega$$

Within this partition, and according to Eq. (A13), the full evolution operator becomes:

$$\{U_{\mathrm{II}}^{0\{1\}}(t)\} = \exp\{-i[H^{00}]t/\hbar\}\{U^{\mathrm{IP}}(t)\}$$

or:

$$\{U_{\mathrm{II}}^{0\{1\}}(t)\} = \exp\{-i[\omega^{\circ} - \beta^{2}\omega]t\}\{U^{\mathrm{IP}}(t)\} \tag{C1}$$

The interaction picture time evolution operator is a solution of the Schrödinger equation:

$$i\hbar\ \partial\{U^{\mathrm{IP}}(t)\}/\partial t = H^{\mathrm{IP}}\{U^{\mathrm{IP}}(t)\}$$

with, according to Eq. (A15):

$$H^{\mathrm{IP}} = \exp\{i\{H^{00}\}t/\hbar\}H\ \exp\{-i\{H^{00}\}t/\hbar\} = H$$

since H^{00} is a scalar.

2. Resolution of the Schrödinger Equation within the Normal Ordering Formalism

In the interaction picture, we have to solve the Schrödinger equation of a driven oscillator given by:

$$i\ \partial\{U^{\mathrm{IP}}(t, a^{\dagger}, a)\}/\partial t = \omega[a^{\dagger}a + \beta(a^{\dagger} + a)]\{U^{\mathrm{IP}}(t, a^{\dagger}, a)\}$$

with the boundary condition $U(0) = 1$.

With the normal ordering formalism [see Appendix A, Eqs. (A28), (A29), and (A30)] the Schrödinger equation governing the dynamics of the normal ordered time evolution operator leads to a scalar partial derivative equation:

$$Ni(\partial\{U^{IP}(t, \alpha^*, \alpha)\}/\partial t)^n = \omega N\{\alpha^*[\alpha + \partial/\partial\alpha^*]\{U^{IP}(t, \alpha^*, \alpha)\}^n$$
$$+ \beta[\alpha^* + \alpha + \partial/\partial\alpha^*]\{U^{IP}(t, a^*, a)\}^n\} \quad \text{(C2)}$$

where N is the normal ordering operator.

Now we perform on the scalar function, the transform:

$$\{U^{IP}(t)\}^n = \exp\{G(t)^n\}$$

with, of course, the boundary condition $G(0) = 0$. Then Eq. (C2) leads to:

$$i(\partial G^n/\partial t) = \omega\{\alpha^*\alpha + \alpha^*(\partial G^n/\partial\alpha)$$
$$+ [\beta\alpha^* + \beta\alpha] + \beta(\partial G^n/\partial\alpha)\} \quad \text{(C3)}$$

Again the solution of this last equation may be sought in the form:

$$G^n(t) = X(t) + B(t)\alpha + C(t)\alpha^* + D(t)\alpha^*\alpha \quad \text{(C4)}$$

Let us perform the partial derivatives of $G(t)$ with respect to t, α, and α^*. Then by identification of the partial derivatives involved in (C3) with the coefficients resulting from the partial derivatives of Eq. (C4), one finds, respectively:

$$i(\partial X/\partial t) = \beta C(t)\omega$$
$$i(\partial B/\partial t) = \beta[1 + D(t)]\omega$$
$$i(\partial C/\partial t) = [C(t) + \beta]\omega$$
$$i(\partial D/\partial t) = [1 + D(t)]\omega$$

Solving this set of equations, owing to the boundary conditions resulting from $G(0) = 0$:

$$X(0) = B(0) = C(0) = D(0) = 0$$

one obtains the following results:

$$X(t) = \beta^2\{\exp[-i\omega t] - 1\} + i\beta^2\omega t \tag{C5}$$

$$C(t) = \beta\{\exp[-i\omega t] - 1\} \tag{C6}$$

$$B(t) = \beta\{\exp[-i\omega t] - 1\} \tag{C7}$$

$$D(t) = \exp[-i\omega t] - 1 \tag{C8}$$

Then coming back to the time evolution operator, one obtains:

$$\{U^{IP}(t)\}^n = N\{\exp[X(t)]\,\exp[C(t)\alpha^*]\,\exp[D(t)\alpha^*\alpha]\,\exp[B(t)\alpha]\} \tag{C9}$$

Again, passing from the scalars to the operators, the evolution operator appears to be:

$$\{U^{IP}(t)\}^n = \exp[X(t)]\,\exp\{C(t)a^\dagger\}N\{\exp\{D(t)\alpha^*\alpha\}\}\,\exp\{B(t)a\} \tag{C10}$$

Or, in view of Eq. (C8) giving $D(t)$

$$\{U^{IP}(t)\} = \exp[X(t)]\,\exp[C(t)a^\dagger]N\{\exp[\exp\{-i\omega t\} - 1]\alpha^*\alpha]\}\,\exp[B(t)a] \tag{C11}$$

Again, in view of Eq. (A31), one may write:

$$N\{\exp\{\exp\{-i\omega t\} - 1\}\alpha^*\alpha\}\} = \exp\{-i\omega a^\dagger at\}$$

3. The Time Evolution Operator

a. Some Transformations. As a consequence, the time evolution operator (C11) takes the form:

$$\{U^{IP}(t)\} = \exp\{X(t)\}\,\exp\{C(t)a^\dagger\}\,\exp\{-i\omega ta^\dagger a\}\,\exp\{B(t)a\} \tag{C12}$$

Next inserting after $\exp\{B(t)a\}$ the unity operator as given by $1 = \exp\{i\omega ta^\dagger a\}\,\exp\{-i\omega ta^\dagger a\}$, the above equation yields:

$$\begin{aligned}
\{U^{IP}(t)\} = {} & \exp\{X(t)\}\,\exp\{C(t)a^\dagger\}\,\exp\{-i\omega ta^\dagger a\} \\
& \times \exp\{B(t)a\}\,\exp\{i\omega ta^\dagger a\}\,\exp\{-i\omega ta^\dagger a\}
\end{aligned} \tag{C13}$$

Now, applying theorem (A21) of Appendix A, this equation becomes:

$$\{U^{\mathrm{IP}}(t)\} = \exp\{X(t)\}\ \exp\{C(t)a^\dagger\}$$
$$\times \exp\{B(t)\ \exp\{i\omega t\}a\}\ \exp\{-i\omega t a^\dagger a\} \tag{C14}$$

It may be remarked that, according to Eqs. (C6) and (C7):

$$B(t)\ \exp\{i\omega t\} = -C^*(t) \tag{C15}$$

so that the time evolution operator becomes:

$$\{U^{\mathrm{IP}}(t)\} = \exp\{X(t)\}\ \exp\{C(t)a^\dagger\}\ \exp\{-C^*(t)a\}\ \exp\{-i\omega t a^\dagger a\} \tag{C16}$$

Now changing the notation with:

$$\{\Phi^0(t)\} = C(t) = \beta\{\exp\{-i\omega t\} - 1\} \tag{C17}$$

the time evolution operator becomes:

$$\{U^{\mathrm{IP}}(t)\} = \exp\{X(t)\}\ \exp\{\{\Phi^0(t)\}a^\dagger\}\ \exp\{-\{\Phi^0(t)\}^*a\}\ \exp\{-i\omega t a^\dagger a\} \tag{C18}$$

b.   An Expression of the Time Evolution Operator. Besides, if we note that Eq. (C5) giving $X(t)$ may be written:

$$X(t) = \beta^2[\cos(\omega t) - 1 - i\ \sin(\omega t) + i\omega t] \tag{C19}$$

the time evolution operator becomes:

$$\{U^{\mathrm{IP}}(t)\} = \exp\{\beta^2[\cos(\omega t) - 1]\}\ \exp\{-i\beta^2\ \sin(\omega t)\}$$
$$\times \exp\{i\beta^2\omega t\}\ \exp\{\Phi^0(t)a^\dagger\}$$
$$\times \exp\{-\Phi^{0*}(t)a\}\ \exp\{-i\omega t a^\dagger a\} \tag{C20}$$

Finally, we may use the definition (A32) of the translation operator and the Glauber theorem (A18) to obtain:

$$\exp\{\Phi^0(t)a^\dagger\}\ \exp\{-\Phi^{0*}(t)a\} = \exp\{\tfrac{1}{2}|\Phi^0(t)|^2\}\{A(\Phi^0(t))\} \tag{C21}$$

Thus, owing to Eq. (C17) this leads to:

$$|\Phi^0(t)|^2 = -2\beta^2[\cos(\omega t) - 1] \tag{C22}$$

The time evolution operator takes the final result:

$$\{U^{IP}(t)\} = \exp\{-i\beta^2 \sin(\omega t)\} \exp\{i\beta^2 \omega t\}\{A(\Phi^0(t))\} \exp\{-i\omega t a^\dagger a\} \quad \text{(C23)}$$

c. *Another Suitable Expressions of the Time Evolution Operator.* Note that we may also insert $1 = \exp\{-i\omega t a^\dagger a\} \exp\{i\omega t a^\dagger a\}$ in Eq. (C12) in another fashion to give:

$$\{U^{IP}(t)\} = \exp\{X(t)\} \exp\{-i\omega t a^\dagger a\} \exp\{i\omega t a^\dagger a\}$$
$$\times \exp\{C(t)a^\dagger\} \exp\{-i\omega t a^\dagger a\} \exp\{B(t)a\} \quad \text{(C24)}$$

Moreover using theorem (A21) of Appendix A, one finds:

$$\{U^{IP}(t)\} = \exp\{X(t)\} \exp\{-i\omega t a^\dagger a\} \exp\{C(t)a^\dagger \exp\{i\omega t\}\} \exp\{B(t)a\} \quad \text{(C25)}$$

Next, in view of Eq. (C6) giving $C(t)$, we get:

$$\{U^{IP}(t)\} = \exp\{X(t)\} \exp\{-i\omega t a^\dagger a\} \exp\{\beta\{\exp\{-i\omega t\} - 1\}a^\dagger$$
$$\times \exp\{i\omega t\}\} \exp\{B(t)a\} \quad \text{(C26)}$$

or, in view of Eqs. (C7) and (C17):

$$\{U^{IP}(t)\} = \exp\{X(t)\} \exp\{-i\omega t a^\dagger a\} \exp\{-\Phi^{0*}(t)a^\dagger\} \exp\{\Phi^0(t)a\} \quad \text{(C27)}$$

and given Eqs. (C5), (C22), and (A45), this leads to:

$$\{U^{IP}(t)\} = \exp\{-i\beta^2 \sin(\omega t)\}$$
$$\times \exp\{i\beta^2 \omega t\} \exp\{-i\omega t a^\dagger a\}\{A(-\Phi^{0*}(t))\} \quad \text{(C28)}$$

4. *The Full-Time Evolution Operator Generated by the Hamiltonian (47)*

According to Eqs. (C1), (C23), and (C28), the full-time evolution operator generated by Eq. (47) may be written equivalently as follows:

$$\{U^{0\{1\}}_{II}(t)\} = \exp\{-i[\omega^\circ - \beta^2\omega]t\} \exp\{-i\beta^2 \sin(\omega t)\}$$
$$\times \exp\{i\beta^2 \omega t\}\{A(\Phi^0(t))\} \exp\{-i\omega t a^\dagger a\} \quad \text{(C29)}$$
$$\{U^{0\{1\}}_{II}(t)\} = \exp\{-i[\omega^\circ - \beta^2\omega]t\} \exp\{-i\beta^2 \sin(\omega t)\}$$
$$\times \exp\{i\beta^2 \omega t\} \exp\{-i\omega t a^\dagger a\}\{A(-\Phi^0(t))^*\} \quad \text{(C30)}$$

It may be of interest to perform on the starting Hamiltonian (47), that is, the first equation of this appendix, the following partition:

$$\{H_{\mathrm{II}}^{0\{1\}}\} = H^0 + \hbar\omega[\beta(a^\dagger + a) - \beta^2] + \hbar\omega^\circ \tag{C31}$$

$$H^0 = \hbar\omega a^\dagger a \tag{C32}$$

Then the full-time evolution operator may be written according to Eqs. (A12) and (A13):

$$\{U_{\mathrm{II}}^{0\{1\}}(t)\} = \exp\{-iH^0 t/\hbar\}\{U_{\mathrm{II}}^{0\mathrm{IP}}(t)\}$$

or:

$$\{U_{\mathrm{II}}^{0\{1\}}(t)\} = \exp\{-i\omega t a^\dagger a\}\{U_{\mathrm{II}}^{0\mathrm{IP}}(t)\} \tag{C33}$$

Thus by identification with (C30), one obtains the new interaction picture of the full-time evolution operator.

$$\{U_{\mathrm{II}}^{0\mathrm{IP}}(t)\} = \exp\{-i\beta^2 \, \sin(\omega t)\} \, \exp\{i\beta^2 \omega t\}\{A(-\Phi^{0*}(t))\} \tag{C34}$$

APPENDIX D

Calculation of the Thermal Average of the Translation Operator

Let us look at the translation operator (A32). Owing to the Glauber theorem (A18), it may be written:

$$\{A(\Phi(t))\} = \exp\{\Phi(t)a^\dagger - \Phi(t)^*a\}$$
$$= \exp\{\Phi(t)a^\dagger\} \, \exp\{-\Phi(t)^*a\} \, \exp\{-\tfrac{1}{2}[\Phi(t)a^\dagger, -\Phi(t)^*a]\} \tag{D1}$$

with $[a, a^\dagger] = 1$, and where $\Phi(t)$ is a scalar.

$$[\Phi(t)a^\dagger, -\Phi(t)^*a] = -|\Phi(t)|^2[a^\dagger, a] = |\Phi(t)|^2$$

As a consequence, the translation operator becomes:

$$\{A(\Phi(t))\} = \exp(\Phi(t)a^\dagger\} \, \exp\{-\Phi(t)^*a\} \, \exp\{\tfrac{1}{2}|\Phi(t)|^2\}$$

Now let us look at the thermal average of the translation operator performed on the Boltzmann operator. Owing to the above equation it is:

$$\varepsilon \, \mathrm{tr}\{\exp\{-\lambda a^\dagger a\}\{A(\Phi(t))\}\} = \varepsilon \, \mathrm{tr}\{\exp\{-\lambda a^\dagger a\} \exp\{\Phi(t)a^\dagger\}$$
$$\times \exp\{-\Phi(t)^*a\}\} \exp\{\tfrac{1}{2}|\Phi(t)|^2\} \quad \text{(D2)}$$

Applying the theorem (A23) of Appendix A to the right-hand side thermal average, one obtains:

$$\varepsilon \, \mathrm{tr}\{\exp\{-\lambda a^\dagger a\} \exp\{\Phi(t)a^\dagger\} \exp\{-\Phi(t)^*a\}\}$$
$$= \langle 0|\langle 0| \exp\{-\Phi(t)^*[(1 + \langle n\rangle)^{1/2}a + \langle n\rangle^{1/2}b^\dagger]\}$$
$$\times \exp\{\Phi(t)[(1 + \langle n\rangle)^{1/2}a^\dagger + \langle n\rangle^{1/2}b]\}|0\rangle|0\rangle$$

Here, $|0\rangle$ is the ground state of the harmonic oscillator. $b^\dagger$ and b are bosons in the same way as $a^\dagger$ and a and $\langle n\rangle$ is the thermal average of the number occupation operator $a^\dagger a$. Now, since the creation and annihilation operators $a^\dagger$ and $b^\dagger$ or a and b are acting on different spaces, the above equation may be written:

$$\varepsilon \, \mathrm{tr}\{\exp\{-\lambda a^\dagger a\} \exp\{\Phi(t)a^\dagger\} \exp\{-\Phi(t)^*a\}\}$$
$$= \langle 0| \exp\{-\Phi(t)^*[1 + \langle n\rangle]^{1/2}a\} \exp\{\Phi(t)[1 + \langle n\rangle]^{1/2}a^\dagger\}|0\rangle$$
$$\times \langle 0| \exp\{-\Phi(t)^*\langle n\rangle^{1/2}b^\dagger\} \exp\{\Phi(t)\langle n\rangle^{1/2}b\}|0\rangle$$

Now, expanding the four right-hand side exponentials, one obtains:

$$\varepsilon \, \mathrm{tr}\{\exp\{-\lambda a^\dagger a\} \exp\{\Phi(t)a^\dagger\} \exp\{-\Phi(t)^*a\}\}$$
$$= \langle 0| \sum\sum ((-\Phi(t)^*)^j[1 + \langle n\rangle]^{(1/2)j}a^j/j!)(\Phi(t)^k[1 + \langle n\rangle]^{(1/2)k}a^{\dagger k}/k!)|0\rangle$$
$$\times \langle 0| \sum\sum ((-\Phi(t)^*)^j\langle n\rangle^{(1/2)j}b^{\dagger j}/j!)(\Phi(t)^k\langle n\rangle^{(1/2)k}b^k/k!)|0\rangle$$

That leads after rearranging:

$$\varepsilon \, \mathrm{tr}\{\exp\{-\lambda a^\dagger a\} \exp\{\Phi(t)a^\dagger\} \exp\{-\Phi(t)^*a\}\}$$
$$= \sum\sum |\Phi(t)|^{j+k}(-1)^j[1 + \langle n\rangle]^{(1/2)[j+k]}\langle 0|a^j a^{\dagger k}|0\rangle/(j!\,k!)$$
$$\times \sum\sum |\Phi(t)|^{j+k}(-1)^j\langle n\rangle^{(1/2)[j+k]}\langle 0|b^{\dagger j}b^k|0\rangle/(j!\,k!)$$

Now, as a consequence of the properties of the boson operator, one may write:

$$a^j a^{\dagger k} |0\rangle = k!/\sqrt{(k-j)!} \, |k-j\rangle$$

$$b^{\dagger j} b^k |0\rangle = \delta_{k,\,0} |0\rangle$$

Then the above equation becomes:

$$\varepsilon \, \text{tr}\{\exp\{-\lambda a^\dagger a\} \, \exp\{\Phi(t)a^\dagger\} \, \exp\{-\Phi(t)^* a\}\}$$

$$= \sum \sum |\Phi(t)|^{j+k} (-1)^k [1 + 2\langle n\rangle]^{(1/2)[j+k]}$$

$$\times \langle 0 | k-j\rangle\langle 0 | 0\rangle (k!/\sqrt{(k-j)!})/(j!\,k!)$$

That leads to:

$$\varepsilon \, \text{tr}\{\exp\{-\lambda a^\dagger a\} \, \exp\{\Phi(t)a^\dagger\} \, \exp\{-\Phi(t)^* a\}\} = \sum |\Phi(t)|^{2j} (-1)^j [1 + 2\langle n\rangle]^j/j!$$

Finally, owing to the expansion property of the exponential, one obtains the final result

$$\varepsilon \, \text{tr}\{\exp\{-\lambda a^\dagger a\} \, \exp\{\Phi(t)a^\dagger\} \, \exp\{-\Phi(t)^* a\}\} = \exp\{-|\Phi(t)|^2 [1 + 2\langle n\rangle]\}$$

As a consequence, in view of Eq. (D2), the thermal average of the translation operator is

$$\varepsilon \, \text{tr}\{\exp\{-\lambda a^\dagger a\}\{A(\Phi(t))\}\} = \exp\{-\tfrac{1}{2}|\Phi(t)|^2 [1 + 2\langle n\rangle]\} \qquad \text{(D3)}$$

APPENDIX E

Calculation of the Product of Two Translation Operators

Let us calculate the product of two translation operators defined by Eq. (A32):

$$A(\beta) = \exp\{a^\dagger \beta - a\beta^*\}$$

$$A(\Phi) = \exp\{a^\dagger \Phi - a\Phi^*\}$$

where β and Φ are scalars and with $[a, a^\dagger] = 1$. The product is of course:

$$A(\beta)A(\Phi) = \exp\{a^\dagger \beta - a\beta^*\} \, \exp\{a^\dagger \Phi - a\Phi^*\}$$

Now we may observe this last equation may also be written with the Glauber theorem (A18):

$$A(\beta)A(\Phi) = \exp\{(a^\dagger\beta - a\beta^*) + (a^\dagger\Phi - a\Phi^*)\}$$
$$\times \exp\{\tfrac{1}{2}[(a^\dagger\beta - a\beta^*), (a^\dagger\Phi - a\Phi^*)]\}$$

Then by expansion of the commutator involved in the second exponential of the right-hand side, we get:

$$A(\beta)A(\Phi) = \exp\{(a^\dagger\beta - a\beta^*) + (a^\dagger\Phi - a\Phi^*)\} \exp\{\tfrac{1}{2}[\beta\Phi^* - \beta^*\Phi]\}$$

Besides, suppressing the brackets and changing in the first exponential of this last equation, the position of the bosons, we find:

$$A(\beta)A(\Phi) = \exp\{a^\dagger(\beta + \Phi) - a(\beta^* + \Phi^*)\} \exp\{\tfrac{1}{2}[\beta\Phi^* - \beta^*\Phi]\}$$

Again, since we have by definition of the translation operator:

$$A(\beta + \Phi) = \exp\{a^\dagger(\beta + \Phi) - a(\beta^* + \Phi^*)\}$$

we obtain the required result:

$$A(\beta)A(\Phi) = A(\beta + \Phi) \exp\{\tfrac{1}{2}[\beta\Phi^* - \beta^*\Phi]\}$$

APPENDIX F

Half-width of the Spectrum

Consider the thermal average at time t of the pulsation of the high frequency stretching mode:

$$\langle \omega^\circ(t) \rangle = \mathrm{tr}\{\{\rho_{\mathrm{II}}^{\{1\}}(t)\}(\{H_{\mathrm{II}}^{\{1\}}\} - \{H_{\mathrm{II}}^{\{0\}}\})/\hbar\} \tag{F1}$$

$\{H_{\mathrm{II}}^{\{0\}}\}$ and $\{H_{\mathrm{II}}^{\{1\}}\}$ are given by Eqs. (46) and (47) of the main text:

$$\{H_{\mathrm{II}}^{\{0\}}\} = a^\dagger a\hbar\omega^{\circ\circ}$$
$$\{H_{\mathrm{II}}^{\{1\}}\} = a^\dagger a\hbar\omega^{\circ\circ} + \alpha^\circ(a^\dagger + a)\hbar\omega^{\circ\circ} - \alpha^{\circ 2}\hbar\omega^{\circ\circ} + \hbar\omega^\circ \tag{F2}$$

$\{\rho_{\mathrm{II}}(t)\}$ is given by Eq. (122), that is, according to the following equations:

$$\{\rho_{\mathrm{II}}^{\{1\}}(t)\} = \varepsilon \exp\{-\lambda(a^\dagger - \Phi_{\mathrm{II}}(t))(a - \Phi_{\mathrm{II}}(t)^*)\}$$
$$\{\Phi_{\mathrm{II}}(t)\} = \beta[\exp\{-\gamma t/2\} \exp\{-i\omega^{\circ\circ}t\} - 1]$$

$$\lambda = \hbar\omega^{\circ\circ}/(k_B T), \qquad \varepsilon = 1 - \exp\{-\lambda\}$$

$$\beta = \alpha^{\circ}(2\omega^{\circ\circ 2} + i\gamma\omega^{\circ\circ})/[2(\omega^{\circ\circ 2} + \gamma^2/4)]$$

In a similar way, the second moment of the spectrum is:

$$\langle \omega^{\circ 2}(t) \rangle = \mathrm{tr}\{\{\rho_{II}^{\{1\}}(t)\}[(\{H_{II}^{\{1\}}\} - \{H_{II}^{\{0\}}\})/\hbar]^2\} \qquad (F3)$$

Moreover, the half-width of the spectrum is:

$$\Delta\omega = [\langle \omega^{\circ 2}(t) \rangle - \langle \omega^{\circ}(t) \rangle^2]^{1/2}$$

Hence, from Eqs. (F1) to (F3), we obtain:

$$\Delta\omega = \alpha^{\circ}\omega^{\circ\circ}[\mathrm{tr}\{\{\rho_{II}^{\{1\}}(t)\}(a^{\dagger} + a + \eta)^2\} - [\mathrm{tr}\{\{\rho_{II}^{\{1\}}(t)\}(a^{\dagger} + a + \eta)\}]^2]^{1/2}$$

with $\eta = \omega^{\circ}/\omega^{\circ\circ} - \alpha^{\circ}$. Now, we may observe that according to Eq. (A44) the density operator may be expressed in terms of the Boltzmann operator and of the translation operator given by Eq. (129):

$$\{\rho_{II}^{\{1\}}(t)\} = \varepsilon\{A(\Phi_{II}(t))\} \exp\{-\lambda a^{\dagger}a\}\{A(\Phi_{II}(t))\}^{\dagger}$$

As a consequence, the first trace involved on the right-hand side of the equation giving $\Delta\omega$, may be written:

$$\mathrm{tr}\{\{\rho_{II}^{\{1\}}(t)\}(a^{\dagger} + a + \eta)\} = \varepsilon\,\mathrm{tr}\{\{A(\Phi_{II}(t))\} \exp\{-\lambda a^{\dagger}a\}\{A(\Phi_{II}(t))\}^{\dagger}(a^{\dagger} + a + \eta)\}$$

Then using the invariance of the trace with respect to a circular permutation, we find:

$$\mathrm{tr}\{\{\rho_{II}^{\{1\}}(t)\}(a^{\dagger} + a + \eta)\} = \varepsilon\,\mathrm{tr}\{\exp\{-\lambda a^{\dagger}a\}\{A(\Phi_{II}(t))\}^{\dagger}(a^{\dagger} + a + \eta)\{A(\Phi_{II}(t))\}\}$$

Moreover, using Eq. (A44) of Appendix A, we get:

$$\mathrm{tr}\{\{\rho_{II}^{\{1\}}(t)\}(a^{\dagger} + a + \eta)\} = \varepsilon\,\mathrm{tr}\{\exp\{-\lambda a^{\dagger}a\}(a^{\dagger} + \Phi_{II}(t) + a + \Phi_{II}(t)^* + \eta)\}$$

performing the trace, one obtains:

$$\mathrm{tr}\{\{\rho_{II}^{\{1\}}(t)\}(a^{\dagger} + a + \eta)\} = [\{\Phi_{II}(t)\} + \{\Phi_{II}(t)^*\} + \eta]$$

In a similar way, it is possible to show that:

$$\text{tr}\{\{\rho_{II}^{(1)}(t)\}(a^\dagger + a + \eta)^2\} = \varepsilon \, \text{tr}\{\exp\{-\lambda a^\dagger a\}[a^\dagger + \Phi_{II}(t) + a + \Phi_{II}(t)^* + \eta]^2\}$$

and then, performing the trace, one finds:

$$\text{tr}\{\{\rho_{II}^{(1)}(t)\}(a^\dagger + a + \eta)^2\} = [\{\Phi_{II}(t)\} + \{\Phi_{II}(t)^*\} + \eta]^2 + [2\langle n \rangle + 1]$$

where $\langle n \rangle$ is the mean occupation number at temperature T given by the equations:

$$\langle n \rangle = \varepsilon \, \text{tr}\{\exp\{-\lambda a^\dagger a\} a^\dagger a\}$$

$$\langle n \rangle = 1/[\exp\{\hbar\omega^{\circ\circ}/(k_B T)\} - 1]$$

As a consequence, the width of the band is:

$$\Delta\omega = \alpha^\circ \omega^{\circ\circ} [2\langle n \rangle + 1]^{1/2}$$

or: (F4)

$$\Delta\omega = \alpha^\circ \omega^{\circ\circ} \coth^{1/2}(\hbar\omega^{\circ\circ}/(2k_B T))$$

APPENDIX G

Some Details of the Rösch and Ratner Calculations

1. Proof of the Rösch and Ratner Autocorrelation Function (357)

a.   The Interaction Picture. Let us look at the time evolution operator solution of the Schrödinger equation involving the full Hamiltonian (349):

$$i\hbar(\partial U_{Rat}(t)/\partial t) = \{H_{Rat}\}\{U_{Rat}(t)\} \tag{G1}$$

By aid of the partition (349) of this Hamiltonian, and considering the part $\{H_{III\,damp}\}$ as the perturbation, this time evolution operator in the inter-action picture is given by:

$$\{U_{Rat}(t)\} = \{U_{Rat}^{unp}(t)\}\{U_{Rat}^{IP}(t)\} \tag{G2}$$

Here the unperturbed time evolution operator is:

$$\{U_{Rat}^{unp}(t)\} = \exp\{-i\{H_{III}^0\}t/\hbar\} \tag{G3}$$

whereas the IP time evolution operator is given by the Schrödinger equation:

$$i\hbar(\partial U_{\text{Rat}}^{\text{IP}}(t)/\partial t) = \{H_{\text{III damp}}^{\text{IP}}\}\{U_{\text{Rat}}^{\text{IP}}(t)\}$$

where the perturbation Hamiltonian in the interaction picture is given [see Eq. (A17)] by:

$$\{H_{\text{III damp}}^{\text{IP}}\} = \exp\{i\{H_{\text{III}}^{0}\}t/\hbar\}\{H_{\text{III damp}}\}\exp\{-i\{H_{\text{III}}^{0}\}t/\hbar\}$$

b. *The Autocorrelation Function.* The autocorrelation function of the hydrogen bond is, according to Eq. (356), given by

$$\{G_{\text{Rat}}(t)\} = |\mu_{10}|^{2}\varepsilon \; \text{tr}'\{\exp\{-\lambda a^{\dagger}a\}\,|\{0\}\rangle\langle\{1\}|\{A^{\{1\}}(\alpha^{\circ})\}^{\dagger}$$
$$\times \langle\{U_{\text{Rat}}(t)\}^{\dagger}\rangle\{A^{\{1\}}(\alpha^{\circ})\}\,|\{1\}\rangle\langle\{0\}|\langle\{U_{\text{Rat}}(t)\}\rangle\}$$

where the trace must be performed on both the bases $\{|\{k\}\rangle\}$ and $\{|n\rangle\}$. In view of the expression (G2) and (G3) of the time evolution operator, this function may be written after performing the trace on the basis $\{|\{k\}\rangle\}$:

$$\{G_{\text{Rat}}(t)\} = |\mu_{10}|^{2}\varepsilon \; \text{tr}\{\exp\{-\lambda a^{\dagger}a\}\sum\langle\{k\}|\{0\}\rangle\langle\{1\}|\{A^{\{1\}}(\alpha^{\circ})\}^{\dagger}$$
$$\times \langle\{U_{\text{Rat}}^{\text{IP}}(t)\}^{\dagger}\rangle\exp\{i\{H_{\text{III}}^{0}\}t/\hbar\}\{A^{\{1\}}(\alpha^{\circ})\}\,|\{1\}\rangle\langle\{0\}|$$
$$\times \exp\{-i\{H_{\text{III}}^{0}\}t/\hbar\}\langle\{U_{\text{Rat}}^{\text{IP}}(t)\}\rangle\,|\{k\}\rangle\} \tag{G4}$$

Again, in view of the expression (350) of the Hamiltonian involved in the phase transformation, this equation leads, after simplifications, to:

$$\{G_{\text{Rat}}(t)\} = |\mu_{10}|^{2}\varepsilon \; \text{tr}\{\exp\{-\lambda a^{\dagger}a\}\langle\{1\}|\{A^{\{1\}}(\alpha^{\circ})\}^{\dagger}\langle\{U_{\text{Rat}}^{\text{IP}}(t)\}^{\dagger}\rangle$$
$$\times \exp\{i\{H_{\text{III}}^{0\{1\}}\}t/\hbar\}\{A^{\{1\}}(\alpha^{\circ})\}\,|\{1\}\rangle\langle\{0\}|$$
$$\times \exp\{-i\{H_{\text{III}}^{0\{0\}}\}t/\hbar\}\langle\{U_{\text{Rat}}^{\text{IP}}(t)\}\rangle\,|\{0\}\rangle\}$$

At last, owing to Eq. (351), giving the expression of the effective Hamiltonian, the above equation leads to:

$$\{G_{\text{Rat}}(t)\} = |\mu_{10}|^{2}\varepsilon \; \text{tr}\{\exp\{-\lambda a^{\dagger}a\}\langle\{1\}|\{A^{\{1\}}(\alpha^{\circ})\}^{\dagger}\langle\{U_{\text{Rat}}^{\text{IP}}(t)\}^{\dagger}\rangle$$
$$\times \exp\{ia^{\dagger}a\omega^{\circ\circ}t\}\{A^{\{1\}}(\alpha^{\circ})\}\,|\{1\}\rangle\langle\{0\}|$$
$$\times \exp\{-ia^{\dagger}a\omega^{\circ\circ}t\}\langle\{U_{\text{Rat}}^{\text{IP}}(t)\}\rangle\,|\{0\}\rangle\}$$
$$\times \exp\{i(\omega^{\circ} - 2\alpha^{\circ2}\omega^{\circ\circ})t\}$$

Moreover, using the fact that the translation and the boson operators do commute with $|\{0\}\rangle$ and $|\{1\}\rangle$, we get:

$$\{G_{\text{Rat}}(t)\} = |\mu_{10}|^2 \varepsilon \, \text{tr}\{\exp\{-\lambda a^\dagger a\}\{A^{\{1\}}(\alpha^\circ)\}^\dagger\langle\{1\}|\langle\{U_{\text{Rat}}^{\text{IP}}(t)\}^\dagger\rangle|\{1\}\rangle$$
$$\times \exp\{ia^\dagger a\omega^{\circ\circ}t\}\{A^{\{1\}}(\alpha^\circ)\} \exp\{-ia^\dagger a\omega^{\circ\circ}t\}\langle\{0\}|$$
$$\times \langle\{U_{\text{Rat}}^{\text{IP}}(t)\}\rangle|\{0\}\rangle\} \exp\{i(\omega^\circ - 2\alpha^{\circ 2}\omega^{\circ\circ})t\}$$

Besides, the unitary transformation involving the translation operator appearing in this equation gives in view of theorem (A21) of Appendix A:

$$\exp\{ia^\dagger a\omega^{\circ\circ}t\}\{A^{\{1\}}(\alpha^\circ)\} \exp\{-ia^\dagger a\omega^{\circ\circ}t\} = \{A^{\{1\}}(\Phi_{\text{III}}^0(t)^*)\}$$

where the argument of the time-dependent translation operator is given by:

$$\{\Phi_{\text{III}}^0(t)\} = \alpha^\circ \exp\{-i\omega^{\circ\circ}t\}$$

As a consequence, the autocorrelation function becomes:

$$\{G_{\text{Rat}}(t)\} = |\mu_{10}|^2 \exp\{i(\omega^\circ - 2\alpha^{\circ 2}\omega^{\circ\circ})t\}$$
$$\times \varepsilon \, \text{tr}\{\exp\{-\lambda a^\dagger a\}\langle\{1\}|\langle\{U_{\text{Rat}}^{\text{IP}}(t)\}^\dagger\rangle|\{1\}\rangle\langle\{0\}|$$
$$\times \langle\{U_{\text{Rat}}^{\text{IP}}(t)\}\rangle|\{0\}\rangle\{A^{\{1\}}(\alpha^\circ)\}^\dagger\{A^{\{1\}}(\Phi_{\text{III}}^0(t)^*)\}\} \qquad \text{(G5)}$$

2. Diagonal Matrix Elements of the IP Evolution Operator

In order to find the explicit expression of the damped autocorrelation function (358), it is required to find the thermal average of the IP time evolution operator appearing in this equation.

$$\{U_{\text{Rat}}^{\text{IP}}(t)\} = \exp\left\{-i \int_0^t d\tau \, \{H_{\text{III damp}}^{\text{IP}}(\tau)\}\right\} \qquad \text{(G6)}$$

In the interaction picture according to Eq. (A15) of Appendix A, the term involved in the exponential is:

$$\{H_{\text{III damp}}^{\text{IP}}(\tau)\} = \{U_{\text{Rat}}^{\text{unp}}(\tau)\}^\dagger\{H_{\text{III damp}}\}\{U_{\text{Rat}}^{\text{unp}}(\tau)\}$$

The IP evolution operator becomes therefore:

$$\{U_{\text{Rat}}^{\text{IP}}(t)\} = \exp\left\{-i \int_0^t d\tau \, \{U_{\text{Rat}}^{\text{unp}}(\tau)\}^\dagger\{H_{\text{III damp}}\}\{U_{\text{Rat}}^{\text{unp}}(\tau)\}\right\}$$

Or, in view of Eq. (353):

$$\{U_{\text{Rat}}^{\text{IP}}(t)\} = \exp\left\{-i \sum \sum \mu_{kl} \int_0^t d\tau \, \{E^0(\tau)\}\{U_{\text{Rat}}^{\text{unp}}(\tau)\}^\dagger \right.$$
$$\left. \times \{A(k\alpha^\circ)\} \,|\, \{k\}\rangle\langle\{l\}\,|\, \{A(l\alpha^\circ)\}^\dagger \{U_{\text{Rat}}^{\text{unp}}(\tau)\} \right\}$$

Then, using Eq. (G3):

$$\{U_{\text{Rat}}^{\text{IP}}(t)\} = \exp\left\{-i \sum \sum \mu_{kl} \int_0^t d\tau \, \{E^0(\tau)\} \, \exp\{i\{H_{\text{III}}^0\}\tau/\hbar\} \right.$$
$$\left. \times \{A(k\alpha^\circ)\} \,|\, \{k\}\rangle\langle\{l\}\,|\, \{A(l\alpha^\circ)\}^\dagger \, \exp\{-i\{H_{\text{III}}^0\}\tau/\hbar\} \right\}$$

Again, using the expansion (350) of Hamiltonian $\{H_{\text{III}}^0\}$ in terms of the effective Hamiltonians $\{H_{\text{III}}^{0\{k\}}\}$, one gets:

$$\{U_{\text{Rat}}^{\text{IP}}(t)\} = \exp\left\{-i \sum \sum \mu_{kl} \int_0^t d\tau \, \{E^0(\tau)\} \, \exp\{i\{H_{\text{III}}^{0\{k\}}\}\tau/\hbar\} \right.$$
$$\left. \times \{A(k\alpha^\circ)\} |\, \{k\}\rangle\langle\{l\}\,|\, \{A(l\alpha^\circ)\}^\dagger \, \exp\{-i\{H_{\text{III}}^{0\{l\}}\}\tau/\hbar\} \right\}$$

Now we may observe that:

$$\exp\{i\{H_{\text{III}}^{0\{k\}}\}\tau/\hbar\}\{A(k\alpha^\circ)\} \,|\, \{k\}\rangle\langle\{l\}\,|\, \{A(l\alpha^\circ)\}^\dagger \, \exp\{-i\{H_{\text{III}}^{0\{l\}}\}\tau/\hbar\}$$
$$= \exp\{ia^\dagger a\omega^{\circ\circ}t\}\{A(k\alpha^\circ)\} \,|\, \{k\}\rangle\langle\{l\}\,|\, \{A(l\alpha^\circ)\}^\dagger$$
$$\times \exp\{-ia^\dagger a\omega^{\circ\circ}t\} \, \exp\{i\omega_{kl}\tau\}$$
$$\omega_{kl} = (k - l)\omega^\circ - [k(k + 1) - l(l + 1)]\alpha^{\circ 2}\omega^{\circ\circ}$$

Besides, inserting the unity operator $\mathbf{1} = \exp\{-ia^\dagger a\omega^{\circ\circ}t\} \, \exp\{ia^\dagger a\omega^{\circ\circ}t\}$, one obtains:

$$\exp\{i\{H_{\text{III}}^{0\{k\}}\}\tau/\hbar\}\{A(k\alpha^\circ)\} \,|\, \{k\}\rangle\langle\{l\}\,|\, \{A(l\alpha^\circ)\}^\dagger \, \exp\{-i\{H_{\text{III}}^{0\{l\}}\}\tau/\hbar\}$$
$$= \exp\{ia^\dagger a\omega^{\circ\circ}t\}\{A(k\alpha^\circ)\} \,|\, \{k\}\rangle \, \exp\{-ia^\dagger a\omega^{\circ\circ}t\}$$
$$\times \exp\{ia^\dagger a\omega^{\circ\circ}t\}\langle\{l\}\,|\, \{A(l\alpha^\circ)\}^\dagger \, \exp\{-ia^\dagger a\omega^{\circ\circ}t\} \, \exp\{i\omega_{kl}\tau\}$$

This may also be written:

$$\exp\{i\{H_{III}^{0\{k\}}\}\tau/\hbar\}\{A(k\alpha^\circ)\} \,|\,\{k\}\rangle\langle\{l\}\,|\,\{A(l\alpha^\circ)\}^\dagger \exp\{-i\{H_{III}^{0\{l\}}\}\tau/\hbar\}$$
$$= \exp\{ia^\dagger a\omega^{\circ\circ}t\}\{A(k\alpha^\circ)\} \exp\{-ia^\dagger a\omega^{\circ\circ}t\}\,|\,\{k\}\rangle\langle\{l\}\,|$$
$$\times \exp\{ia^\dagger a\omega^{\circ\circ}t\}\{A^\dagger(l\alpha^\circ)\} \exp\{-ia^\dagger a\omega^{\circ\circ}t\} \exp\{i\omega_{kl}\tau\}$$

Again, using theorem (A21) of Appendix A, one obtains:

$$\exp\{i\{H_{III}^{0\{k\}}\}\tau/\hbar\}\{A(k\alpha^\circ)\} \,|\,\{k\}\rangle\langle\{l\}\,|\,\{A(l\alpha^\circ)\}^\dagger \exp\{-i\{H_{III}^{0\{l\}}\}\tau/\hbar\}$$
$$= \{A(k\Phi_{III}^0(\tau)^*)\} \,|\,\{k\}\rangle\langle\{l\}\,|\,\{A(l\Phi_{III}^0(\tau)^*)\}^\dagger \exp\{i\omega_{kl}\tau\}$$

As a consequence, the IP time evolution operator becomes:

$$\{U_{Rat}^{IP}(t)\} = \exp\left\{-i\sum\sum\mu_{kl}\int_0^t \tau\{E^0(\tau)\}\exp\{i\omega_{kl}\tau\}\right.$$
$$\left.\times \{A(k\Phi_{III}^0(\tau)^*)\} \,|\,\{k\}\rangle\langle\{l\}\,|\,\{A^\dagger(l\Phi_{III}^0(\tau)^*)\}\right\} \qquad \text{(G7)}$$

3. Thermal Average Value of the IP Exponential Operator Involved in Eq. (364)

We start from Eq. (364):

$$\langle\{U^{IP\{r,\,r\}}(t)\}\rangle = \langle\{r\}\,|\,\exp\left\{-\sum\sum\sum\sum\mu_{kl}\mu_{pq}\int_0^t\int_0^{\tau_1} d\tau_1\,d\tau_2\right.$$
$$\times \exp\{i[\omega_{kl}\tau_1 + \omega_{pq}\tau_2]\}\langle E^0(\tau_1)E^0(\tau_2)\rangle$$
$$\times \varepsilon\,\mathrm{tr}\{\exp\{-\lambda a^\dagger a\}\{A(k\Phi_{III}^0(\tau_1))\} \,|\,\{k\}\rangle\langle\{l\}\,|\,\{A(l\Phi_{III}^0(\tau_1))\}^\dagger$$
$$\left.\times \{A(p\Phi_{III}^0(\tau_2)^*)\} \,|\,\{p\}\rangle\langle\{q\}\,|\,\{A(q\Phi_{III}^0(\tau_2)^*)\}^\dagger\right\}\,|\,\{r\}\rangle \qquad \text{(G8)}$$

Now, we may observe that, using the fact that the eigenstates of the fast mode commute with the effective translation operator and that the thermal average may be performed on the Boltzmann operator of the slow mode, the thermal average involved in the last equation may be written,

$$\langle\{A(k\Phi_{III}^0(\tau_1))\} \,|\,\{k\}\rangle\langle\{l\}\,|\,\{A(l\Phi_{III}^0(\tau_1)^*)\}^\dagger$$
$$\times \{A(p\Phi_{III}^0(\tau_2))\} \,|\,\{p\}\rangle\langle\{q\}\,|\,\{A(q\Phi_{III}^0(\tau_2)^*)\}^\dagger\rangle$$
$$= \varepsilon\,\mathrm{tr}\{\exp\{-\lambda a^\dagger a\}\{A(k\Phi_{III}^0(\tau_1)^*)\}\langle\{l\}\,|\,\{p\}\rangle\,|\,\{k\}\rangle$$
$$\times \langle\{q\}\,|\,\{A(l\Phi_{III}^0(\tau_1)^*)\}^\dagger\{A(p\Phi_{III}^0(\tau_2)^*)\}\{A(q\Phi_{III}^0(\tau_2)^*)\}^\dagger\}$$

As a consequence, one obtains, after simplification using the orthogonality of the eigenkets of the slow mode:

$$
\langle\{U^{IP\{r,\,r\}}(t)\}\rangle = \langle\{r\}\,|\,\exp\Big\{-i\sum\sum\sum \mu_{kl}\,\mu_{pq}\int_0^t\int_0^{\tau_1} d\tau_1\,d\tau_2
$$

$$
\times\,\exp\{i[\omega_{kl}\,\tau_1+\omega_{pq}\,\tau_2]\}\langle E^0(\tau_1)E^0(\tau_2)\rangle
$$

$$
\times\,\varepsilon\,\mathrm{tr}\{\exp\{-\lambda a^\dagger a\}\{A(k\Phi^0_{III}(\tau_1)^*)\}\,|\,\{k\}\rangle\{A(l\Phi^0_{III}(\tau_1)^*)\}^\dagger
$$

$$
\times\,\{A(l\Phi^0_{III}(\tau_2)^*)\}\langle\{q\}\,|\,\{A(q\Phi^0_{III}(\tau_2)^*)\}^\dagger\Big\}\,|\,\{r\}\rangle
$$

Using the orthogonality of the eigenkets of the slow mode once, this reduces to:

$$
\langle\{U^{IP\{r,\,r\}}(t)\}\rangle = \exp\Big\{-i\sum|\mu_{rl}|^2\int_0^t\int_0^{\tau_1} d\tau_1\,d\tau_2
$$

$$
\times\,\exp\{i\omega_{rl}(\tau_1-\tau_2)\}\langle E^0(\tau_1)E^0(\tau_2)\rangle
$$

$$
\times\,\varepsilon\,\mathrm{tr}\{\exp\{-\lambda a^\dagger a\}\{A(r\Phi^0_{III}(\tau_1)^*)\}
$$

$$
\times\,\{A(l\Phi^0_{III}(\tau_1)^*)\}^\dagger\{A(l\Phi^0_{III}(\tau_2)^*)\}\{A(r\Phi^0_{III}(\tau_2)^*)\}^\dagger\}\Big\}
$$

Again, using the properties of the translation operators:

$$
\langle\{U^{IP\{r,\,r\}}(t)\} = \exp\Big\{-\sum|\mu_{rl}|^2\int_0^t\int_0^{\tau_1} d\tau_1\,d\tau_2
$$

$$
\times\,\exp\{i\omega_{rl}(\tau_1-\tau_2)\}\langle E^0(\tau_1)E^0(\tau_2)\rangle
$$

$$
\times\,\varepsilon\,\mathrm{tr}\{\exp\{-\lambda a^\dagger a\}\{A(r\Phi^0_{III}(\tau_1)^*)\}
$$

$$
\times\,\{A(-l\Phi^0_{III}(\tau_1)^*)\}\{A(l\Phi^0_{III}(\tau_2)^*)\}\{A(-r\Phi^0_{III}(\tau_2)^*)\}\}\Big\}
$$

Besides, since the autocorrelation functions involved in the above expression are stationary, the integrand is a function of the difference between the two arguments, $\tau=\tau_1-\tau_2$. The rectangular region of integration can be divided into two triangular regions; after coordinate transformation and quadrature the above equation reduces to:

$$\langle\{U^{\text{IP}(r,\,r)}(t)\}\rangle = \exp\left\{-\sum|\mu_{rl}|^2\int_0^t d\tau\,(t-\tau)\right.$$
$$\times\,\exp\{i\omega_{rl}(\tau_1-\tau_2)\}\{\langle E^0(\tau)E^0(0)\rangle + \langle E^0(-\tau)E^0(0)\rangle\}$$
$$\times\,\varepsilon\,\text{tr}\{\exp\{-\lambda a^\dagger a\}\{A(r\Phi^0_{\text{III}}(0)^*)\}\{A(-l\Phi^0_{\text{III}}(0)^*)\}$$
$$\times\,\{A(l\Phi^0_{\text{III}}(\tau)^*)\}\{A(-r\Phi^0_{\text{III}}(\tau)^*)\}\bigg\}$$

Note that, owing to Eq. (A45), we may write, respectively:

$$\{A(r\Phi^0_{\text{III}}(0)^*)\}\{A(-l\Phi^0_{\text{III}}(0)^*)\} = \{A([r-l]\Phi^0_{\text{III}}(0)^*)\}$$
$$\{A(l\Phi^0_{\text{III}}(\tau)^*)\}\{A(-r\Phi^0_{\text{III}}(\tau)^*)\} = \{A([l-r]\Phi^0_{\text{III}}(\tau)^*)\}$$

Moreover, according to the same theorem (A45), the product of the right-hand side translation operators is:

$$\{A([r-l]\Phi^0_{\text{III}}(0)^*)\}\{A([r-l]\Phi^0_{\text{III}}(\tau)^*)\}$$
$$= \{A([r-l][\Phi^0_{\text{III}}(0)^* - \Phi^0_{\text{III}}(\tau)^*])\}\,\exp\{i[r-l]^2\,|\alpha^\circ|^2\,\sin(\omega^{\circ\circ}\tau)\}$$

Again, in view of Eq. (D3), the partial trace on the right-hand side translation operator of the last equation is:

$$\varepsilon\,\text{tr}\{\exp\{-\lambda a^\dagger a\}\{A([r-l][\Phi^0_{\text{III}}(0)^* - \Phi^0_{\text{III}}(\tau)^*])\}\}$$
$$= \exp\{[r-l]^2\,|\alpha^\circ|^2(1+2\langle n\rangle)[\cos(\omega^{\circ\circ}\tau)-1]\}$$

Here $\langle n\rangle$ is the thermal average of the number occupation operator given by Eq. (H90). Then, using the above equations, one obtains finally:

$$\langle\{U^{\text{IP}(r,\,r)}(t)\}\rangle = \exp\left\{-\sum|\mu_{rl}|^2\int_0^t d\tau\,(t-\tau)\,\exp\{i\omega_{rl}(\tau_1-\tau_2)\}\right.$$
$$\times\,\{\langle E^0(\tau)E^0(0)\rangle + \langle E^0(-\tau)E^0(0)\rangle\}$$
$$\times\,\exp\{[r-l]^2\,|\alpha^\circ|^2(1+2\langle n\rangle)[\cos(\omega^{\circ\circ}\tau)-1]\}$$
$$\times\,\exp\{i[r-l]^2\,|\alpha^\circ|^2\,\sin(\omega^{\circ\circ}\tau)\}\bigg\} \qquad\text{(G9)}$$

APPENDIX H

The Classical and Quantum Relaxation of the Harmonic Oscillator

In this appendix we give some general information on the connection between the classical and quantum relaxation of the harmonic oscillator. That is, we clarify the physics at the basis of the classical and quantum autocorrelation function of the dipole moment operator of the high-frequency stretching mode of the hydrogen bond that is used in the main text. We shall look first at the classical approach of the damped harmonic oscillator, which is the basis of the Robertson (1978) and Sakun (1985) models and, second, we shall consider the quantum study that is the basis of the Boulil et al. model (1988, 1994).

1. The Classical Approach of the Damped Harmonic Oscillator

Now let us look at the classical approach of the damped harmonic oscillator. For that, we shall refer extensively to Coffey (1985). We shall recall the connection between the Langevin method, the diffusion equation.

a. The Ornstein–Uhlenbeck Approach within the Langevin Method and the Corresponding Fokker–Planck Equation. Consider the Ornstein–Uhlenbeck analysis (1930) of a classical damped harmonic oscillator of reduced mass m and angular frequency ω. The authors used to obtain the dynamics of the position q of the oscillator, the Langevin method. In terms of the velocity v, the Langevin equation is:

$$M\,dv/dt + M\gamma v = F(t) + F^0 \tag{H1}$$

Here M is the reduced mass of the oscillator, γ the damping parameter, F^0 the applied force, and $F(t)$ a stochastic force acting on the q coordinate and reflecting the action of the medium. (Note that in the present appendix, q plays the role of Q in the main text since it is the slow mode that must be described by a Brownian operator.) The velocity and the applied force are given, respectively, by:

$$v = dq/dt \tag{H2}$$

$$F^0 = -\partial U(q)/\partial q \tag{H3}$$

For a harmonic oscillator, the potential $U(q)$ is:

$$U(q) = \tfrac{1}{2}M\omega^2 q^2 \tag{H4}$$

For the damped harmonic oscillator, the Langevin equation is therefore:

$$M \, d^2q/dt^2 + M\gamma \, dq/dt = F(t) - M\omega^2 q$$

In the Langevin theory the fluctuating force is considered as having a zero average value and a Dirac delta autocorrelation function and obeys the equations:

$$\langle F(t) \rangle = 0 \tag{H5}$$

$$\langle F(t)F(0) \rangle = \langle F(0)^2 \rangle \{\delta(t)\} \tag{H6}$$

$$\langle F(0)^2 \rangle = 2k_B T\gamma M \tag{H7}$$

Consider the underdamped situation corresponding to $\omega > \gamma$. Within the Langevin formalism, Ornstein and Uhlenbeck (1930) and later Wang and Uhlenbeck (1945) were able to obtain for the underdamped and over-damped situations the following results that hold for positive times:

$$\langle q(t) \rangle = 0$$

$$\langle q(t)q(0) \rangle = \langle q^2 \rangle \exp\{-\gamma t/2\}\{\cos(\omega_{und} t) + [\gamma/(2\omega_{und})] \sin(\omega_{und} t)\} \tag{H8}$$

$$\omega_{und}^2 = [\omega^2 - \gamma^2/4]$$

$$\langle q(t)q(0) \rangle = \langle q^2 \rangle \exp\{-\gamma t/2\}$$

$$\times \{\cosh(\omega_{over} t) + [\gamma/(2\omega_{over} t)] \sinh(\omega_{over} t)\} \tag{H9}$$

$$\omega_{over}^2 = -[\omega^2 - \gamma^2/4]$$

The authors did not show how the probability distribution of the displacements for the oscillator may be obtained by aid of a diffusion equation. This was performed by Klein, who showed that the phase-space distribution function:

$$\{f(q, v, t)\} = \{f(q, v, t \,|\, q^0, v^0, 0)\} \tag{H10}$$

where q^0 and v^0 are the initial values of the position and velocity at $t = 0$, may be obtained by the following equation, which is known as the Fokker–Planck–Kramers equation:

$$(\partial f(q, v, t)/\partial t) = -v(\partial f(q, v, t)/\partial q)$$

$$+ (1/M)(\partial U(q)/\partial q)(\partial f(q, v, t)/\partial v)$$

$$+ \gamma\{\partial/\partial v\}\{(vf(q, v, t)) + [k_B T/M](\partial f(q, v, t)/\partial v)\} \tag{H11}$$

The solution of this equation allows one to perform average velocities that are the same as those obtained from the Langevin equation (H1) averaged on the stochastic force and subject to the conditions (H5)–(H7) and leads therefore to the same autocorrelation function of the q coordinate.

b.   Connection with the Boltzmann Equation. It must be observed that Eq. (H11) may be written as a special case of the Boltzmann equation, according to:

$$(\partial f(q, v, t)/\partial t) + v(\partial f(q, v, t)/\partial q) - (1/M)(\partial U(q)/\partial q)(\partial f(q, v, t)/\partial v)$$
$$= D\{f(q, v, t)\}/Dt \quad \text{(H12)}$$

in which the effect of collisions is represented by the term:

$$D\{f(q, v, t)\}/Dt = \gamma(\partial/\partial v)\{(vf(q, v, t)) + [k_B T/M](\partial f(q, v, t)/\partial v)\} \quad \text{(H13)}$$

Recall that the Boltzmann equation is for a system involving very high-freedom degrees, an expression of the Liouville equation:

$$(\partial f(q, v, t)/\partial t) = \{H, f(q, v, t)\} \quad \text{(H14)}$$

in which the right-hand side term is the Poisson bracket given by:

$$\{H, f(q, v, t)\} = \sum (\partial f(q, v, t)/\partial p_i)(\partial H/\partial q_i) - (\partial f(q, v, t)/\partial p_i)(\partial H/\partial q_i) \quad \text{(H15)}$$

where q_i and p_i are, respectively, the position and the conjugate momentum of the particle i in the phase space.

c.   Special Situation When Inertial Effects Are Neglected. It may be of interest to observe that if the force does not change appreciably over distance of the order of $k_B T/m\gamma$, and if we are only interested in times that are great with respect to the inverse of the damping parameter (i.e., $T > 1/\gamma$), this implies that as time passes, the distribution of velocity v becomes of the Maxwell–Boltzmann type, while that of the q coordinate remains still away of it. In other words, an order exists for reaching the equilibrium state: First the velocity and then the position. From this viewpoint, that is, when the above approximation may be verified, the inertial effect can be neglected in both the Langevin equation (H1) and the Fokker–Planck–Kramers equation (H11). Then the Langevin equation (H1) reduces to:

$$M\gamma v + \partial U(q)/\partial q = F(t) \quad \text{(H16)}$$

whereas the Fokker–Planck–Kramers equation (H11) transforms into the following equation governing the distribution function in the coordinate space:

$$(\partial f(q, t)/\partial t) = (1/M\gamma)(\partial((\partial U(q)/\partial q)f(q, t))/\partial q) + (k_B T/M\gamma)(\partial^2 f(q, t)/\partial q^2)$$
(H17)

In Eq. (H17), named the Smoluchowski equation, the conditional probability is given by:

$$\{f(q, t)\} = \{f(q, t \mid q^0, 0)\} \tag{H18}$$

At least, it may be remarked that the Smoluchowski equation reduces in the absence of potential to the Einstein equation for Brownian motions:

$$(\partial f(q, t)/\partial t) = D(\partial^2 f(q, t)/\partial q^2) \tag{H19}$$

with the Einstein diffusion constant given by:

$$D = k_B T/M\gamma \tag{H20}$$

d. *The Solution Fokker–Planck Equation for Velocity in the Absence of Force with and without Neglecting the Inertia.* By solving Eq. (H11), in which the potential $U(q)$ is zero, and integrating the resulting expression for the distribution function over all the velocities, one finds:

$$\{f(q, t \mid q^0, 0)\} = \{M\gamma^2/[2\pi k_B T(2\gamma t - 3 + 4 \exp\{-\gamma t\} - \exp\{-2\gamma t\})]\}^{1/2}$$
$$\times \exp\{-M\gamma^2[q - q^0 - v^0(1 - \exp\{-\gamma t\})/\gamma]^2/$$
$$[2k_B T(2\gamma t - 3 + 4 \exp\{-\gamma t\} - \exp\{-2\gamma t\})]\} \quad \text{(H21)}$$

On the other hand, when one neglects the inertia effect, it must be observed that the Fokker–Planck equation for the distribution function of velocities in the absence of force F is:

$$\{\partial f(v, t)/\partial t\} = \gamma\{\partial/\partial v\}\{(vf(v, t)) + (k_B T/M)(\partial f(v, t)/\partial v)\} \tag{H22}$$

Appearing in this equation is the conditional probability given by:

$$\{f(v, t)\} = f(v, t \mid v^0, 0) \tag{H23}$$

The solution of this Fokker–Planck equation is:

$$\{f(v, t \mid v^0, 0)\} = \{M/[2\pi k_B T(1 - \exp\{-2\gamma t\})]\}^{1/2}$$

$$\times \exp\{-M[v - v^0 \exp\{-\gamma t\}]^2/[2k_B T(1 - \exp\{-2\gamma t\})]\}$$

(H24)

Note that, as required, this equation reduces for infinite time to the Maxwell–Boltzmann distribution:

$$\{f(v, \infty \mid v^0, 0)\} = \{M/[2\pi k_B T]\}^{1/2} \exp\{-Mv^2/[2k_B T]\} \quad (H25)$$

e. *The Generalized Langevin Equation.* Now consider the generalized Langevin equation of a Brownian oscillator of mass m governing the velocity v, when the stochastic force is such that its correlation function is not delta peaked but is involving a finite time correlation. This generalized Langevin equation, which may be compared to the Langevin equation (H1), is, defining by Γ the memory function:

$$M \, d^2q/dt^2 + M \int_0^t \{\Gamma(t - t')\}\{dq(t')/dt\} \, dt' = F(t) - M\omega^2 q$$

$$M \, d^2q/dt^2 + (M/k_B T) \int_0^t \{F(t - t')F(0)\}\{dq(t')/dt\} \, dt' = F(t) - M\omega^2 q \quad (H26)$$

$F(t)$ is the stochastic force obeying the following equations:

$$\langle F(t)F(0)\rangle = \langle F(0)^2\rangle \exp\{-t/\tau_c^S\} \quad (H27)$$

$$\langle F(t)\rangle = 0 \quad (H28)$$

where τ_c^S is the correlation time of the generalized Langevin force.

It may be observed that when the correlation time of the stochastic force reaches zero, according to Eq. (H27) the autocorrelation function of the stochastic force reduces to (H6) and (H7) of the crude Langevin approach:

$$\langle F(t)F(0)\rangle = 2\langle F(0)^2\rangle \, \delta(t) = 2M\gamma k_B T \quad (H29)$$

Note that in such a situation, the corresponding diffusion equation in the phase space is the generalized Fokker–Planck–Kramers equation:

$$(\partial f(q, v, t)/\partial t) = -v(\partial f(q, v, t)/\partial q) + (1/M)(\partial U(q)/\partial q)(\partial f(q, v, t)/\partial v)$$

$$+ \Gamma(t)[(\partial(vf(q, v, t))/\partial v) + [k_B T/M](\partial^2 f(q, v, t)/\partial v^2)] \quad (H30)$$

2. The Basic Equations of the Quantum Approach of the Driven Damped Harmonic Oscillator

Now let us look at the quantum approach of the driven damped quantum harmonic oscillator starting from a coherent state at any temperature. For this section we shall refer to the Louisell works (1965, 1973). We shall give shortly the connection between the starting von Neumann–Liouville equation and the equation governing the reduced density operator of the driven damped harmonic oscillator. The full solution of this equation is given by Louisell and Walker and is used in the Boulil et al. autocorrelation function and the corresponding perturbative solution expressed by a master equation, which within the normal ordering technique and passing to scalars leads to a Fokker–Planck equation that is similar to that of a Brownian free particle.

The Hamiltonian of a driven damped harmonic oscillator may be written, neglecting the zero-point energies, according to the following equations:

$$H = H^{00} + H^0 + H_{\text{int}} + H_\theta \tag{H31}$$

$$H^{00} = a^\dagger a \hbar \omega$$

$$H^0 = \alpha^\circ [a^\dagger + a] \hbar \omega \tag{H32}$$

$$H_{\text{int}} = \sum \{b_k^\dagger a \kappa_k + b_k a^\dagger \kappa_k\} \hbar \tag{H33}$$

$$H_\theta = \sum \{b_k^\dagger b_k \hbar \omega_k\} \tag{H34}$$

H^0 is the Hamiltonian of the driven harmonic oscillator, H_θ is that of the thermal bath considered as an infinity of harmonic oscillators, the frequencies of which ω_k vary continuously, whereas H_{int} is the interaction Hamiltonian between the driven oscillator and the thermal bath; a and $a^\dagger$ are the bosons of the driven harmonic oscillator, b and $b^\dagger$ those of the thermal bath. At last, α° is the strength of the term driving the harmonic oscillator and κ_k are the parameters characterizing the strength of the coupling between the driven harmonic oscillator and the thermal bath.

It is assumed that at initial time the density operator $\{\rho(0)\}$ of the driven harmonic oscillator is that of a coherent state similar to that given by Eq. (82) of the main text and characterizing the density operator of the slow mode in representation $\{III\}$:

$$\{\rho(0)\} = \varepsilon \exp\{-\lambda(a^\dagger - \alpha_\circ)(a - \alpha_\circ)\} \tag{H35}$$

$$\lambda = \hbar \omega / (k_B T) \tag{H36}$$

$$\varepsilon = (1 - \exp\{-\lambda\}) \tag{H37}$$

Note that one must distinguish between α° appearing in Eq. (H32) and $\alpha_{\circ}$ appearing in Eq. (H35). Moreover, in the following we shall consider a special situation of the initial conditions that may be written according to the equations:

$$\{\rho(0)\} = |\alpha_{\circ}\rangle\langle\alpha_{\circ}|$$
$$a|\alpha_{\circ}\rangle = \alpha_{\circ}|\alpha_{\circ}\rangle \tag{H38}$$

where $|\alpha_{\circ}\rangle$ is a coherent state and we have used for the last equation, Eq. (A46).

The dynamics of the reduced density operator governing the dynamics of the oscillator embedded in the thermal bath is given by the von Neumann–Liouville equation:

$$i\hbar(\partial\rho(t)/\partial t) = \mathrm{tr}_{\theta}[H, \{\rho_{\mathrm{tot}}(t)\}] \tag{H39}$$

in which the right-hand side involves a partial trace over the density operator of the thermal bath of the commutator of the total Hamiltonian (H29) with the full density operator $\{\rho_{\mathrm{tot}}(t)\}$.

$$[H, \{\rho_{\mathrm{tot}}(t)\}] = H\{\rho_{\mathrm{tot}}(t)\} - \{\rho_{\mathrm{tot}}(t)\}H \tag{H40}$$

This quantum von Neumann–Liouville equation may be connected with the classical Liouville equation (H14) governing the dynamics of the distribution function in the phase space, the quantum commutator (H39) playing the role of the Poisson bracket (H15), and the diagonal matrix elements of the density operator, that of the distribution function $f(q, v, t)$.

The full density operator appearing on the right-hand side of Eq. (H39) is given by:

$$\{\rho_{\mathrm{tot}}(0)\} = \rho(0)\rho_{\theta} \tag{H41}$$

Here ρ_{θ} is the density operator of the thermal bath that does not depend on time, as required for a thermal bath. It is taken as a product of the Boltzmann operators of the oscillators of the thermal bath and thus is defined by the following equations.

$$\rho_{\theta} = \prod \varepsilon_k \exp\{-\lambda_k(b_k^{\dagger} b_k)\} \tag{H42}$$

$$\lambda_k = \hbar\omega_k/(k_{\mathrm{B}} T) \tag{H43}$$

$$\varepsilon_k = (1 - \exp\{-\lambda_k\}) \tag{H44}$$

It may be of interest to observe that the present model is similar to another one in which there are a very high number of quantum harmonic oscillators that are linearly coupled. The first starts with the initial situation given by a coherent state [i.e., by Eq. (H38)], leading to a dynamical situation where the entropy of the system viewed as a lack of knowledge on the details of the model reaches progressively a stationary maximum value, from which the energy repartition becomes that of a Boltzmann distribution (Blaise et al., 1994).

3. The Main Lines of the Direct Resolution of the Liouville Equation by Louisell and Walker (1965)

a. The Liouville Equation within the Normal Ordering Formalism. The first step in the solution of the Liouville equation (H39) is to express the full density operator at time t in normal order and to apply the N^{-1} operator by aid of Eq. (A28) of Appendix A:

$$N^{-1}\{\rho_{\text{tot}}^{(n)}(t)\} = \{\rho_{\text{tot}}^{(n)}(\zeta, t)\} \tag{H45}$$

where ζ is a scalar which will be defined later.

The Liouville equation (H39) may therefore be written in normal order with Eq. (A29). Owing to the fact that there are two bosons for describing the damped harmonic oscillator and $2N$ bosons to describe the N oscillators simulating the thermal bath, this Liouville equation may be written:

$$\{\partial\rho_{\text{tot}}^{(n)}(\zeta, t)/\partial t\} = \{[\mathbf{N}]^t[\partial/\partial\zeta] + [\zeta]^t[\mathbf{M}][\partial/\partial\zeta]\}\{\rho_{\text{tot}}^{(n)}(\zeta, t)\} \tag{H46}$$

$[\zeta]$, $[\partial/\partial\zeta]$, and $[\mathbf{N}]$ are three column vectors of order $2(N + 1)$, the transposes of which are, respectively:

$$[\zeta]^t = [[\alpha]^t, [\beta]^t]$$
$$[\partial/\partial\zeta]^t = [[\partial/\partial\alpha]^t, [\partial/\partial\beta]^t] \tag{H47}$$
$$[\mathbf{N}]^t = [[\alpha^\circ]^t, [\mathbf{0}]^t]$$

$[\alpha]$, $[\partial/\partial\alpha]$, and $[\alpha^\circ]$ are three column vectors of dimension 2 characterizing the damped oscillator, the transposes of which are:

$$[\alpha]^t = [\alpha^*, \alpha] \tag{H48}$$
$$[\partial/\partial\alpha]^t = [\partial/\partial\alpha^*, \partial/\partial\alpha] \tag{H49}$$
$$[\alpha^\circ]^t = [\alpha^\circ, -\alpha^\circ] \tag{H50}$$

$[\beta]$, $[\partial/\partial\beta]$, and $[0]$ are three column vectors of dimension $2N$, these $2N$ dimensions being related to the N oscillators of the thermal bath, the transposes of which are, respectively:

$$[\beta]^t = [\beta_1^*, \beta_1, \ldots, \beta_k^*, \beta_k, \ldots, \beta_N^*, \beta_N]$$

$$[\partial/\partial\beta]^t = [\partial/\partial\beta_1^*, \partial/\partial\beta_1, \ldots, \partial/\partial\beta_k^*, \partial/\partial\beta_k, \ldots, \partial/\partial\beta_N^*, \partial/\partial\beta_N] \quad \text{(H51)}$$

$$[0]^t = [0, 0, \ldots, 0, 0, \ldots, 0, 0]$$

Besides, $[\mathbf{M}]$ is a $(2N + 2) \times (2N + 2)$ square matrix that may be written formally:

$$[\mathbf{M}] = \begin{vmatrix} \mathbf{M}_{11} & \mathbf{M}_{12} \\ \mathbf{M}_{21} & \mathbf{M}_{22} \end{vmatrix} \quad \text{(H52)}$$

Here appear four matrices that are, respectively, of dimension 2×2, $2 \times 2N$, $2N \times 2$, and $2N \times 2N$, the last one being diagonal. The first ones are:

$$\mathbf{M}_{11} = \begin{vmatrix} \omega & 0 \\ 0 & -\omega \end{vmatrix}, \quad \mathbf{M}_{22} = \begin{vmatrix} \omega_1 & & 0 \\ & \ddots & \\ 0 & & -\omega_N \end{vmatrix} \quad \text{(H53)}$$

$$\mathbf{M}_{12} = \begin{vmatrix} \kappa_1 & 0 & \kappa_2 & 0 & \cdots & \kappa_N & 0 \\ 0 & -\kappa_1^* & 0 & \kappa_2^* & \cdots & 0 & -\kappa_N^* \end{vmatrix} \quad \text{(H54)}$$

b. The Fourier Transform of the Liouville Equation. Now let us perform the Fourier transform of the scalar distribution function according to:

$$\{F_{\text{tot}}(\xi, t)\} = \int \cdots \int \cdots \int \{\rho_{\text{tot}}^{(n)}(\zeta, t)\} \exp\{-i[\xi]^t \cdot [\zeta]\} \, d[\zeta] \quad \text{(H55)}$$

$[\xi]^t$ is a line vector of dimension $2N + 2$ given by:

$$[\xi]^t = [[\pi]^t [\sigma]^t] \quad \text{(H56)}$$

$[\pi]^t$ and $[\sigma]^t$ are, respectively, line vectors of dimensions 2 and $2N$ given by:

$$[\pi]^t = [\pi^*, \pi] \quad \text{(H57)}$$

$$[\sigma]^t = [\sigma_1^*, \sigma_1, \ldots, \sigma_k^*, \sigma_k, \ldots, \sigma_N^*, \sigma_N] \quad \text{(H58)}$$

$d[\zeta]$ is the short notation:

$$d[\zeta] = d\alpha \, d\alpha^* \, d\beta_1 \, d\beta_1^* \cdots d\beta_k \, d\beta_k^* \cdots d\beta_N \, d\beta_N^*/(4\pi)^{N+1}$$

In terms of this Fourier transform, the Liouville equation (H46) becomes after some rearrangement:

$$i \, \partial\{F_{\text{tot}}(\xi, t)\}/\partial t = i[\mathbf{N}]'[\sigma]\{F_{\text{tot}}(\xi, t)\} - [\partial/\partial\sigma]'\{F_{\text{tot}}(\xi, t)\}[\mathbf{M}][\sigma] \quad \text{(H59)}$$

The solution of the differential equation (H59) may be written formally as:

$$\{F_{\text{tot}}(\xi, t)\} = \exp\{-[i[\mathbf{Q}(t)]'[\xi] + [\xi]'[\mathbf{P}][\xi]]\} \quad \text{(H60)}$$

$[\mathbf{P}]$ is a $(2N + 2) \times (2N + 2)$ square matrix given by:

$$[\mathbf{P}] = \begin{vmatrix} \mathbf{P}_{11} & \mathbf{P}_{12} \\ \mathbf{P}_{21} & \mathbf{P}_{22} \end{vmatrix} \quad \text{(H61)}$$

where appear four matrices $\mathbf{P}_{11}$, $\mathbf{P}_{12}$, $\mathbf{P}_{21}$ and $\mathbf{P}_{22}$, which are, respectively, of dimension 2×2, $2 \times 2N$, $2N \times 2$, and $2N \times 2N$. Besides $[\mathbf{Q}(t)]$ is a line vector of dimension $2(1 + N)$, the transpose of which is:

$$[\mathbf{Q}(t)]' = [[\mathbf{Q}_{11}(t)]', [\mathbf{Q}_{12}(t)]'] \quad \text{(H62)}$$

$[\mathbf{Q}_{11}(t)]$ and $[\mathbf{Q}_{12}(t)]$ are, respectively, column vectors of dimension 2 and $2 \times N$, the elements of which are unknown at time t, and at initial time are given by:

$$[\mathbf{Q}_{11}(0)]' = [\alpha_\circ^*, \alpha_\circ] \quad \text{(H63)}$$

$$[\mathbf{Q}_{12}(0)]' = [0, 0, \ldots, 0, 0, \ldots, 0, 0] \quad \text{(H64)}$$

By derivation of Eq. (H60) with respect to time and identification to the Liouville equation (H59), one obtains, respectively:

$$\partial[\mathbf{P}]/\partial t = i\{[\mathbf{P}][\mathbf{M}] + [\mathbf{M}]'[\mathbf{P}]\} \quad \text{(H65)}$$

$$\partial[\mathbf{Q}]/\partial t = i[\mathbf{Q}][\mathbf{M}] + i[\mathbf{N}] \quad \text{(H66)}$$

According to Eqs. (H61) and (H62), the last equation (H66) appears to be given by:

$$i \, \partial[\mathbf{Q}_{11}]/\partial t = -[\mathbf{Q}_{11}][\mathbf{M}_{11}] - [\mathbf{Q}_{12}]'[\mathbf{M}_{21}] - [\alpha^\circ] \quad \text{(H67)}$$

$$i \, \partial[\mathbf{Q}_{12}]/\partial t = -[\mathbf{Q}_{11}][\mathbf{M}_{12}] - [\mathbf{Q}_{12}][\mathbf{M}_{22}] \quad \text{(H68)}$$

These equations must be solved owing to the boundary conditions (H63) and (H64). Recall that $[\mathbf{Q}_{12}]$ and $[\mathbf{M}_{12}]$ are, respectively, a vector of dimension $2N$ and a matrix of dimension $2 \times 2N$, where N is the number of oscillators simulating the thermal bath that must be considered for this reason as infinite.

c. *The Wigner–Weisskopf Approximation.* Next, solving Eqs. (H67) and (H68), with the Laplace transform method, and using, during the procedure, the Wigner–Weisskopf approximation, which is usual for systems as the present one with infinite freedom degrees, Louisell and Walker found:

$$[\mathbf{Q}_{11}]^t = [\Phi(t)^*, \, \Phi(t)]$$

where the two components of this vector may be obtained by:

$$\{\Phi(t)\} = \alpha_\circ \exp\{-\gamma t/2\} \exp\{-i(\omega^{\circ\circ} + \Delta)t\}$$
$$- i\alpha^\circ \omega^{\circ\circ} \int_0^t \exp\{-\gamma t'/2\} \exp\{-i(\omega^{\circ\circ} + \Delta)t'\} \, dt' \quad \text{(H69)}$$

Here γ is the damping parameter related to the strength of the coupling of the slow mode with the thermal bath,

$$\gamma = 2\pi \sum |\kappa_k|^2 \, \delta(\omega_k - \omega^{\circ\circ}) \quad \text{(H70)}$$

Recall that, since the coupling of the oscillator with the bath is weak, one has $\gamma < \omega^{\circ\circ}$, which is the underdamped condition. Besides, Δ is an angular frequency shift given by the Cauchy principal part

$$\Delta = \mathrm{P} \sum |\kappa_k|^2/(\omega_k - \omega^{\circ\circ})$$

For a quantum driven damped harmonic oscillator starting from an equilibrium Boltzmann distribution, which corresponds to the quantum approach of the indirect relaxation of the hydrogen bond in representation $\{II\}$, that is, when $\alpha_\circ = 0$ and $\alpha^\circ \neq 0$, the above equation reduces, after neglecting Δ to:

$$\{\Phi_{II}(t)\} = -i\alpha^\circ \omega^{\circ\circ} \int_0^t \exp\{-\gamma t'/2\} \exp\{-i(\omega^{\circ\circ})t'\} \, dt' \quad \text{(H71)}$$

By integration one obtains:

$$\{\Phi_{II}(t)\} = \beta[\exp\{-\gamma t/2\} \exp\{-i(\omega^{\circ\circ})t\} - 1] \quad \text{(H72)}$$

where β is a nondimensional effective anharmonic coupling parameter given by:

$$\beta = \alpha^{\circ}(2\omega^{\circ\circ2} + i\gamma\omega^{\circ\circ})/[2(\omega^{\circ\circ2} + \gamma^2/4)] \tag{H73}$$

Note that the absolute value of β which is of interest is:

$$|\beta| = \alpha^{\circ}(4\omega^{\circ\circ4} + \gamma^2\omega^{\circ\circ2})^{1/2}/[2(\omega^{\circ\circ2} + \gamma^2/4)] \tag{H74}$$

For a free quantum damped harmonic oscillator starting from an initial coherent state, which corresponds to the quantum approach of the indirect relaxation of the hydrogen bond within representation {II}, that is, when $\alpha_{\circ} \neq 0$ and $\alpha^{\circ} = 0$, Eq. (H69), neglecting the Lamb shift, reduces to:

$$\{\Phi_{\text{III}}(t)\} - \alpha_{\circ} \exp\{-\gamma t/2\} \exp\{-i\omega^{\circ\circ}t\} \tag{H75}$$

d. *Obtaining the Reduced Density Operator Using the Partial Trace over the Thermal Bath Variables.* Now, returning to the Fourier transform (H55); it may be written explicitly, owing to Eq. (H56):

$$\{F_{\text{tot}}(\xi, t)\} = \{F_{\text{tot}}(\pi, \sigma, t)\} \tag{H76}$$

As we shall see, its knowledge, when the scalar variables of the thermal bath are zero (i.e., when $\sigma = 0$), will allow us to obtain the normal ordered distribution function corresponding to the reduced density operator of the oscillator and obtained by performing a partial trace over the thermal bath:

$$\{\rho^{(n)}(\alpha, t)\} = \text{tr}_{\theta}\{\rho_{\text{tot}}^{(n)}(\alpha, \beta, t)\} \tag{H77}$$

As a matter of fact, according to an important theorem given in Louisell's book (1973, p. 148), the trace of an operator function of the bosons may be obtained acording to:

$$\text{tr}\{f^{(n)}(a, a^{\dagger})\} = \{F^{(n)}(0, 0)\}$$

where the right-hand side term is the Fourier transform of the scalar function corresponding to the operator function of the bosons, given by:

$$\{F^{(n)}(\xi, \xi^*)\} = \int\int \{f^{(n)}(\alpha, \alpha^*)\} \exp\{-i[\alpha\xi + \alpha^*\xi^*]\} \, d\alpha \, d\alpha^*$$

in which ξ and ξ^* have to take the value 0. Moreover, suppose that one has to perform a partial trace over the subspace in which some bosons b and $b^\dagger$ are working, according for instance to:

$$\{f^{(n)}(a, a^\dagger)\} = \mathrm{tr}'\{f^{(n)}_{\mathrm{tot}}(a, a^\dagger, b, b^\dagger)\}$$

The same theorem leads to:

$$\mathrm{tr}'\{f^{(n)}_{\mathrm{tot}}(a, a^\dagger, b, b^\dagger)\} = \int \int \{F^{(n)}(\xi, \xi^*, 0, 0)\} \exp\{-i[\alpha\xi + \alpha^*\xi^*]\}\, d\xi\, d\xi^*$$

where the right-hand side term is the Fourier transform of the scalar function corresponding to the operator function of the bosons, given by:

$$\{F^{(n)}(\xi, \xi^*, \zeta, \zeta^*)\} = \int \int \int \int \{f^{(n)}(\alpha, \alpha^*, \beta, \beta^*)\}$$

$$\times \exp\{-i[\alpha\xi + \alpha^*\xi^* + \beta\zeta + \beta^*\zeta^*]\}\, d\alpha\, d\alpha^*\, d\beta\, d\beta^*$$

in which the variables ζ and ζ^* are taken to be zero.

As a consequence, the reduced distribution function obtained by performing a partial trace over the thermal bath, must be got by performing the Fourier transform (H55) in which the variables σ of the thermal bath involved in Eq. (H58) are all taken to be zero:

$$\{\rho^{(n)}(\alpha, t)\} = \int \int \{F_{\mathrm{tot}}(\pi, 0, t)\} \exp\{-[\pi]'[\alpha]\}\, d\pi\, d\pi^* \qquad \text{(H78)}$$

Owing to Eq. (H60), the reduced distribution function may be written, in view of Eqs. (H56)–(H58) and (H62)–(H64):

$$\{F_{\mathrm{tot}}(\pi, 0, t)\} = \exp\{-[\pi]'[\mathbf{P}_{11}][\pi] - i[\mathbf{Q}_{11}(t)]'[\pi]\} \qquad \text{(H79)}$$

where we must recall that $\mathbf{P}_{11}$ is a 2×2 matrix, the four elements of which are a priori different. However, since the Louisell and Walker calculations show that the diagonal matrix elements are approximately zero for weak coupling and that the off elements are equal, it may be suitable to write this matrix as:

$$[\mathbf{P}_{11}] = \tfrac{1}{2}\begin{vmatrix} 0 & 1/\varepsilon \\ 1/\varepsilon & 0 \end{vmatrix} \qquad \text{(H80)}$$

where ε is given by Eq. (H37). According to Louisell and Walker (1965), the solution of Eq. (H78) is, within normal ordering:

$$\{\rho^{(n)}(\alpha, t)\} = \varepsilon \exp\{-\varepsilon(\alpha^* - \Phi(t))(\alpha - \Phi(t)^*)\} \tag{H81}$$

Again, by aid of the operator N, one may pass from the scalar function to the corresponding density operator in the normal form:

$$N\{\rho^{(n)}(\alpha, \alpha^*, t)\} = \{\rho^{(n)}(a^\dagger, a, t)\} \tag{H82}$$

The result is, owing to the expression (H37) of ε, and to theorem (A31) of Appendix A:

$$\{\rho^{(n)}(a^\dagger, a, t)\} = \varepsilon \exp\{-\lambda(a^\dagger - \Phi(t))(a - \Phi^*(t))\} \tag{H83}$$

where $\Phi(t)$ works for $\Phi_{II}(t)$ or $\Phi_{III}(t)$ given, respectively, by Eqs. (H72) and (H75), according to the fact that one is dealing with a driven damped quantum harmonic oscillator starting from a Boltzmann density operator or a free damped quantum harmonic oscillator starting from the density operator of a coherent state at some temperature.

4. Perturbative Treatment of the Liouville Equation

Now let us look at the perturbative solution of the von Neumann–Liouville equation. For that we shall follow Louisell (1973) and Cohen–Tannoudji (1992).

a. The Master Equation. In the interaction picture, with respect to the unperturbed Hamiltonian $H^{00} + H_\theta$, the perturbative resolution of the von Neumann–Liouville equation (H39) up to second order gives:

$$\{\rho^{IP}(t)\} = \{\rho^{IP}(t^0)\}$$

$$+ 1/(i\hbar) \int_{t^0}^{t} \text{tr}_\theta[\{H_{\text{int}}^{IP}(t' - t^0)\}, \{\rho^{IP}(t^0)\rho_\theta\}] \, dt'$$

$$+ [1/(i\hbar)^2] \int_{t^0}^{t} \int_{t^0}^{t'} \text{tr}_\theta[\{H_{\text{int}}^{IP}(t' - t^0)\}, [\{H_{\text{int}}^{IP}(t'' - t^0)\}, \{\rho^{IP}(t^0)\rho_\theta\}]] \, dt' \, dt''$$

$$\tag{H84}$$

Again, performing the trace over the thermal bath, one obtains:

$$
\{\rho^{\mathrm{IP}}(t)\} - \{\rho^{\mathrm{IP}}(t^0)\} = \int_{t^0}^{t} [a^\dagger \rho(t^0) - \rho(t^0) a^\dagger] \langle F^{(+)}(t' - t^0) \rangle \, dt'
$$

$$
+ \int_{t^0}^{t} \int_{t^0}^{t'} [a\rho(t^0)a^\dagger - a^\dagger a \rho(t^0)]
$$

$$
\times \langle F^{(+)}(t' - t^0) F^{(-)}(t'' - t^0) \rangle \exp\{-i\omega^{\circ\circ}(t' - t'')\} \, dt' \, d''
$$

$$
+ \int_{t^0}^{t} \int_{t^0}^{t'} [a^\dagger \rho(t^0) a - \rho(t^0) a a^\dagger] \langle F^{(-)}(t'' - t^0) F^{(+)}(t' - t^0) \rangle
$$

$$
\times \exp\{-i\omega^{\circ\circ}(t' - t'')\} \, dt' \, dt'' + \text{H.c.} \qquad \text{(H85)}
$$

Here, the thermal averages are given, respectively, by:

$$
\langle F^{(+)}(t' - t^0) \rangle = \mathrm{tr}_\theta \{ \sum \kappa_j \{ \rho_\theta \, b_j^{\mathrm{IP}} \} \} \exp\{ -i\omega_j(t' - t^0) \}
$$

$$
\langle F^{(-)}(t' - t^0) F^{(+)}(t'' - t^0) \rangle = \mathrm{tr}_\theta \{ \sum |\kappa_j|^2 \{ \rho_\theta \, b_j^{\mathrm{IP}\dagger} b_j^{\mathrm{IP}} \} \} \exp\{ -i\omega_j(t' - t'') \}
$$

$$
\text{(H86)}
$$

$$
\langle F^{(-)}(t' - t^0) F^{(+)}(t'' - t^0) \rangle = \mathrm{tr}_\theta \{ \sum |\kappa_j|^2 \{ \rho_\theta \, b_j^{\mathrm{IP}} b_j^{\mathrm{IP}\dagger} \} \} \exp\{ -i\omega_j(t' - t'') \}
$$

$$
\text{(H87)}
$$

Performing the thermal average, one finds:

$$
\langle F^{(+)}(t' - t^0) \rangle = 0 \qquad \text{(H88)}
$$

$$
\langle F^{(+)}(t' - t^0) F^{(-)}(t'' - t^0) \rangle = \sum |\kappa_j|_j^2 \{ \langle n \rangle + 1 \} \exp\{ -i\omega_j(t' - t'') \}
$$

$$
\langle F^{(-)}(t'' - t^0) F^{(+)}(t' - t^0) \rangle = \sum |\kappa_j|^2 \langle n \rangle \exp\{ -i\omega_j(t' - t'') \} \qquad \text{(H89)}
$$

In this last equation, $\langle n \rangle$ is the mean number occupation at temperature T of the oscillator that is given by:

$$
\langle n \rangle = 1/[\exp\{\hbar\omega/(k_{\mathrm{B}} T)\} - 1] \qquad \text{(H90)}
$$

It must be emphasized that there is an analogy between Eqs. (H88) and (H89) governing the quantum statistical force damping the quantum oscillator and the Langevin equations (H5) and (H6) governing the stochastic force acting on the classical Brownian oscillator. Again, taking $t - t^0$ as a small interval Δt, one may perform a change in the time variables and extend the limit of integration of the right integral according to:

$$\int_0^t \int_0^{t'} \{ \ \} \, dt' \, dt'' = (1/\Delta t) \int_0^\infty d(t - t'') \int_0^{t + \Delta t} \{ \ \} \, dt'$$

The result of the integrations gives the rate of variation of the reduced density operator of the driven damped coherent state in the interaction picture with respect only to H^{00} because we have performed the partial trace on the thermal bath:

$$\Delta \rho^{IP} / \Delta t = \{ \rho^{IP}(t) \} - \{ \rho^{IP}(t^0) \} / \Delta t$$
$$= -i\alpha^\circ \{ [a^\dagger, \rho^{IP}(t)] + [a, \rho^{IP}(t)] \}$$
$$+ \tfrac{1}{2}\gamma \{ 2a\rho^{IP}(t)a^\dagger - a^\dagger a \rho^{IP}(t) - \rho^{IP}(t)a^\dagger a \}$$
$$+ \gamma \langle n \rangle \{ a^\dagger \rho^{IP}(t)a + a\rho^{IP}(t)a^\dagger - a^\dagger a \rho^{IP}(t) - \rho^{IP}(t)aa^\dagger \} \quad \text{(H91)}$$

with γ given by Eq. (H70). Finally, taking Δt as infinitesimal and returning to the Schrödinger picture, one obtains the master equation.

$$\{ \partial \rho(t) / \partial t \} = -i\omega [a^\dagger a, \rho(t)] - i\alpha^\circ \{ [a^\dagger, \rho(t)] + [a, \rho(t)] \}$$
$$+ \tfrac{1}{2}\gamma \{ 2a\rho(t)a^\dagger - a^\dagger a\rho(t) - \rho(t)a^\dagger a \} + \gamma \langle n \rangle$$
$$\times \{ a^\dagger \rho(t)a + a\rho(t)a^\dagger - a^\dagger a\rho(t) - \rho(t)aa^\dagger \} \quad \text{(H92)}$$

b. *The Fokker–Planck Equation.* Equation (H92) may be transformed in antinormal form using the following relations on any operator in antinormal form, which may be obtained by Eqs. (A29) and (A30):

$$[\rho^{(a)}, a] = -(\partial \rho^{(a)} / \partial a^\dagger) \quad \text{and} \quad [\rho^{(a)}, a^\dagger] = -(\partial \rho^{(a)} / \partial a) \quad \text{(H93)}$$

which lead to:

$$\{ \partial \rho^{(a)}(a^\dagger, a, t) / \partial t \} = -i\omega^{\circ\circ} \{ a(\partial \rho^{(a)}(t) / \partial a) - (\partial \rho^{(a)}(t) / \partial a^\dagger)a^\dagger \}$$
$$+ i\alpha^\circ \{ (\partial \rho^{(a)}(t) / \partial a) - (\partial \rho^{(a)}(t) / \partial a^\dagger) \}$$
$$+ \tfrac{1}{2}\gamma \{ \partial / \partial a [a\rho^{(a)}(t)] + \partial / \partial a^\dagger [\rho^{(a)}(t)a^\dagger] \}$$
$$+ \gamma \langle n \rangle (\partial^2 \rho^{(a)}(t) / \partial a \, \partial a^\dagger) \quad \text{(H94)}$$

Of course, in antinormal form the initial situation (H38) is:

$$\{ \rho^{(a)}(0) \} = | \alpha_\circ \rangle \langle \alpha_\circ | \quad \text{(H95)}$$

Now let us define the complex scalar distribution function f according to:

$$\{f(\alpha, \alpha^*, t)\} = \{\rho^{(a)}(\alpha, \alpha^*, t)\} \tag{H96}$$

Then the distribution function obeys the Fokker–Planck equation:

$$\{\partial f(\alpha, \alpha^*, t)/\partial t\} = i\omega^{\circ\circ}\{[\alpha \ \partial f(t)/\partial \alpha] - [\alpha^* \ \partial f(t)/\partial \alpha^*]\}$$
$$+ i\alpha^\circ\{[\partial f(t)/\partial \alpha] - [\partial f(t)/\partial \alpha^*]\}$$
$$+ \tfrac{1}{2}\gamma\{\partial/\partial\alpha[\alpha f(t)] + \partial/\partial\alpha^*[\alpha^* f(t)]\}$$
$$+ \gamma\langle n\rangle[\partial^2 f(t)/\partial\alpha \ \partial\alpha^*] \tag{H97}$$

with the initial situation

$$\{f(\alpha, \alpha^*, 0)\} = \delta(\alpha - \alpha_\circ)\,\delta(\alpha^* - \alpha_\circ^*) \tag{H98}$$

The distribution function satisfying the Fokker–Planck equation (H97) owing to the initial situation (H98) is:

$$\{f(\alpha, \alpha^*, t)\} = \{f(\alpha, \alpha^*, t \,|\, \alpha_\circ, \alpha_\circ^*, 0)\} \tag{H99}$$

It may be observed that when the harmonic oscillator is damped but not driven, the driven parameter α° is zero so that the Fokker–Planck equation (H97) reduces to:

$$\{\partial f(\alpha, \alpha^*, t)\,|\, \partial t\} = i\omega^{\circ\circ}\{[\alpha \ \partial f(t)/\partial\alpha] - [\alpha^* \ \partial f(t)/\partial\alpha^*]\}$$
$$+ \tfrac{1}{2}\gamma\{\partial/\partial\alpha[\alpha f(t)] + \partial/\partial\alpha^*[\alpha^* f(t)]\}$$
$$+ \gamma\langle n\rangle[\partial^2 f(t)/\partial\alpha \ \partial\alpha^*] \tag{H100}$$

Then, when the temperature is high, $k_B T > \hbar\omega^{\circ\circ}$, the mean number occupation is:

$$\langle n\rangle \approx k_B T/\hbar\omega^{\circ\circ} \tag{H101}$$

Thus the distribution function (H99), which is the solution of the Fokker–Planck equation (H100), takes the form:

$$\{f(\alpha, t \,|\, \alpha_\circ, 0)\} = \{\hbar\omega^{\circ\circ}/[\pi k_B T(1 - \exp\{-\gamma t\})]\}^{1/2}$$
$$\times \exp\{-|\alpha \exp\{-i\omega^{\circ\circ}t\}$$
$$- \alpha_\circ \exp\{-\tfrac{1}{2}\gamma t\}|^2 \hbar\omega^{\circ\circ}/[k_B T(1 - \exp\{-\gamma t\})]\} \tag{H102}$$

Now one may recall that this expression of the distribution function is within the normal ordering formalism and that it governs the dynamics of the irreversible behavior of the scalars, which are directly related to the impulsion operator, and thus to the velocity operator of the oscillator. Keeping that in mind, it may be of interest to compare this distribution function to the classical one (H24) governing the velocity of a Brownian free particle. As it appears, the equations have the same structure, the complex distribution function $\{f(\alpha, t \,|\, \alpha_\circ, 0)\}$ playing in (H102) the role of the distribution function $f(v, t \,|\, v^0, 0)$ in (H24).

APPENDIX I

Extraction of the Reduced Time Evolution Operator of the Driven Damped Quantum Harmonic Oscillator

It is shown in Appendix H, (see Eqs (H72) and (H23)), that the reduced density operator of the driven quantum harmonic oscillator is given by:

$$\{\rho_{II}^{\{1\}}(t)\} = \varepsilon \exp\{-\lambda(a^\dagger - \Phi_{II}(t))(a - \Phi_{II}^*(t))\}$$

with respectively, according to Eqs. (H36), (H37), (H72), and (H73):

$$\lambda = \hbar\omega^{\circ\circ}/(k_B T) \qquad \varepsilon = 1 - \exp\{-\lambda\}$$

$$\{\Phi_{II}(t)\} = \beta[\exp\{-\gamma t/2\}\exp\{-i\omega^{\circ\circ}t\} - 1]$$

$$\beta = \alpha^\circ(2\omega^{\circ\circ 2} + i\gamma\omega^{\circ\circ})/[2(\omega^{\circ\circ 2} + \gamma^2/4)]$$

Observe that the above density operator may be considered as the phase transformation on the Boltzmann operator, according to Eq. (128) (see section B1b):

$$\{\rho_{II}^{\{1\}}(t)\} = \varepsilon\{A^{\{1\}}(\Phi_{II}(t))\}\exp\{-\lambda a^\dagger a\}\{A^{\{1\}}(\Phi_{II}(t))\}^\dagger$$

This phase transformation involves the translation operator defined by:

$$\{A^{\{1\}}(\Phi_{II}(t))\} = \exp\{\Phi_{II}(t)a^\dagger - \Phi_{II}^*(t)a\}$$

Note that the reduced density operator at time t of the driven quantum harmonic oscillator may be more generally given by the following phase transformation using time evolution operator (Blaise et al., 1992):

$$\{\rho_{II}^{\{1\}}(t)\} = \{U_{II}^{\{1\}}(t)\}\exp\{-\lambda a^\dagger a\}\{U_{II}^{\{1\}}(t)\}^\dagger$$

The time evolution operator must be a scalar phase factor times the known translation operator, and times the exponential of an operator that commutes with the Boltzmann operator, i.e. with the number occupation. Thus we may write:

$$\{U_{\text{II}}^{\{1\}}(t)\} = \exp\{-i\zeta(t)\}\{A^{\{1\}}(\Phi_{\text{II}}(t))\} \exp\{-i\xi(t)\}$$

where ζ is a scalar and ξ an operator which commutes with $a^\dagger a$. Our purpose is to find $\exp\{-i\zeta(t)\}$ and $\exp\{-i\xi(t)\}$.

Now, return some moment to the situation in the absence of damping, i.e. to the behavior of the driven damped quantum harmonic oscillator. The corresponding time evolution operator, that is given by Eq. (C29) may be written by analogy as follows:

$$\{U_{\text{II}}^{0\{1\}}(t)\} = \exp\{-i\zeta^0(t)\}\{A^{\{1\}}(\Phi_{\text{II}}^0(t))\} \exp\{-i\xi^0(t)\}$$

with, respectively:

$$\{A^{\{1\}}(\Phi_{\text{II}}^0(t))\} = \exp(\Phi_{\text{II}}^0(t)a^\dagger - \Phi_{\text{II}}^{0*}(t)a\}$$

$$\{\Phi_{\text{II}}^0(t)\} = \alpha^\circ[\exp\{-i\omega^{\circ\circ}t\} - 1] \tag{I1}$$

$$\exp\{-i\zeta^0(t)\} = \exp\{-i(\omega^\circ - \alpha^{\circ 2}\omega^{\circ\circ})t\} \exp\{-i\alpha^{\circ 2} \sin(\omega^{\circ\circ}t)\} \tag{I2}$$

$$\exp\{-i\xi^0(t)\} = \exp\{-ia^\dagger a\omega^{\circ\circ}t\} \tag{I3}$$

Note, that when the damping is missing, i.e. when $\gamma \to 0$, $\beta(\gamma \to 0) = \alpha^\circ$, then the argument of the translation operator of the driven damped quantum harmonic oscillator reduces to that involved in the corresponding undamped situation given by Eq. (94). At the inverse, one may say that when we pass from the undamped to the underdamped situation, one has the following transformation:

$$\{\Phi_{\text{II}}^0(t)\} \to \{\Phi_{\text{II}}(t)\}$$
$$\alpha^\circ[\exp\{-i\omega^{\circ\circ}t\} - 1] \to \beta[\exp\{-\gamma t/2\} \exp\{-i\omega^{\circ\circ}t\} - 1] \tag{I4}$$

In the presence of damping, the simple phase factor $\exp\{i\xi^0(t)\}$ that would disappear in the interaction picture (see Eq. (C34)), cannot be affected by the damping, since it is dealing with the dynamics of a free quantum harmonic oscillator, so that one may write for the damped situation:

$$\exp\{-i\xi(t)\} = \exp\{-i\xi^0(t)\} = \exp\{-ia^\dagger a\omega^{\circ\circ}t\} \tag{I5}$$

Now, look at the phase factor $\zeta(t)$ involved in the damped time evolution operator one may assume that when we are passing from the undamped to the damped one, the sine term involved in the expression (I2) of $\zeta^0(t)$ is affected by the damping in the same way as the cosine term in the translation operator when one passes from the argument in the absence of damping, to the corresponding one in the damped situation. That leads therefore to multiply the sine term by $\exp\{-\gamma t/2\}$ and to replace α by β and thus to write:

$$\exp\{-i\zeta(t)\} = \exp\{-i(\omega^\circ - |\beta|^2\omega^{\circ\circ})t\}\ \exp\{-i|\beta|^2\ \sin(\omega^{\circ\circ}t)\ \exp\{-\gamma t/2\}\}$$

Therefore, the time evolution operator in the presence of damping may be written:

$$\{U_{\mathrm{II}}^{\{1\}}(t)\} = \{A^{\{1\}}(\Phi_{\mathrm{II}}(t))\}\ \exp\{-ia^\dagger a\omega^{\circ\circ}t\}\ \exp\{-i(\omega^\circ - \alpha^{\circ 2}\omega^{\circ\circ})t\}$$

$$\times\ \exp\{i(|\beta|^2\omega^{\circ\circ})t\}\ \exp\{-i|\beta|^2\ \exp\{-\gamma t/2\}\ \sin(\omega^{\circ\circ}t)\} \quad (I6)$$

APPENDIX J

Time Evolution Operator $\{U(T)\}$ of the Driven Quantum Harmonic Oscillator and Related Functions

1. Time Evolution Operator of the Undamped Oscillator: Hamiltonian

$$\{U_{\mathrm{II}}^{0\{1\}}(t)\} = \exp\{-i(\omega^\circ - \alpha^{\circ 2}\omega^{\circ\circ})t\}\ \exp\{i\alpha^{\circ 2}\omega^{\circ\circ}t\}$$

$$\times\ \exp\{-i\alpha^{\circ 2}\ \sin(\omega^{\circ\circ}t)\}\{A^{\{1\}}(\Phi_{\mathrm{II}}^0(t))\}$$

$$\times\ \exp\{-ia^\dagger a\omega^{\circ\circ}t\}$$

$$\{U_{\mathrm{II}}^{0\{1\}}(t)\} = \exp\{-i(\omega^\circ - \alpha^{\circ 2}\omega^{\circ\circ})t\}\ \exp\{i\alpha^{\circ 2}\omega^{\circ\circ}t\}$$

$$\times\ \exp\{-i\alpha^{\circ 2}\ \sin(\omega^{\circ\circ}t)\}$$

$$\times\ \exp\{-ia^\dagger a\omega^{\circ\circ}t\}\{A^{\{1\}}(-\Phi_{\mathrm{II}}^0(t)^*)\}$$

$$\{U_{\mathrm{II}}^{0\{1\}}(t)\}^\dagger = \exp\{i(\omega^\circ - \alpha^{\circ 2}\omega^{\circ\circ})t\}\ \exp\{-i\alpha^2\omega^{\circ\circ}t\}$$

$$\times\ \exp\{i\alpha^2\ \sin(\omega^{\circ\circ}t)\}\{A^{\{1\}}(\Phi_{\mathrm{II}}^0(t))\}$$

$$\times\ \exp\{ia^\dagger a\omega^{\circ\circ}t\}$$

2. Time Evolution Operator of the Damped Oscillator: Hamiltonian

$$\{U_{II}^{\{1\}}(t)\} = \exp\{-i(\omega^\circ - \alpha^{\circ 2}\omega^{\circ\circ})t\} \, \exp\{i(|\beta|^2\omega^{\circ\circ})t\}$$
$$\times \exp\{-i|\beta|^2 \exp\{-\gamma t/2\} \, \sin(\omega^{\circ\circ}t)\}\{A^{\{1\}}(\Phi_{II}(t))\}$$
$$\times \exp\{-ia^\dagger a\omega^{\circ\circ}t\}$$

$$\{U_{II}^{\{1\}}(t)\} = \exp\{-i(\omega^\circ - \alpha^{\circ 2}\omega^{\circ\circ})t\} \, \exp\{i(|\beta|^2\omega^{\circ\circ})t\}$$
$$\times \exp\{-i|\beta|^2 \exp\{-\gamma t/2\} \, \sin(\omega^{\circ\circ}t)\}$$
$$\times \exp\{-ia^\dagger a\omega^{\circ\circ}t\}\{A^{\{1\}}(-\Phi_{II}(t)^*)\}$$

$$\{U_{II}^{\{1\}}(t)\}^\dagger = \exp\{i(\omega^\circ - \alpha^{\circ 2}\omega^{\circ\circ})t\} \, \exp\{-i|\beta|^2\omega^{\circ\circ}t\}$$
$$\times \exp\{i|\beta|^2 \exp\{-\gamma t/2\} \, \sin(\omega^{\circ\circ}t)\}$$
$$\times \{A^{\{1\}}(\Phi_{II}(t))\} \, \exp\{ia^\dagger a\omega^{\circ\circ}t\}$$

3. Translation Operators $\{A(\Phi)\}$ Involved in Representations $\{II\}$ and $\{III\}$

$$\{A(\Phi)\} = \exp\{\tfrac{1}{2}(\Phi a^\dagger - \Phi^* a)\} = \{A(-\Phi)\}^\dagger$$

$$\{A^{\{1\}}(\Phi_{III}(0))\}\{A^{\{1\}}(-\Phi_{III}(t))\} = \exp\{-i\alpha^{\circ 2} \exp\{-\gamma t/2\} \sin(\omega^{\circ\circ}t)\}\{A^{\{1\}}(-\Phi_{II}(t))\}$$

4. The Arguments Φ of the Translation Operators with and without Damping

without damping

$$\{\Phi_{III}^0(t)\} = \alpha^\circ \, \exp\{-i\omega t\}$$
$$\{\Phi_{II}^0(t)\} = \alpha^\circ\{\exp\{-i\omega t\} - 1\}$$

with damping

$$\{\Phi_{III}(t)\} = \alpha^\circ \, \exp\{-\gamma t/2\} \, \exp\{-i\omega^{\circ\circ}t\}$$
$$\{\Phi_{II}(t)\} = \beta[\exp\{-\gamma t/2\} \, \exp\{-i\omega^{\circ\circ}t\} - 1]$$
$$\beta = \alpha^\circ(2\omega^{\circ\circ 2} + i\gamma\omega^{\circ\circ})/[2(\omega^{\circ\circ 2} + \gamma^2/4)]$$

$\omega^{\circ\circ}$ is the angular frequency of the slow mode γ is the damping parameter of the slow mode.

α° is the adimensional anharmonic coupling between the slow and fast mode, i.e. the adimensional driving parameter.

$a^\dagger$ and a are, respectively, the creation and annihilation operators.

AN EFFECTIVE HAMILTONIAN TO TREAT ADIABATIC AND NONADIABATIC EFFECTS IN THE ROTATIONAL AND VIBRATIONAL SPECTRA OF DIATOMIC MOLECULES

R. M. HERMAN

Department of Physics, Pennsylvania State University, University Park, PA

J. F. OGILVIE*

Department of Chemistry, Oregon State University Corvallis, OR

CONTENTS

I. INTRODUCTION

According to Bohr's law, the frequency of a spectral transition is proportional to the difference of energies of two states of the active species. For transitions of free diatomic molecules in absorption or emission in the

* Permanent address: Centre for Experimental and Constructive Mathematics, Simon Fraser University, Burnaby, BC V5A 1S6 Canada

Advances in Chemical Physics, Volume 103, Edited by I. Prigogine and Stuart A. Rice.
ISBN 0-471-24752-9   © 1998 John Wiley & Sons, Inc.

infrared region, or in Raman scattering with comparable spectral shifts, narrow spectral lines are customarily assigned to transitions between discrete eigenstates associated with varied extents of rotational and vibrational energies. For such transitions, the eigenenergies of a diatomic molecule in the absence of externally applied electric or magnetic field, expressed as spectral terms in wavenumber unit, were represented systematically by Dunham in 1932 according to the formula [1]

$$\tilde{E}_{vJ} = \sum_{k=0} \sum_{l=0} Y_{kl}(v + \tfrac{1}{2})^k [J(J + 1)]^l \tag{1}$$

that is applicable to states with small values of vibrational quantum number v (i.e. vibrational states well below the dissociation limit) for molecules with contributions of neither net electronic spin nor orbital angular momentum for any isotopic variant to the total angular momentum; thus the quantum number J for the latter quantity reflects essentially only the rotational angular momentum. This formula is equivalent to then prevailing empirical notation involving spectral parameters such as ω_e, $\omega_e x_e$, B_e, α_e etc. which were coefficients in expansions in series of the arguments $v + \tfrac{1}{2}$ and $J(J + 1)$. (Alternative expansions with arguments v and $(J + \tfrac{1}{2})^2$ were also applied in the early literature [2].) Dunham's expressions for the coefficients Y_{kl} were considerably refined compared with previous expressions.

By means of methods [3] involving phase integrals that he developed, Dunham [1] related these term coefficients Y_{kl} to parameters describing the internuclear potential energy. He showed [1] that Eq. (1) is consistent with the eigenstates of a rotating oscillator that has a function for potential energy of the form

$$U(x) = hca_0 x^2 \left(1 + \sum_{j=1} a_j x^j\right) \tag{2}$$

in which x is a reduced internuclear displacement

$$x = (R - R_e)/R_e \tag{3}$$

in terms of instantaneous, R, and equilibrium, R_e, internuclear distances. For application to molecular spectra this oscillator corresponds to a molecule composed of two atoms (point masses). Dunham thus decomposed Y_{kl} into a sum of contributions,

$$Y_{kl} = Y_{kl}^{(0)} + Y_{kl}^{(2)} + Y_{kl}^{(4)} + \cdots \tag{4}$$

in which the superscripts indicate orders of expansion in JBKW theory [3], and developed expressions for various contributions in terms of parameters B_e and $\omega_e = (4B_e a_0)^{1/2}$, and a_j in Eq. (2); the rapidly decreasing magnitudes of successive contributions $Y_{kl}^{(2n)}$ generally assure acceptable convergence. Herman and Short extended [4] the notation of Eq. (1) to include first-order perturbations that led to the incorporation of further term coefficients Z_{kl} in a form equivalent to

$$\tilde{E}_{vJ} = \sum_{k=0} \sum_{l=0} (Y_{kl} + Z_{kl})(v + \tfrac{1}{2})^k [J(J+1)]^l \tag{5}$$

These quantities Z_{kl} essentially specify expectation values of functions of internuclear separation. For a neutral molecule AB that dissociates into atoms A and B of disparate atomic number and having masses M_a and M_b, Ogilvie [5] extended this form to

$$\tilde{E}_{vJ} = \sum_{k=0} \sum_{l=0} (Y_{kl} + Z_{kl}^{v;\,a} + Z_{kl}^{v;\,b} + Z_{kl}^{r;\,a} + Z_{kl}^{r;\,b})(v + \tfrac{1}{2})^k [J(J+1)]^l \tag{6}$$

Here the primary term coefficients Y_{kl} contain an implicit dependence on atomic reduced mass $\mu_{at} = M_a M_b/(M_a + M_b)$, whereas the auxiliary coefficients Z_{kl}, which represent small terms arising through deviations from Born–Oppenheimer behaviour, depend on masses of atoms A or B separately; the latter quantities are divided into those associated with vibrational–rotational effects, with superscript v, and those that pertain to further rotational effects, with superscript r.

An alternative representation of vibration–rotational terms of any isotopic variant σ of a diatomic molecular species, still within the same electronic state of class $^1\sum^+$ or 0^+, is

$$\tilde{E}_{vJ}^{\sigma} = \sum_{k=0} \sum_{l=0} U_{kl} \mu_{at}^{-[(1/2)k+l]} [1 + m_e(\Delta_{kl}^a/M_a + \Delta_{kl}^b/M_b)](v + \tfrac{1}{2})^k [J(J+1)]^l$$

$$\tag{7}$$

which has an empirical origin [6] that was subsequently given a theoretical basis [7]; in this relation the coefficients U_{kl} and $\Delta_{kl}^{a;\,b}$ are formally independent of mass, as all dependence on mass required to reproduce transition wavenumbers of multiple isotopic variants of a particular diatomic molecular species is embodied in the atomic reduced mass and in atomic masses of a particular isotopic variant.

Because ratios of spectral parameters of isotopic molecules deviate from appropriate ratios of their reduced atomic masses, the approximation treated theoretically by Born and Oppenheimer [8], and later by others [9], is corrected to account for the fact that molecules consist not merely of structureless atoms according to the classical model but of nuclei and electrons; when one is concerned with rotational and vibrational spectra, one must consider the fact that electrons follow imperfectly the nuclei in their rotations and vibrations about the centre of molecular mass. van Vleck [10] presented the first analysis of adiabatic and nonadiabatic effects that are hence associated with partial failure of the Born–Oppenheimer model. Adiabatic effects reflect the fact that the internuclear potential energy depends not only on the static coulombic interactions but also, through reactions of nuclear motions, on electronic motions in an attempt to conserve momentum, and hence depend upon inverse nuclear masses. Nonadiabatic effects were taken to signify interactions between the electronic state of interest, generally the ground electronic state, and other (excited) states, induced by rotational and vibrational motions of the nuclei, described respectively as nonadiabatic rotational and vibrational effects that depend upon the actual nuclear velocities. In his treatment [10] van Vleck associated a further radial function, $\sum_j k_j x^j$, with adiabatic effects and produced expressions containing coefficients k_j for a few contributions to spectra terms $\tilde{E}_{vJ}$.

As molecular spectra became available with increased resolution and for multiple isotopic variants, Herman and Asgharian [11] undertook a renewed theoretical analysis of observable phenomena attributed to these adiabatic and nonadiabatic effects; for this purpose they derived from an exact Hamiltonian for nuclei and electrons in an electronic state $^1\sum$ an effective Hamiltonian for nuclear motion that was more nearly complete than that in van Vleck's treatment (which suffers from cumbersome notation and difficulty of direct application to molecular spectra recorded under contemporary conditions). Subsequent refinements of this theory by Bunker [12] and Watson [13] (and their coworkers) yielded additional connections between theoretical quantities, implied in adiabatic and nonadiabatic effects, and the empirical parameters in Eq. (7) for instance.

As molecular spectra become increasingly abundant and precisely measured with relentlessly expanding sophistication of experiments to record these spectra (of not only electrically neutral diatomic molecules but also molecular ions), additional development of theory within a comprehensive and systematic treatment is required to forge relationships between isotopic variants of a particular molecular species: this task we take as the objective of our present work. Our approach is to develop an effective molecular Hamiltonian and to demonstrate how functions therein relate through their isotopic dependence to several molecular properties (e.g. functions for the electric and magnetic moments) that are, at present, or may be, in the near future, measured for many diatomic compounds. Our analysis shows that frequencies or wavenumbers of (centres of) spectral lines that are measured for rotational and vibrational transitions of a given free diatomic species in multiple isotopic variants can provide information about the radial function for the electric dipolar moment; information taken to govern intensities of rotational and vibrational transitions in absorption or emission thus exists implicitly within data of merely frequency type.

II. DERIVATION OF AN EFFECTIVE HAMILTONIAN

In considering the Hamiltonian of a diatomic molecule, we neglect magnetic effects characteristic of an electronic state of class other than $^1\Sigma^+$ or 0^+. We outline here the principal steps of the required derivation as a basis for a succeeding treatment of particular terms of this Hamiltonian. The operator for kinetic energy of multiple particles in the Schrödinger equation is proportional to

$$\sum_i \nabla_i'^2/m_e + \sum_k \nabla_k'^2/m_k \tag{8}$$

in which m_k is the mass of nucleus k (k denotes nucleus either a or b of the diatomic molecule AB), and m_e is the electronic mass; i sums over n electrons and k over nuclei A and B; x_i' and x_k' are laboratory coordinates of the appropriate particles. Let the coordinates of the centre of nuclear mass relative to the nuclei be

$$\mathbf{X}_{CNM} = \left[\sum_k m_k \mathbf{x}_k'\right]\Big/(m_a + m_b) \tag{9}$$

in which $m_a + m_b$ is the total nuclear mass. The coordinates of particles relative to the centre of nuclear mass (rather than to the centre of molecular mass) are

$$\begin{aligned}
\mathbf{x}_i &= \mathbf{x}_i' - \mathbf{X}_{CNM} \\
\mathbf{x}_k &= \mathbf{x}_k' - \mathbf{X}_{CNM}
\end{aligned} \tag{10}$$

We denote the relative nuclear separation by

$$\mathbf{x} = \mathbf{x}_a' - \mathbf{x}_b' = \mathbf{x}_a - \mathbf{x}_b \tag{11}$$

Through standard manipulation we obtain

$$\sum_k \nabla_k'^2/m_k = \nabla^2/\mu + \nabla_{CNM}^2/(m_a + m_b) \tag{12}$$

in which μ is the nuclear reduced mass. Through standard means as in the reduction of an atomic Hamiltonian, we proceed to write

$$\sum_i \nabla_i'^2/m_e + \nabla_{CNM}^2/(m_a/m_b) = \sum_i \nabla_i^2/m_e + \left(\sum_i \nabla_i\right)^2 \bigg/ (m_a + m_b) + \nabla_{CMM}^2/M \tag{13}$$

in which M is the total molecular mass, with $\mathbf{X}_{CMM}$ being the position of the centre of molecular mass; the second term on the right side provides the conventional effects of mass polarization and reduced mass. Consistent with our objectives both latter effects can be treated perturbatively, ultimately to appear as effects in the adiabatic potential energy that depend on the separate nuclear masses, even though the effects of reduced mass could have been included exactly. The operator for total kinetic energy becomes proportional to

$$\sum_i \nabla_i^2/m_e + \left(\sum_i \nabla_i\right)^2 \bigg/ (m_a + m_b) + \nabla^2/\mu + \nabla_{CMM}^2/M \tag{14}$$

The motion of the centre of molecular mass, in the latter term, readily factors out, leaving in the frame of the centre of mass only the preceding three terms to be actively considered.

At this point we transform to spherical coordinates as far as the relative nuclear coordinates are concerned. When we take

$$\psi = R^{-1}\Phi(\mathbf{x}_i, \mathbf{x}) \tag{15}$$

the Schrödinger equation becomes

$$\left[-\tfrac{1}{2}\hbar^2\left(\sum_i \frac{\nabla_i^2}{m_e} + \frac{(\sum_i \nabla_i)^2}{(m_a + m_b)} + \frac{1}{\mu}\left(\frac{\partial^2}{\partial R^2} - \frac{J^2}{R^2\hbar^2}\right)\right) + V(\mathbf{x}_i, \mathbf{x})\right]\Phi_{\lambda;\, vJM} = E_{\lambda;\, vJM}\Phi_{\lambda;\, vJM} \tag{16}$$

with

$$\mathbf{J} = -i\hbar\mathbf{x} \times \nabla \tag{17}$$

λ denotes an electronic state, and v, J and M are conventional vibrational and rotational quantum numbers. The instantaneous value of $\mathbf{J}$ is invariably perpendicular to an instantaneous displacement $\mathbf{x}$, as is indicated by the fact that the product $\mathbf{x} \cdot \mathbf{J}$ is identically zero. At each point on a unit sphere for a unit vector $\hat{\mathbf{x}}$ for a relative nuclear displacement, we define mutually perpendicular angular displacements that correspond to that in θ and an orthogonal angle χ; θ has its standard meaning in spherical polar coordinates, and χ would be measured along a great circle passing through $\mathbf{x}$ perpendicular to θ at that point. This definition is equivalent to a local system of coordinates with orthogonal rectangular unit vectors $\hat{\mathbf{x}}$, $\hat{\boldsymbol{\theta}}$ and $\hat{\boldsymbol{\chi}}$. Local to the point in question, $\mathbf{J}$ is thereby represented conventionally [14]. Then

$$\mathbf{J} = -(\hat{R} \times \hat{\boldsymbol{\theta}})i\hbar\partial/\partial\theta - (\hat{R} \times \hat{\boldsymbol{\chi}})i\hbar\partial/\partial\chi \tag{18}$$

with

$$J^2 = -\hbar^2(\partial^2/\partial\theta^2 + \partial^2/\partial\chi^2) \tag{19}$$

So defined, the relationships between θ and χ automatically yield the form of J^2 that is familiar in spherical polar coordinates. This property is fundamental in considering mixed terms of form $\langle\lambda|\mathbf{V}|0\rangle \cdot \mathbf{V}$, and their meaning.

Based on the Schrödinger equation and according to a conventional Born–Oppenheimer separation of Φ, we directly recover the adiabatic potential energy from Eq. (16),

$$U_{ad}(R) = -(\tfrac{1}{2}\hbar^2/\mu)\langle\Phi_{BO}^{el}(\mathbf{x}_i; \mathbf{x})|(\partial^2/\partial R^2 - J^2/(\hbar^2 R^2))|\Phi_{BO}^{el}(\mathbf{x}_i; \mathbf{x})\rangle$$

$$- \{\tfrac{1}{2}\hbar^2/(m_a + m_b)\}\langle\Phi_{BO}^{el}(\mathbf{x}_i; \mathbf{x})|\left(\sum_i \nabla_i\right)^2|\Phi_{BO}^{el}(\mathbf{x}_i; \mathbf{x})\rangle \tag{20}$$

which supplements the ordinary function $U_{BO}(R)$ for the Born–Oppenheimer potential energy governing nuclear motions.

The nonadiabatic terms are divided according to the symmetry of the molecule into one involving $\partial^2/\partial R^2$ and two orthogonal terms, exhibited in J^2:

$$H_{nad}^{eff} = -\frac{\hbar^4}{\mu^2}\left[\sum_{\lambda \neq 0} \frac{\langle 0|\partial/\partial R|\lambda\rangle\langle\lambda|\partial/\partial R|0\rangle}{E_{\lambda 0}} \frac{\partial^2}{\partial R^2} \right.$$

$$\left. -\frac{1}{R^2}\sum_{\lambda \neq 0} \frac{\langle 0|\partial/\partial\theta|\lambda\rangle\langle\lambda|\partial/\partial\theta|0\rangle}{E_{\lambda 0}} \frac{J^2}{\hbar^2 R^2}\right] \tag{21}$$

The operator $\partial/\partial\theta$ pertains to the angular momentum of electrons relative to the nuclear centre of mass. As far as forming matrix elements is concerned, the effect of rotating the nuclear frame through an angle $\partial\theta$ is identical to the effect of holding the nuclear coordinates fixed while making derivatives $-\partial/\partial\theta$ for all electrons; for this reason the nuclear derivatives are equal and opposite in their effect to the sum of electronic derivatives in generating a rotational inertial correction. These terms in Eq. (21) thus lead [11] to a term $(m_e/m_p)g_2^{el}(R)$ for the vibrational inertial correction and to a corresponding term $(m_e/m_p)g_1^{el}(R)$ for the rotational inertial correction, which also equals that part of a magnetogyric-ratio function of R arising from all electrons (not merely 'valence' electrons); here m_p denotes the mass of the proton, which enters because of the traditional form of the rotational g factor. If one were to ascribe effects of non-rigid atoms to only the part of the electronic distribution not near a nucleus and to find that the latter electrons lead rather than lag nuclear motion, as a negative total molecular magnetogyric ratio would indicate, possibly the electrons involved would be distributed in space quite extended beyond the less massive nucleus; in that case the electronic contribution to the magnetic moment would be larger than if the electrons were more concentrated between the nuclei. If the less massive atomic centre serves as the negative pole of the electric dipole, this condition might suffice to generate also such effects.

The first term of Eq. (21) arises through the nonadiabatic perturbation which, taken to second order, is more rigorously written

$$\Delta E_{vJ}^{nad} = -(\hbar^2/\mu)^2 \sum_\lambda \sum_{v'} \{\langle 0|\partial/\partial R|\lambda\rangle\langle vJM|\partial/\partial R|v'JM\rangle\langle\lambda|\partial/\partial R|0\rangle$$
$$\times \langle v'JM|\partial/\partial R|vJM\rangle/[E_{\lambda 0} + (E_{v'J}^0 - E_{vJ}^0)]\} \qquad (22)$$

As long as

$$(E_{v'J}^0 - E_{vJ}^0) \ll E_{\lambda 0} \qquad (23)$$

which is an alternative statement of the Born–Oppenheimer approximation,* the above expression becomes approximately

$$\Delta E_{vJ}^{nad} \cong -(\hbar^2/\mu)^2 \sum_{\lambda \neq 0} \langle 0|\partial/\partial R|\lambda\rangle\langle\lambda|\partial/\partial R|0\rangle\langle vJM|\partial^2/\partial R^2|vJM\rangle/E_{\lambda 0}$$

$$(24)$$

Within the approximation according to inequality (23), perturbation theory in higher orders reveals that ΔE_{vJ}^{nad} is not given simply according to Eq. (24); instead one should consider the operator given by the first term of Eq. (21) to be part of an effective Hamiltonian governing the nuclear degrees of

---

* Retaining the first term in the Taylor series involving vibrational energies generates perturbational corrections proportional to $(m_e/m_p)^{1/2}$, but they are extremely small.

freedom that must, itself, be regarded as a perturbing influence on nuclear dynamics and, in that sense, can enter as perturbations of higher order. We thus write

$$\{[-\tfrac{1}{2}\hbar^2/\mu][1 + m_e g_2^{el}(R)/m_p]\, \partial^2/\partial R^2 + [1 + m_e g_1^{el}(R)/m_p]J^2\hbar^2/(2\mu R^2)$$
$$+ U_{BO}(R) + U_{ad}(R)\}\psi_{vJM}^{nuc}(R) = E_{vJ}\,\psi_{vJM}^{nuc}(R) \quad (25)$$

The forms of correction terms of the radial second derivative are considerably simpler in the present work than in past treatments [11,15,16] because R^{-1} is factored in the original wave function. In previous work [5,15] a term equivalent to $g_2(R)$ was included in the radial differential operator in a form

$$\partial/\partial R[1 + m_e g_2(R)/m_p]\partial/\partial R \quad (26)$$

Between their Eqs. (11) and (14), Bunker and Moss explained [16] how proceeding between this form and that of Eq. (25) yields further corrections containing a ratio of electronic to a nuclear mass to powers only greater than the first power relative to $U_{BO}(R)$, which can be safely ignored; the latter corrections are thus consistent with those calculated in the present treatment. In that sense whether $g_2(R)$ appears in the form of Eq. (25) or expression (26) above is immaterial. An application of hypervirial perturbation theory may be practicable, however, only with the form in expression (26); that theory [17] applied to the latter form [15] allows the derivation of expressions that are currently used in analyses of molecular spectra [5]. Whether the form of the corrections shown in Eq. (25) leads to a more practical form than the one achieved elsewhere [15] is uncertain.

In conclusion of this section we note the apparent division of terms into effects of $1/\mu$ and $1/(m_a + m_b)$, rather than of $1/m_a$ and $1/m_b$. This form is deceptive because effects of the nuclear derivatives $\mathbf{V}$ on the electronic wave functions $\Phi_{BO}^{el}(\mathbf{x}_i;\mathbf{x})$ depend upon the position of the centre of nuclear mass. Noting that $\sum_i \mathbf{V}_i$ has the same effect on Φ_{BO}^{el} as $-\mathbf{V}_{CNM}$, one expresses the adiabatic potential in the form

$$U_{ad}(R) = -(\tfrac{1}{2}\hbar^2)\langle\Phi_{BO}^{el}\,|\,\mathbf{V}_{CNM}^2/(m_a + m_b) + \mathbf{V}^2/\mu\,|\,\Phi_{BO}^{el}\rangle \quad (27)$$

that, through standard manipulation, is equivalent to

$$U_{ad}(R) = -(\tfrac{1}{2}\hbar^2)\langle\Phi_{BO}^{el}\,|\,\mathbf{V}_a^2/m_a + \mathbf{V}_b^2/m_b\,|\,\Phi_{BO}^{el}\rangle \quad (28)$$

Hence $U_{ad}(R)$ becomes written exactly in terms of m_a^{-1} and m_b^{-1}, thus in the spirit of Eq. (7), with otherwise constant functional dependences upon R.

We see below that H_{nad}^{eff} separates into operators that depend on m_a^{-1} and m_b^{-1}, in accordance with results of preceding authors [13]; note, however, our comments below regarding molecular ions. These separations are remarkable, in view of the mathematical primacy of μ^{-1} and $(m_a + m_b)^{-1}$ in the mathematical expression of the problem. The resolution of all effects of deviations from Born–Oppenheimer behaviour for electrically neutral molecules to terms in m_a^{-1} and m_b^{-1} implies that a simpler description of the entire problem might result from direct use of an operator for nuclear
kinetic energy in terms of $(\mathbf{V}_a^2/m_a + \mathbf{V}_b^2/m_b)$, rather than of $[\mathbf{V}^2/\mu + \mathbf{V}_{CNM}^2/(m_a + m_b)]$, but there seems no obvious procedure to follow for that purpose.

III. EFFECTS OF ISOTOPIC SUBSTITUTION

A. Derivation of Electronic Quantities

The problem of transformations between isotopic variants is complicated. To begin we define the reduced differential operators

$$\mathbf{V} \equiv \hat{R} \, \partial/\partial R + R^{-1}[\hat{\theta} \, \partial/\partial\theta + \hat{\chi} \, \partial/\partial\chi] \tag{29}$$

with

$$\mathbf{V}^2 = \partial^2/\partial R^2 + R^{-2}[\partial^2/\partial\theta^2 + \partial^2/\partial\chi^2] \tag{30}$$

The total operator for kinetic energy of the normal molecular species is

$$T = -\tfrac{1}{2}\hbar^2\left[\sum_i \mathbf{V}_i^2/m_e + \left(\sum_i \mathbf{V}_i\right)^2 \bigg/ (m_a + m_b) + \mathbf{V}^2/\mu + \mathbf{V}_{CMM}^2/M\right] \tag{31}$$

in which the notation conforms to that above. The centre of mass of an isotopic variant, henceforth denoted by primed quantities (superseding preceding usage in relation to laboratory coordinates), is shifted by a displacement $\kappa\mathbf{x}$ from that of the original species,

$$\mathbf{X}'_{CMM} = \mathbf{X}_{CMM} + \kappa\mathbf{x} \tag{32}$$

in which, in various useful forms,

$$\begin{aligned}
\kappa &= \tfrac{1}{2}[(m'_a - m'_b)/(m'_a + m'_b) - (m_a - m_b)/(m_a + m_b)] \\
&= m'_a/(m'_a + m'_b) - m_a/(m_a + m_b) \\
&= m_b/(m_a + m_b) - m'_b/(m'_a + m'_b) \\
&= (m'_a m_b - m'_b m_a)/[(m'_a + m'_b)(m_a + m_b)]
\end{aligned} \tag{33}$$

with small effects due to masses of electrons neglected. Although the use of atomic and molecular masses instead of nuclear masses might be preferable, the derivation thereby becomes extremely difficult. For electron i of an isotopic variant the coordinates become correspondingly altered to

$$\mathbf{x}'_i = \mathbf{x}_i - \kappa \mathbf{x} \tag{34}$$

whereas

$$\mathbf{x}' = \mathbf{x}_a - \mathbf{x}_b \tag{35}$$

remains unaltered.

$$\mathbf{x}' = \mathbf{x} \tag{36}$$

From Eqs. (32), (34) and (36) we have

$$\nabla'_i = \nabla_i \tag{37}$$

$$\nabla'_{CMM} = \nabla_{CMM} \tag{38}$$

$$\nabla' = \nabla + \kappa \sum_i \nabla_i - \kappa \nabla_{CMM} \tag{39}$$

For an isotopic variant in terms of coordinates of the original species, we thus find the kinetic energy, beginning with an expression for T' analogous to Eq. (31):

$$T' = -\tfrac{1}{2}\hbar^2 \left[\sum_i \nabla_i^2/m_e + \left(\sum_i \nabla_i \right)^2 \Big/ (m'_a + m'_b) \right.$$
$$\left. + \left(\nabla + \kappa \sum_i \nabla_i - \kappa \nabla_{CMM} \right)^2 \Big/ \mu' + \nabla_{CMM}^2/M' \right] \tag{40}$$

To establish from first principles quantities such as $g_{1,2}(R)'$ in terms of $g_{1,2}(R)$, electronic positional moments etc. this relation is crucial; one must proceed by expressing the dynamics of an isotopic variant in terms of those of the original molecule: for that purpose the latter relation is essential. At this point we note that the motion of $\mathbf{X}_{CMM}$ is not totally separated: there remains some inertial coupling through ∇ and other dynamic variables, but these coupling operators can be chosen to have zero expectation values.

Certain terms in this operator for kinetic energy warrant explanation. The operator $\sum_i \nabla_i^2/m_e$ primarily determines the electronic states

$\Phi_{BO;\,\lambda}^{el}(\mathbf{x}_i;\mathbf{x})$ and the function representing the Born–Oppenheimer potential energy. The term $(\sum_i \mathbf{V}_i)^2/(m_a' + m_b')$ likewise contains the mass-reducing effect for electrons and the standard mass polarization terms $\sim \mathbf{V}_i \cdot \mathbf{V}_j$ for $i \neq j$. These terms affect the radial potential energy for nuclear motions, producing the equivalent of a contribution to the adiabatic potential energy dependent on total nuclear mass $m_a' + m_b'$. $\mathbf{V}^2/\mu'$ provides the kinetic energy for nuclear motions, an adiabatic potential energy dependent on the reduced mass, and functions g_1 and g_2 for the original molecule reduced according to the ratio μ/μ', leading to nonadiabatic energy terms, in turn reduced according to $(\mu/\mu')^2$. Other contributions to $\mathbf{V}^2$ energies arise from cross terms of two types. The terms in $1/\mu'$ that involve $\mathbf{V}$ are

$$-\tfrac{1}{2}\hbar^2\left(\mathbf{V}^2 + 2\kappa\mathbf{V}\cdot\sum_i \mathbf{V}_i - 2\kappa\mathbf{V}\cdot\mathbf{V}_{CMM}\right)\Big/\mu' \tag{41}$$

The latter term generates within an isotopic variant an inertial coupling between internal nuclear motions and the centre of mass of the original molecule. As previously mentioned, when one chooses the approximate eigenstate to have no dependence upon $\mathbf{X}_{CMM}$, this term vanishes in first order. In second order there appears a small effect proportional to the translational energy of the centre of mass. An eventual reversion to the coordinates and propagation constant of the centre of mass of the isotopic variant eliminates entirely not only this term but also terms involving $\mathbf{V}_{CMM} \cdot \sum_i \mathbf{V}_i$, giving the normal translational energy of the entire isotopically invariant molecule.

The remaining important effects in second order that depend on κ are represented by

$$-(\kappa\hbar^2/\mu')^2 \sum_\lambda \sum_\alpha \langle 0|\sum_i \mathbf{V}_{i\alpha}|\lambda\rangle\langle\lambda|\sum_i \mathbf{V}_{i\alpha}|0\rangle\mathbf{V}_\alpha^2/E_{\lambda 0} \tag{42}$$

(in which α is an index that denotes a component R, θ or χ) which supplements the familiar cross terms,

$$-(\kappa/\mu')\sum_\alpha \mathbf{V}_\alpha \cdot \sum_i \mathbf{V}_{i\alpha} \tag{43}$$

taken to second order in Eq. (21), and

$$-\kappa(\hbar^2/\mu'^2) \sum_\lambda \sum_\alpha \left\{\langle 0|\mathbf{V}_\alpha|\lambda\rangle\langle\lambda|\sum_i \mathbf{V}_{i\alpha}|0\rangle + \langle 0|\sum_i \mathbf{V}_{i\alpha}|\lambda\rangle\langle\lambda|\mathbf{V}_\alpha|0\rangle\right\}\mathbf{V}_\alpha^2/E_{\lambda 0}$$

$$\tag{44}$$

that results from mixed perturbations of second order arising from the terms $-\frac{1}{2}\hbar^2 \mathbf{V}_\alpha^2/\mu'$ and $-(\kappa\hbar^2/\mu')\mathbf{V}_\alpha \cdot \sum_i \mathbf{V}_{i\alpha}$ of Eq. (40). A less obvious factor two that is present arises because

$$\hbar^2 \mathbf{V}_\alpha^2 [\Phi_{BO;\ \lambda}^{el}(\mathbf{x}_i;\ \mathbf{x})\psi_{vJM;\ \lambda}^{nuc}(\mathbf{x})] \tag{45}$$

contains the cross term $2\mathbf{V}_\alpha\Phi \cdot \mathbf{V}_\alpha\psi$. As

$$\mathbf{V}_i = -[H, \mathbf{x}_i]/(m_e/\hbar^2) \tag{46}$$

the former term, expression (42), becomes

$$m_e(\kappa\hbar/\mu')^2 \sum_\alpha \langle 0 | \sum_i \mathbf{V}_{i\alpha} \cdot \sum_j x_{j\alpha} | 0 \rangle \mathbf{V}_\alpha^2 \tag{47}$$

or

$$\tfrac{1}{2}m_e(\kappa\hbar/\mu')^2 \sum_\alpha \langle 0 | \left[\sum_i \mathbf{V}_{i\alpha}, \sum_j x_{j\alpha}\right] | 0 \rangle \mathbf{V}_\alpha^2 = \tfrac{1}{2}m_e(\kappa\hbar/\mu')^2(Z_a + Z_b)\mathbf{V}^2 \tag{48}$$

in which $Z_a + Z_b$ is the total number of electrons; this term serves as a sort of transformation of principal axes of the total mass of electrons, and combines with the ordinary nonadiabatic correction for this case. The latter term, expression (44), becomes reduced with the aid of expression (45) to the form

$$\kappa m_e(\hbar/\mu')^2 \sum_\lambda \sum_\alpha \left\{ \langle 0 | \mathbf{V}_\alpha | \lambda \rangle \langle \lambda | \sum_i x_{i\alpha} | 0 \rangle - \langle 0 | \sum_i x_{i\alpha} | \lambda \rangle \langle \lambda | \mathbf{V}_\alpha | 0 \rangle \right\} \mathbf{V}_\alpha^2 \tag{49}$$

By orthogonality

$$\mathbf{V}_\alpha \langle 0 | \lambda \rangle = 0 = \langle 0 | \mathbf{V}_\alpha | \lambda \rangle + \langle \lambda | \mathbf{V}_\alpha | 0 \rangle^* \tag{50}$$

Hence expression (44) becomes, through closure,

$$-\kappa m_e(\hbar/\mu')^2 \sum_\alpha \left\{ \langle \Phi_{BO}^{el} | \sum_i x_{i\alpha} | \mathbf{V}_\alpha \Phi_{BO}^{el} \rangle + \langle \mathbf{V}_\alpha \Phi_{BO}^{el} | \sum_i x_{i\alpha} | \Phi_{BO}^{el} \rangle \right\} \mathbf{V}_\alpha^2 \tag{51}$$

which is equivalent to

$$-\kappa m_e(\hbar/\mu')^2 \sum_\alpha \left(\mathbf{V}_\alpha \langle 0 | \sum_i x_{i\alpha} | 0 \rangle \right) \mathbf{V}_\alpha^2 \tag{52}$$

For α representing a direction perpendicular to $\hat{R}$, $\langle 0 | \sum_i x_{i\alpha} | 0 \rangle$ is of course zero for any diatomic molecule because charge is symmetrically distributed about the direction $\hat{R}$. Nevertheless an operation $\partial \langle 0 | \sum_i x_{i\alpha} | 0 \rangle / \partial \theta$ implies a mathematical possibility of virtual differential displacements of the entire molecule away from $\hat{R}$ in the direction $\hat{\theta}$, so that

$$\partial \langle 0 | \sum_i x_{i\alpha} | 0 \rangle / \partial \theta = \langle 0 | \sum_i x_{iR} | 0 \rangle \, \delta_{\alpha, \theta} \tag{53}$$

with analogous arguments for $\partial \langle 0 | \sum_i x_{i\alpha} | 0 \rangle / \partial \chi$. As the positional moment is related to the total electronic part of the electric dipolar moment $p^{el}(R)$ through

$$p^{el}(R) = -e \langle 0 | \sum_i x_{iR} | 0 \rangle \tag{54}$$

with $-e$ as the charge on an electron, expression (52) becomes

$$\kappa (\hbar/\mu')^2 (m_e/e) [(dp^{el}(R)/dR) \, \partial^2/\partial R^2 - p^{el}(R) J^2 / (\hbar^2 R^3)] \tag{55}$$

Collecting all terms dependent on ∇^2, we find

$$\begin{aligned} T = -(\tfrac{1}{2}\hbar^2/\mu') \Bigg[& 1 - (2\kappa m_e/\mu' e) dp^{el}(R)/dR - \kappa^2 (Z_a + Z_b) m_e/\mu' \\ & + (2\hbar^2/\mu') \sum_\lambda \langle 0 | \partial/\partial R | \lambda \rangle \langle \lambda | \partial/\partial R | 0 \rangle / E_{\lambda 0} \Bigg] \partial^2/\partial R^2 \\ & + [1/(2\mu' R^2)] \Bigg[1 - (2\kappa m_e/(\mu' eR)) p^{el}(R) - \kappa^2 (Z_a + Z_b) m_e/\mu' \\ & + (2\hbar^2/(\mu' R^2)) \sum_\lambda \langle 0 | \partial/\partial \theta | \lambda \rangle \langle \lambda | \partial/\partial \theta | 0 \rangle / E_{\lambda 0} \Bigg] J^2 \end{aligned} \tag{56}$$

Because in the latter equation we identify second-order sums as

$$(\mu/\mu')(m_e/m_p) g^{el}_{1,2}(R) \tag{57}$$

and the bracketed terms themselves are

$$[1 + m_e g^{el}_{1,2}(R)'/m_p] \tag{58}$$

we obtain the important results

$$g_1^{el}(R)' = \mu g_1^{el}(R)/\mu' - m_p \kappa^2 (Z_a + Z_b)/\mu' - (2\kappa m_p/e\mu')p^{el}(R)/R \quad (59)$$

and

$$g_2^{el}(R)' = \mu g_2^{el}(R)/\mu' - m_p \kappa^2 (Z_a + Z_b)/\mu' - (2\kappa m_p/e\mu')dp^{el}(R)/dR \quad (60)$$

Here $p^{el}(R)$, $g_1^{el}(R)$ and $g_2^{el}(R)$ are due to all electrons, and $\mathbf{p}^{el}(R)$ is reckoned relative to the centre of nuclear mass of the original molecule (not its isotopic variant).

For a molecule that has a total electronic charge $-(Z_a + Z_b)e$ the dipolar moment $p^{el}(R)'$ of an isotopic variant (with respect to the centre of mass as origin) is simply

$$p^{el}(R)' = p^{el}(R) + \kappa e R(Z_a + Z_b) \quad (61)$$

Thus we achieve ultimately the symmetric results

$$\mu' g_1^{el}(R)' = \mu g_1^{el}(R) - \kappa m_p [p^{el}(R) + p^{el}(R)']/(eR) \quad (62)$$

and

$$\mu' g_2^{el}(R)' = \mu g_2^{el}(R) - (\kappa m_p/e)\, d[p^{el}(R) + p^{el}(R)']/dR \quad (63)$$

In the case that one treats a molecular ion, $(Z_a + Z_b)$ becomes replaced throughout by $(Z_a + Z_b - Q)$, in which Q represents the number of electrons removed from the neutral molecule to form the molecular cation. Thereby Eqs. (62) and (63) are readily established for that case.

B.   Conversion from Electronic to Molecular Quantities

We proceed to show how the total magnetogyric function $g_1(R)$, the corresponding total vibrational inertial function $g_2(R)$ and a total function $\mathbf{p}(R)$ for the electric dipolar moment, together with the reduced mass μ_{at} for neutral atoms A and B, are to be used instead of the purely electronic contributions, as an elaboration of earlier Eqs. (35)–(37) [11]. The inverse of the reduced mass of two neutral atoms is

$$1/\mu_{at} = [(m_a + Z_a m_e) + (m_b + Z_b m_e)]/[(m_a + Z_a m_e)(m_b + Z_b m_e)] \quad (64)$$

As both m_a and $m_b \gg m_e$,

$$1/\mu_{at} \cong [1 - m_e(Z_a m_b/m_a + Z_b m_a/m_b)/(m_a + m_b)]/\mu \quad (65)$$

The magnetic dipole due to only the nuclei is

$$|\mathbf{m}^{nuc}| = \tfrac{1}{2}(Z_a R_a v_a + Z_b R_b v_b)e/m_e = \tfrac{1}{2}[(Z_a m_b^2 + Z_b m_a^2)/(m_a + m_b)^2]eRv/m_e \quad (66)$$

in which we envisage circular orbits with $\mathbf{v}$ being the velocity associated with the vector $\mathbf{R}$; an orbit of other or general shape makes no difference. As $\mathbf{J} = \mu\mathbf{R} \times \mathbf{v}$,

$$\mathbf{m}^{nuc} = \tfrac{1}{2}e[(Z_a m_b^2 + Z_b m_a^2)/(m_a + m_b)^2]\mathbf{J}/\mu$$
$$= (\tfrac{1}{2}e\hbar/m_p)g_1^{nuc}(R)\mathbf{J}/\hbar \quad (67)$$

with

$$g_1^{nuc}(R) = g_1^{nuc} = m_p(Z_a m_b/m_a + Z_b m_a/m_b)/(m_a + m_b) \quad (68)$$

for the nuclear contribution to $g_1(R)$. Hence, by comparison of Eqs. (65) and (68) we see that for either the original species or its isotopic variant the kinetic energy of rotation becomes

$$[1 + m_e g_1^{el}(R)/m_p]J^2/(2\mu R^2)$$
$$= [1 + m_e g_1(R)/m_p - m_e(Z_a m_b/m_a$$
$$+ Z_b m_a/m_b)/(m_a + m_b)]J^2/(2\mu R^2)$$
$$\cong [1 + m_e g_1(R)/m_p]J^2/(2\mu_{at} R^2) \quad (69)$$

This result agrees with the previously reported expressions [11] involving $g_1(R)$, the total molecular magnetogyric ratio. A similar derivation applies to $g_2(R)$, the total vibrational $g(R)$-function, which is a measure of the vibrational inertia of each nucleus, with its electrons, were they to move rigidly within the molecule, less the actual vibrational inertia.

Because the total functions $g_{1,2}(R)$ are important, we reexamine Eq. (59) for transformations of $g_1(R)$ on isotopic substitution, knowing that transformations of $g_2(R)$ operate analogously. Regarding the nuclear terms one can easily show from Eq. (68), the second and third expressions of κ in Eq.

(33) and the expression for the nuclear contribution to the electric dipolar moment,

$$|\mathbf{p}^{nuc}(R)| = eR(Z_a m_b - Z_b m_a)/(m_a + m_b) \tag{70}$$

following some algebraic manipulation, that

$$\mu'g_1^{nuc}(R)' = \mu g_1^{nuc}(R) - \kappa m_p[p^{nuc}(R) + p^{nuc}(R)']/(eR) \tag{71}$$

which is entirely analogous to Eq. (62) for the function $g_1^{el}(R)'$. Adding Eqs. (62) and (70) we achieve the important result

$$\mu'g_1(R)' = \mu g_1(R) - \kappa m_p[p(R) + p(R)']/(eR) \tag{72}$$

with the analogous expression for $g_2(R)$

$$\mu'g_2(R)' = \mu g_2(R) - (\kappa m_p/e)\, d[p(R) + p(R)']/dR \tag{73}$$

valid for both neutral and ionic species. In the case of a diatomic species having electric neutrality, $p(R) = p(R)'$, with resulting simplification of these equations, but to include the ionic cases we leave them in the present form.

The total operator for the effective kinetic energy is thus

$$T = -(\tfrac{1}{2}\hbar^2/\mu_{at})[1 + m_e g_2(R)/m_p]\, \partial^2/\partial R^2 + [1/(2\mu_{at} R^2)][1 + m_e g_1(R)/m_p]J^2 \tag{74}$$

for which the relations between the functions $g_{1,2}(R)'$ for an isotopic variant in terms of properties of the original species are given in Eqs. (72) and (73). To achieve the total effective Hamiltonian, one adds to the operator for kinetic energy the potential energy $U^{BO}(R) + U^{ad}(R)$ as presented above.

C.   Dependences of $g_{1,2}(R)$ upon m_a^{-1} and m_b^{-1} for Neutral Molecules

Elsewhere [13] appears the argument that one can separate $g_{1,2}(R)$ into functions dependent on m_a^{-1} and m_b^{-1} that are otherwise independent of nuclear mass and that serve to allow one readily to establish connections among spectra of isotopic variants. To verify this condition, we work with $g_1(R)$, which can, we assume, be cast into the form

$$\mu g_1(R) = m_p[g_1^{irr}(R) - (\mu p(R)/eR)(m_a^{-1} - m_b^{-1})] \tag{75}$$

Because $\mu^{-1} = m_a^{-1} + m_b^{-1}$, $g_1(R)$ shows the requisite behaviour. One can regard $g_1^{irr}(R)$ in this expression as an irreducible nonadiabatic function (or magnetogyric function) of a sort. Although one can write such an equation for a single isotopic species, the test of validity is to prove its compatibility with Eq. (64) when one transforms between isotopic variants. On taking a difference for two isotopic species (the original species denoted by unprimed quantities and a variant by primed quantities), we find

$$\mu' g_1(R)' - \mu g_1(R) = (m_p\, p(R)/eR)[\mu'(1/m_a' - 1/m_b') - \mu(1/m_a - 1/m_b)] \quad (76)$$

Because

$$\mu'(1/m_a' - 1/m_b') - \mu(1/m_a - 1/m_b)$$
$$= (m_b' - m_a')/(m_b' + m_a') - (m_b - m_a)/(m_a + m_b)$$
$$= -2\kappa \quad (77)$$

according to the first of relations (33), Eq. (72) is directly confirmed and Eq. (75) is valid for all neutral molecules; the argument relies upon knowing that $p(R) = p(R)'$, which is true for neutral molecules but not in general true for molecular ions. Thus we find that

$$g_1(R) = [g_1^{irr}(R) - p(R)/(eR)]m_p/m_a + [g_1^{irr}(R) + p(R)/(eR)]m_p/m_b \quad (78)$$

and, according to analogous arguments,

$$g_2(R) = [g_2^{irr}(R) - (1/e)\, dp(R)/dR]m_p/m_a + [g_2^{irr}(R) + (1/e)\, dp(R)/dR]m_p/m_b$$
$$(79)$$

To understand the meaning of $g_{1,\,2}^{irr}(R)$, or indeed of $(m_p/\mu)g_{1,\,2}^{irr}(R)$, is difficult, because when one considers the second-order sums occurring in Eq. (21) there seems to be no hint as to how such a separation might be implemented. The above arguments seem somewhat fortuitous, a sense that is enhanced by the fact that a relation such as Eq. (75) cannot in general be found for molecular ions.

D. Note on Molecular Ions

Equations (72) and (73) yield transformations of the type

$$\mu_{at}' g_1(R)' = \mu_{at} g_1(R) - \kappa m_p[p(R) + p(R)']/(eR) \quad (80)$$

in which $p(R)$ and $p(R)'$ are the total electric dipolar moments of the original molecule and its isotopic variant, with the origin of coordinates at the centre of the respective molecular mass. If the species of interest is a diatomic molecular ion bearing a net ionic charge Qe, this equation becomes simply

$$\mu'_{at} g_1(R)' = \mu_{at} g_1(R) - 2\kappa m_p p(R)/(eR) + \kappa^2 m_p Q \tag{81}$$

in which $Q = (Z_a + Z_b - n)$, n being the total number of electrons and $p(R)$ the total electric dipolar moment of the original (unprimed) molecular ion. A method to handle this in the expression for $\tilde{E}_{vJ}$ of an isotopic variant σ is to choose a standard isotopic species ($\sigma = 0$); $\kappa_{\sigma 0}$ ($=\kappa$) is measured relative to κ_{00} ($=0$) for that species. We proceed to define a further inertial function $f_1(R)$, equal to $g_1(R)$ for $\sigma = 0$, such that

$$\mu'_{at} f_1(R)' = \mu_{at} f_1(R) - 2\kappa m_p p(R)/(eR) \tag{82}$$

Thereby $f_1(R)'$ transforms in the manner described for neutral molecules with terms dependent on m_a^{-1} and m_b^{-1}. The κ^2 term remains; in factors such as $(1/\mu_{at})[1 + m_e g_1(R)/m_p]$, which defines $1/\mu_{at}^{eff}$, this quantity for an isotopic variant becomes

$$(1/\mu'_{at})[1 + m_e f_1(R)'/m_p + \kappa^2 Q m_e/\mu'_{at}] \tag{83}$$

with correspondingly a factor of the same type involving $f_2(R)'$ accompanying the vibrational operator. The term containing κ^2 becomes viewed as simply leading to overall modification of the effective reduced mass of an isotopic variant through a relation

$$\mu_{at}^{eff'} = \mu'_{at} - \kappa_{\sigma 0}^2 Q m_e \tag{84}$$

In a consistent manner one can regard this adjustment as altering only the Born–Oppenheimer part of the problem; one then analyses the spectrum at first without this correction, and proceeds subsequently to apply this correction to each coefficient $Y_{kl}^{(0)}$ in Eq. (4), according to its dependence on μ_{at}.

IV. EXTRACTION OF MOLECULAR SPECTRAL PROPERTIES

The effective Schrödinger equation for nuclear motion is

$$\{[-\tfrac{1}{2}\hbar^2/\mu_{at}][1 + m_e g_2(R)/m_p] \, \partial^2/\partial R^2 + [1 + m_e g_1(R)/m_p]\hbar^2 J^2/(2\mu_{at} R^2)$$
$$+ U^{BO}(R) + U^{ad}(R)\}\psi_{vJM}^{nuc}(\mathbf{R}) = E_{vJ} \psi_{vJM}^{nuc}(\mathbf{R}) \tag{85}$$

To use this equation in the context of what is known according to either the JBKW approach or perturbation theory, one can divide by the radially dependent factor $[1 + m_e g_2(R)/m_p]$ to yield an effective radial Schrödinger equation

$$\{-\partial^2/\partial R^2 + J(J+1)[1 + m_e g_1(R)/m_p]/[R^2(1 + m_e g_2(R)/m_p)]$$
$$+ [2\mu_{at} U_{BO}(R)/\hbar^2]/[1 + m_e g_2(R)/m_p] + 2\mu_{at} U_{ad}(R)/\hbar^2$$
$$- [2\mu_{at} E_{vJ}^{BO}/\hbar^2]/[1 + m_e g_2(R)/m_p] - 2\mu_{at} \Delta E_{vJ}/\hbar^2\}\psi_{vJM}^{nuc}(\mathbf{R}) = 0 \quad (86)$$

in which ΔE_{vJ} represents corrections to the vibration–rotational energy associated with the presence of $g_1(R)$, $g_2(R)$ and $U_{ad}(R)$ to first order in the ratio (m_e/m_p). Within a linear approximation, with the variable x according to Eq. (3), and with the definitions $a_0 = [d^2 U_{BO}(R)/dR^2]_{R=R_e}/(2hc/R_e^2)$ and the equilibrium rotational parameter $B_e = h/(8\pi^2 c\mu_{at} R_e^2)$, this equation becomes

$$-B_e \, \partial^2\psi_{vJM}^{nuc}/\partial x^2 + \left\{ B_e J(J+1)[1 + m_e(g_1(x) - g_2(x))/m_p]/(1+x)^2 \right.$$

$$+ a_0 x^2\left(1 + \sum_{j=1} a_j x^j\right)[1 - m_e g_2(x)/m_p]$$

$$\left. + U_{ad}(x) - \tilde{E}_{vJ}^{BO}[1 - m_e g_2(x)/m_p] - \Delta\tilde{E}_{vJ} \right\}\psi_{vJM}^{nuc}(x) = 0 \quad (87)$$

in which B_e, $U_{ad}(x)$ here and energy operators as a function of x in all subsequent usage are assumed to have the unit of wavenumber. The latter equation provides a useful means to evaluate analytic eigenvalues in terms of parameters of the various radial functions in the spirit of Dunham's approach [1].

Following previous developments [5,15,17] according to an alternative approach, we assume a neutral diatomic molecule AB, of relative polarity $^-AB^+$, in an electronic ground state of class $^1\Sigma^+$ or 0^+ with no nearby electronically excited state, such that perturbations of measured pure rotational and vibration–rotational spectra are small and homogeneous; the molecule dissociates into neutral atoms A and B having masses M_a and M_b respectively. We assume an effective Hamiltonian for vibration–rotational motion of the nuclei of the form derived above,

$$H^{eff}(x) = -B_e[1 + m_e g_2(x)/m_p] \, d^2/dx^2$$

$$+ U_{BO}(x) + U_{ad}(x) + B_e J(J+1)[1 + m_e g_1(x)/m_p]/(1+x)^2 \quad (88)$$

with $U_{BO}(x)$, $U_{ad}(x)$, $g_1(x)$ and $g_2(x)$ as previously defined. The eigenvalues of this Hamiltonian are expressed as in Eq. (6) [5]. We assume a radial function for electric dipolar moment of the form

$$M(x) = \sum_{j=0} M_j x^j = M_0 + M_1 x + M_2 x^2 + M_3 x^3 + \cdots \tag{89}$$

in which M_0 is the permanent electric dipolar moment of the molecule (the value of the function $p(R)$ at $x = 0$ or $R = R_e$), which governs the intensity of the pure rotational band in absorption or emission, and $M_1 \equiv R_e(dp(R)/dR)_{R=R_e}$, which correspondingly governs the intensity of the fundamental band ($v = 1 \leftarrow v = 0$) in absorption or emission.

An established model [18,19] to treat the nonadiabatic rotational term $g_1(R)$ is to begin with Eq. (78) in which $g_1^{irr}(R)$ represents irreducible nonadiabatic rotational effects, attributed to interactions between the electronic ground state of interest and electronically excited states of class $^1\prod$ or 1 [10], and the other terms represent the effect of the rotating electric dipole, which is consistent with the above analysis of isotopic variants. The total term $g_1(R)$ consequently becomes expressed as

$$g_1(R) = m_p g_1^{irr}(R)/\mu_{at} - (m_p/e)[1/M_a - 1/M_b]p(R)/R \tag{90}$$

with

$$\alpha(R) \equiv g_1(R)m_e/m_p = m_e[\alpha^a(R)/M_a + \alpha^b(R)/M_b] \tag{91}$$

and

$$\begin{aligned}
\alpha^a(R) &= g_1^{irr}(R) - p(R)/(eR) \\
\alpha^b(R) &= g_1^{irr}(R) + p(R)/(eR)
\end{aligned} \tag{92}$$

In terms of the reduced variable x, the latter three relations become

$$\begin{aligned}
\alpha(x) &= m_e[\alpha^a(x)/M_a + \alpha^b(x)/M_b] \\
\alpha^a(x) &= g_1^{irr}(x) - M(x)/[eR_e(1+x)] \equiv \sum_{j=0} t_j^a x^j \\
\alpha^b(x) &= g_1^{irr}(x) + M(x)/[eR_e(1+x)] \equiv \sum_{j=0} t_j^b x^j
\end{aligned} \tag{93}$$

in which $t_j^{a,b}$ are coefficients in established radial functions [5] for nonadiabatic rotational effects.

We postulate an analogous set of relations for nonadiabatic vibrational effects,

$$g_2(R) = m_p g_2^{irr}(R)/\mu_{at} + (m_p/e)[1/M_a - 1/M_b] \, dp(R)/dR \qquad (94)$$

in which $g_2^{irr}(R)$ represents irreducible nonadiabatic effects, attributed to interactions between the electronic ground state of interest and electronically excited states of class $^1\Sigma^+$ or 0^+ [10], and the other terms represent effects of the vibrating dipole. These nonadiabatic effects are likewise partitioned into contributions for the separate nuclei:

$$\beta(R) \equiv g_2(R)m_e/m_p = m_e[\beta^a(R)/M_a + \beta^b(R)/M_b] \qquad (95)$$

with

$$\begin{aligned}
\beta^a(R) &= g_2^{irr}(R) - (1/e) \, dp(R)/dR \\
\beta^b(R) &= g_2^{irr}(R) + (1/e) \, dp(R)/dR
\end{aligned} \qquad (96)$$

In terms of x, these three relations become

$$\begin{aligned}
\beta(x) &= m_p[\beta^a(x)/M_a + \beta^b(x)/M_b] \\
\beta^a(x) &= g_2^{irr}(x) - (1/eR_e) \, dM(x)/dx \equiv \sum_{j=0} s_j^a x^j \\
\beta^b(x) &= g_2^{irr}(x) + (1/eR_e) \, dM(x)/dx \equiv \sum_{j=0} s_j^b x^j
\end{aligned} \qquad (97)$$

in which $s_j^{a,b}$ are coefficients in radial functions [5] to represent nonadiabatic vibrational effects.

By equating sets of coefficients of x to a common exponent, we obtain the results, using the coefficients $t_j^{a,b}$,

$$M_0 = \tfrac{1}{2}eR_e(t_0^b - t_0^a) \qquad (98)$$

$$g_1^{irr}(x = 0) = \tfrac{1}{2}(t_0^a + t_0^b) \qquad (99)$$

$$g_1(x = 0) = m_p(t_0^a/M_a + t_0^b/M_b) \qquad (100)$$

$$M_1 - M_0 = \tfrac{1}{2}eR_e(t_1^b - t_1^a) \qquad (101)$$

$$dg_1^{irr}(x)/dx \Big|_{x=0} = \tfrac{1}{2}(t_1^a + t_1^b) \qquad (102)$$

$$dg_1(x)/dx \Big|_{x=0} = m_p(t_1^a/M_a + t_1^b/M_b) \qquad (103)$$

and the corresponding results with the coefficients $s_j^{a,\,b}$,

$$M_1 = \tfrac{1}{2}eR_e(s_0^b - s_0^a) \tag{104}$$

$$g_2^{irr}(x = 0) = \tfrac{1}{2}(s_0^a + s_0^b) \tag{105}$$

$$g_2(x = 0) = m_p(s_0^a/M_a + s_0^b/M_b) \tag{106}$$

$$M_2 = \tfrac{1}{4}eR_e(s_1^b - s_1^a) \tag{107}$$

$$dg_2^{irr}(x)/dx\,\bigg|_{x=0} = \tfrac{1}{2}(s_1^a + s_1^b) \tag{108}$$

$$dg_2(x)/dx\,\bigg|_{x=0} = m_p(s_1^a/M_a + s_1^b/M_b) \tag{109}$$

Combined with further relations analogously derived, these results link expansion coefficients M_j for the electric dipolar moment and the corresponding expansion coefficients of $g_1^{irr}(x)$ and $g_2^{irr}(x)$ to radial coefficients $s_j^{a,\,b}$ and $t_j^{a,\,b}$ fitted during reduction of vibration–rotational spectra of sufficient quality and quantity for multiple isotopic variants [5].

One should bear in mind the following conditions during consideration or use of these results. First, coefficients $s_j^{a,\,b}$ and $t_j^{a,\,b}$ are here defined as factors of x^j (not of z^j with $z \equiv 2(R - R_e)/(R + R_e)$ [20], as in recent papers [5,21]). Second, a positive value of M_0 implies the polarity $^-AB^+$, contrary to usage elsewhere [21–23]. Third, these relations with radial coefficients for $M(x)$, $g_{1,\,2}(x)$ and $g_{1,\,2}^{irr}(x)$ in terms of $s_0^{a,\,b}$ and $t_j^{a,\,b}$ are independent whether a term for nonadiabatic vibrational effects appears in the Hamiltonian as $d/dx\,[1 + m_e g_2(x)/m_p]\,d/dx$ or as $[1 + m_e g_2(x)/m_p]\,d^2/dx^2$; however, expressions containing $s_j^{a,\,b}$ with $j > 0$ in various contributions $Z_{kl}[v + \tfrac{1}{2}]^k[J(J + 1)]^l$ to molecular eigenvalues in Eq. (6) depend on the location of $g_2(R)$ in the Hamiltonian.

As explained previously [5], for a molecule with sufficient isotopic variants, dependences of frequencies of pure rotational and vibration–rotational transitions on masses of particular nuclei and the extra rotational dependence might be insufficient to allow evaluation of adiabatic, nonadiabatic rotational and nonadiabatic vibrational effects of nuclei of separate types (i.e. A or B). In principle the above relations still preclude, from only finite data of frequency type in spectra of samples without external fields, separate evaluation of the auxiliary radial functions, i.e. the three sets $u_j^{a,\,b}$ (for adiabatic effects in $U_{ad}(x)$), $t_j^{a,\,b}$ and $s_j^{a,\,b}$ for each nucleus. When alternative coefficients in the set M_j, $u_j^{a,\,b}$, and those in $g_{1,\,2}^{irr}(x)$ (or $g_{1,\,2}(x)$) are used, the number of independent unknown inertial functions is decreased from six to five, but the insufficiency remains. If one derives data for $g_1(x)$ from the rotational g factor through measurements of the Zeeman effect and for M_j from measurements of either the Stark effect or spectral intensities, one can apply this information to constrain parameters in the analysis of spectral frequencies. If adiabatic effects are assumed at the outset to be negligible, some parameters in $M(x)$ or $g_{1,2}(x)$ may become overdetermined.

V. APPLICATIONS AND DISCUSSION

Our objective is to examine the nature of adiabatic and nonadiabatic effects in relation to energies of rotational and vibrational states of diatomic molecules of multiple isotopic variants; our work has naturally proceeded in the light of the original analysis [10] and subsequent developments [11–13, 15,16]. Another derivation [24] of adiabatic and nonadiabatic effects in diatomic molecules, as a limiting case of a linear polyatomic molecule and intended to be applied to a triatomic system [25], lacks features of the present formulation related to isotopic substitution and to molecular ions. The combination of an effective Hamiltonian here derived (in Eq. (25) or its extension in Eq. (85) or (87)) and quantum-mechanical derivation [15] of expressions $Z_{kl}^{a;b}$ containing coefficients $s_j^{a,b}$, $t_j^{a,b}$ and $u_j^{a,b}$ of x^j or z^j provides a complete and sufficient theoretical justification of Eq. (7) that originated empirically, in which parameters $\Delta_{kl}^{a;b}$ absorb effects of not only $s_j^{a,b}$, $t_j^{a,b}$ and $u_j^{a,b}$ (or their equivalents) but also $Y_{kl}^{(2)}$ (in Eq. (4)).

A principal achievement of our work is the relation of the isotopic dependence of nonadiabatic effects to the molecular electric dipolar moment and to its derivatives with respect to internuclear distance. According to quantum-mechanical methods we demonstrate how a dependence of vibration–rotational energies on nuclear mass, through isotopic variants, enables one in principle to relate molecular functions for the rotational magnetogyric ratio and for the electric dipolar moment to spectral data in the form purely of frequencies or wavenumbers of pure rotational or vibration–rotational transitions. In practice, because these spectral data have finite number and precision, such deductions of electric and magnetic properties may depend on neglect of adiabatic effects, as discussed elsewhere [26]. In the present work we find no explicit justification of an empirical observation that adiabatic effects have small magnitudes, relative to those of nonadiabatic effects, for somewhat massive molecules (containing nuclei with Z_a, $Z_b > 10$). For even more massive nuclei, the effects of the finite size of nuclei become detectable, for instance as observed for monofluorides of all elements in group 13 (B—Tl) [23]. Our results apply not only to neutral diatomic molecules in electronic states lacking spin or orbital angular momentum but also to molecular ions. Our approach allows for the first time experimental estimations of intrinsic non-adiabatic vibrational effects, in the radial function $g_2(R)$ (or $\beta_j x^j$). In what follows we present for purposes of illustration a few instances of the way in which the relations involving $\alpha(x)$ and $\beta(x)$ may be utilized.

Application of Eqs. (98)–(109) enables one to estimate a few parameters in radial functions of molecules for which values of $s_j^{a,b}$ and $t_j^{a,b}$ exist. For BrCl for which $s_0^{Br} = 1.70 \pm 0.26$, $s_0^{Cl} = 0.846 \pm 0.058$, $t_0^{Br} = -0.507 \pm 0.054$ and $t_0^{Cl} = -0.6463 \pm 0.0108$ [26], by means of Eqs. (98)–(100) we generate several results. From $t_0^{Br, Cl}$ we obtain $M_0 = (-2.38 \pm 0.78) \times 10^{-30}$ C m, $g_1^{irr}(x = 0) = -0.577 \pm 0.055$ and (for $^{79}Br^{35}Cl$) $g_1(x = 0) = -0.0251 \pm 0.0007$; the single experimental result for comparison is $|M_0| = (1.732 \pm 0.007) \times 10^{-30}$ C m [27], for which the magnitudes agree within the uncertainty of our M_0 propagated from the small difference between t_0^{Br} and t_0^{Cl}. Our approach yields directly also the polarity $^+BrCl^-$, expected according to conventional chemical trends [26]. From $s_0^{Br, Cl}$ and Eqs. (104)–(106) we find $M_1 = (-14.6 \pm 4.6) \times 10^{-30}$ C m, $g_2^{irr}(x = 0) = 1.27 \pm 0.27$ and (for $^{79}Br^{35}Cl$) $g_2(x = 0) = 0.0461 \pm 0.0041$; the only pertinent experimental result is the magnitude $|M_1| = (5.42 \pm 0.43) \times 10^{-30}$ C m [28], for which again the uncertainty of our M_1 propagated from the modest difference between s_0^{Br} and s_0^{Cl} hinders meaningful comparison. The negative sign of M_1 implies that $M(x)$ decreases as x increases about $R = R_e$. The omission of $u_j^{Br, Cl}$ as fitting parameters [26] may contribute to any discrepancies beyond the large uncertainties.

Although other instances of the evaluation of $t_0^{a,b}$ from spectral reductions exist such that estimates of M_0 and $g_1(x = 0)$ are practicable [29], there is at present no other case in which only experimental data serve to yield estimates of both $s_0^{a,b}$. To simulate for LiH the results that might emanate from measurement of expectation values of $g_1(x)$ and $M(x)$ in varied vibration–rotational states by means of Zeeman and Stark effects, quantum-chemical calculations of $g_1(x)$ and $M(x)$ enabled estimates of $t_j^{Li, H}$; from data of pure rotational and vibration–rotational spectra were then generated $s_0^{Li} = 0.669 \pm 0.042$ and $s_0^{H} = -0.3011 \pm 0.0079$ [30], among other parameters. By means of Eqs. (104)–(106) we here derive $M_1 = (-12.39 \pm 0.64) \times 10^{-30}$ C m, $g_2^{irr}(x = 0) = 0.184 \pm 0.043$ and (for $^6Li\,^1H$) $g_2(x = 0) = -0.1890 \pm 0.0009$; the former value is roughly consistent with $M_1 = (-11.075\,89 \pm 0.000\,18) \times 10^{-30}$ C m [30] used to generate $t_j^{Li, H}$. For LiH [30] adiabatic effects were not neglected. To allow meaningful application of Eqs. (101)–(103) or (107)–(109) there are at present available values neither of $s_1^{a,b}$ nor of $t_1^{a,b}$ evidently free of interference from neglected values of $u_j^{a,b}$.

Previous analyses of nonadiabatic rotational effects relied on a classical model according to which a magnetic dipolar moment arises from rotation of a rigid electric dipole, with addition ad hoc of a term for other influences equivalent to $g_1^{irr}(x = 0)$. That the isotopic dependence of $g_1(x = 0)$ according to that model is the same as that here derived for $g_1(x)$, in Eq. (78), purely from quantum-mechanical arguments justifies the former estimates of values of M_0 and g_J or $\langle vJ|g_1(x)|vJ\rangle$ [19,29]. Neither a direct quantal derivation nor a classical model is achieved for Eq. (79), which is thus generated entirely by analogy with Eq. (78). Although $g_2(R)$ might be related to the molecular electric quadrupolar moment or to anisotropy of magnetic susceptibility, neither such relation nor indeed a classical model for this vibrational g-function is obvious at present. When trends of magnitudes of this function become available from experimental data, similar to deductions of values of $g_1(x = 0)$ and $g_1^{irr}(x = 0)$ above, correlations with other molecular properties might appear. Calculations of molecular electronic structure that yield theoretical estimates of $g_2(x)$, analogous to those for $g_1(x)$ of LiH [30], would aid understanding the nature of this function.

For a neutral elemental diatomic species, for which the two nuclei have like proton number but unlike mass numbers, pure rotational or vibration–rotational spectra in absorption or emission are weak, although they might be readily measured in spontaneous or stimulated Raman processes. In these cases the coefficients M_j in Eq. (89) for electric dipolar moment are either zero or nearly zero; slight magnitudes of M_j for $j > 0$ may result from adiabatic effects if masses of the two nuclei are not equal, but these magnitudes are unlikely to be comparable with contributions to $g_1(R)$ and $g_2(R)$ from $g_1^{irr}(R)$ and $g_2^{irr}(R)$ respectively. Hence effectively $s_j^a = s_j^b$, $t_j^a = t_j^b$ and $u_j^a = u_j^b$ for all j; evaluations of Z_{kl} in Eq. (4) then yield values of u_j^a and coefficients in the radial functions $g_1^{irr}(x)$ and $g_2^{irr}(x)$, but unknown parameters in three sets remain to be deduced from Z_{kl}^v and Z_{kl}^r. To evaluate both Z_{kl}^v and Z_{kl}^r of a diatomic molecule with the same protonic number of each atomic centre, spectra are fit in terms of $1/\mu_{at}$, not of $1/M_a$ and $1/M_b$ separately.

Quantitative application of information deduced from parameters such as $u_j^{a,b}$, $t_j^{a,b}$ or $s_j^{a,b}$ for adiabatic and nonadiabatic effects requires a knowledge of propagation of error from the original measurements of spectral frequencies and wavenumbers. For instance, if one seeks to evaluate M_0 with a relative uncertainty less than 10%, then because M_0 results from a difference between t_0^b and t_0^a the corresponding uncertainties of these parameters might require to be about 1%. If, as in the case of BrCl, atomic masses are of the order of 50 u, the ratio m_e/M_a is of the order of 10^{-5}; one would need to evaluate $Y_{1,0}$ and $Y_{0,1}$ with relative uncertainties $|\delta Y_{1,0}/Y_{1,0}|$ and $|\delta Y_{0,1}/Y_{0,1}|$ about 10^{-7}, and the frequencies or wavenumbers would require both the latter internal precision and comparable consistency between data from varied sources (for instance, pure rotational transitions from microwave measurements and vibration–rotational transitions from infrared measurements with an interferometer or laser). To ensure maximum precision of auxiliary parameters s_j^a, t_j^a and u_j^a, one should include spectral data from isotopic variants of atomic centre A with atomic masses over a broad range; this condition is obviously fulfilled most readily with hydrogen as A (or B), for which M_a might be roughly 1, 2 or even 3 u—hence varying over a factor two or three, whereas even for bromine, which has a moderate mass among stable nuclides between hydrogen and bismuth, the range is only about 79–81 u corresponding to variation by only a few parts per hundred. To derive precisely defined values of molecular properties through these adiabatic and nonadiabatic effects, the requirement of precisely measured frequencies and wavenumbers of usable transitions tests the capabilities of present conventional instruments, although precisions and absolute accuracies $|\delta\tilde{\nu}/\tilde{\nu}| \sim 2 \times 10^{-10}$ of wavenumbers are attained in the region about 2×10^5 m^{-1} under exceptional conditions at the limit of current metrological standards [31]. As the precision of such measurements on diatomic molecules improves, so does correspondingly the precision of the electric dipolar moment and other properties derived from these frequency data. Processing of such data for molecules of small mass, such as LiH or other hydrides, would eventually require further extension of the theory to take into account terms beyond m_e/M to the first power (relative to adjacent terms) that we consider here.

In the preceding derivation and discussion we confine the development of theory to a case of a molecule in an electronic state having neither net electronic spin nor net orbital angular momentum, and we ignore effects of intrinsic nuclear angular momenta. If the former condition does not hold, there exist magnetic couplings between electronic spin and orbital angular momenta and between electronic spins of two or more net unpaired electrons, but also between these properties and the angular momentum due to rotation of the nuclei about the centre of molecular mass. The radial dependences of these properties can be modeled readily using radial functions having a form of those applied for adiabatic properties in $U_{ad}(x)$, such as involving coefficients $u_j^{q,\,b}$; several instances are summarized elsewhere [32]. The effects of such phenomena on corrections to Born–Oppenheimer behavior have never been fully considered. Our preliminary investigation of the effects of net electronic spin and orbital angular momentum indicates that the results given above for neither the adiabatic contribution $U_{ad}(R)$ nor the nonadiabatic vibrational term $g_2(R)$ become altered in the presence of net electronic spin or orbital angular momentum. For their effects on the nonadiabatic rotational term $g_1(R)$, if for a molecule in an electronic state of class other than $^1\sum^+$ or 0^+ J^2 is replaced throughout by N^2, in which $\mathbf{N}$ denotes the operator for angular momentum of the nuclear frame, our results presented in Sections III and IV remain valid, but the manifestations become complicated. Work on these aspects is in progress.

VI. CONCLUSIONS AND FUTURE DEVELOPMENTS

In this work we extend the theory of molecular spectra to include an analysis of relationships between the adiabatic and nonadiabatic corrections to the Born–Oppenheimer approximation among isotopic variants of diatomic molecules. We indicate extensions of the theory to include molecular ions and molecular states other than $^1\sum^+$ or 0^+.

Our analysis demonstrates how nonadiabatic effects from only frequency data in rotational and vibrational spectra of multiple isotopic variants of a given diatomic molecular species yield useful information about the electric dipolar moment and its derivatives. In absorption or emission the principal determinant of the intensity of pure rotational transitions is the electric dipolar moment itself, whereas the first derivative of that moment governs the intensity of the fundametal vibration–rotational band and the first and higher derivatives govern the intensity of overtone bands. Thus we prove that information about intensities (of spectra in absorption and emission) is latent within frequency data. Such information can enable estimations of concentrations of molecular ions and free radicals or other transients in chemically reacting systems. As frequencies of transitions in pure rotational and vibration–rotational spectra of diatomic molecules become measured with enhanced precision, even within the present range of metrological conditions, our results become increasingly valuable in comprehensive analyses of these data. Further development of the theory to apply to an electronic state other than $^1\sum^+$ or 0^+ is in progress.

ACKNOWLEDGMENT

J. F. Ogilvie thanks Professor J. W. Nibler for his hospitality at Oregon State University during the preparation of this paper.

REFERENCES

1. J. L. Dunham, *Phys. Rev.* **41**, 721–731 (1932).

2. G. Herzberg, *Molecular Spectra and Molecular Structure. I Spectra of Diatomic Molecules*, Second Edition, van Nostrand, Princeton, NJ, U.S.A.

3. J. L. Dunham, *Phys. Rev.* **41**, 713–720 (1932).

4. R. M. Herman and S. Short, *J. Chem. Phys.* **48**, 1266–1272 (1968) and *J. Chem. Phys.* **50**, 572 (1969).

5. J. F. Ogilvie, *J. Phys. At. Mol. Opt. Phys.* **B27**, 47–61 (1994).

6. A. H. M. Ross, R. S. Eng and H. Kildal, *Opt. Commun.* **12**, 433–438 (1974).

7. P. R. Bunker, *J. Mol. Spectrosc.* **68**, 367–371 (1977).

8. M. Born and J. R. Oppenheimer, *Ann. Phys. Leipzig* **84**, 457–484 (1927).

9. F. M. Fernandez, *Phys. Rev.* **A50**, 2953–2959 (1994) and references therein.

10. J. H. van Vleck, *J. Chem. Phys.* **4**, 327–338 (1936).

11. R. M. Herman and A. Asgharian, *J. Mol. Spectrosc.* **45**, 305–324 (1966).

12. P. R. Bunker, *J. Mol. Spectrosc.* **35**, 306–313 (1970).

13. J. K. G. Watson, *J. Mol. Spectrosc.* **80**, 411–421 (1980) and references therein.

14. M. E. Rose, *Elementary Theory of Angular Momentum*, Wiley, New York, U.S.A., p. 15–22.

15. F. M. Fernandez and J. F. Ogilvie, *Chin. J. Phys.* **30**, 177–193, 599 (1992).

16. P. R. Bunker and R. E. Moss, *Mol. Phys.* **33**, 417–424 (1977).

17. F. M. Fernandez and J. F. Ogilvie, *Phys. Rev.* **A42**, 4001–4006 (1990).

18. C. H. Townes and A. S. Schawlow, *Microwave Spectroscopy*, McGraw-Hill, New York, U.S.A. (1955).

19. J. F. Ogilvie, "New tests of models in chemical binding—extra-mechanical effects and molecular properties", in *Fundamental Principles of Molecular Modeling*, W. Gans, A. Amman and J. C. A. Boeyens, editors, Plenum, New York, U.S.A. (1996), and references therein.

20. J. F. Ogilvie, *Proc. Roy. Soc. London*, **A378**, 287–301 (1981), and **A381**, 479 (1982).

21. E. Tiemann and J. F. Ogilvie, *J. Mol. Spectrosc.* **165**, 377–392 (1994).

22. J. F. Ogilvie, *J. Mol. Spectrosc.* **180**, 193–195 (1996).

23. J. F. Ogilvie, H. Uehara and K. Horiai, *J. Chem. Soc. Faraday Trans.* **91**, 3007–3013 (1995) and references therein.

24. J.-L. Teffo, *Mol. Phys.* **78**, 1493–1512 (1993).

25. J.-L. Teffo and J. F. Ogilvie, *Mol. Phys.* **80**, 1507–1524 (1993).

26. J. F. Ogilvie, *J. Chem. Soc. Faraday Trans.* **91**, 3005–3006 (1995).

27. K. P. R. Nair, J. Hoeft and E. Tiemann, *Chem. Phys. Lett.* **58**, 153–158 (1978).

28. W. V. F. Brooks and B. Crawford, *J. Chem. Phys.* **23**, 363–365 (1954).

29. J. F. Ogilvie, *Mol. Phys.* **88**, 1055–1061 (1996).

30. J. F. Ogilvie, J. Oddershede and S. P. A. Sauer, *Chem. Phys. Lett.* **228**, 183–190 (1994).

31. S. Saupe, M. H. Wapplehorst, B. Meyer, W. Urban and A. G. Maki, *J. Mol. Spectrosc.* **175**, 190–197 (1996).

32. J. F. Ogilvie, *Chin. J. Phys.* **26**, 270–281 (1988).

THE ROLE OF THE STOKES PHENOMENON IN NONADIABATIC TRANSITIONS

S. F. C. O'ROURKE, B. S. NESBITT, AND D. S. F. CROTHERS

Theoretical and Computational Physics Research Division, Department of Applied Mathematics and Theoretical Physics, The Queen's University of Belfast, Belfast BT7 1NN, N. Ireland

CONTENTS

I. INTRODUCTION

Essential for the understanding of nonadiabatic collision processes have been two simple two-state models formulated within a semiclassical framework. One is the curve-crossing model developed independently by Landau (1932) using an adiabatic impact parameter treatment. Zener (1932) using a diabatic impact parameter treatment, and by Stueckelberg (1932) using a JWKB semiclassical phase-integral treatment. The second model is the Rosen–Zener noncrossing model (1932). These models represent two

Advances in Chemical Physics, Volume 103, Edited by I. Prigogine and Stuart A. Rice.
ISBN 0-471-24752-9   © 1998 John Wiley & Sons, Inc.

extreme situations in which the coupling between adiabatic states is due solely to a variation of the diabatic energy difference (Landau–Zener–Stueckelberg) or to a variation of the nonadiabatic coupling element (Rosen–Zener).

The power of Stueckelberg's phase-integral formulation is that it readily provides accurate differential cross sections, including quantal interference effects associated with competing classical trajectories. Such effects have been termed Stueckelberg oscillations (Smith, 1969). From this point of view the phase-integral method pioneered by Stueckelberg, reviewed and developed by Crothers (1971), Child (1978), Crothers (1981), Nakamura (1991), and McCarroll and Crothers (1994), and supplemented by the comparison-equation method provides an understanding of the principal mechanisms involved in nonadiabatic charge exchange.

In the JWKB phase-integral method, the channel wave functions are developed in terms of the classical adiabatic actions and their first-order quantal corrections. Nonadiabatic transitions are associated with transition points, the complex adiabatic degeneracies, and are the physical manifestation of the Stokes phenomenon, in which the coefficient of the purely exponentially subdominant term changes abruptly by an amount proportional to the dominant coefficient and the particular Stokes constant. This occurs at the so-called Stokes line emanating from the transition point.

In the case of the one-transition-point problem, analyticity determines the Stokes constant and guarantees symmetry and unitarity. Two-transition-points pose more of a problem. The phase of the Stokes constant is determined by solving a comparison second-order differential equation exactly in terms of parabolic cylinder functions and by developing their asymptotic expansions in the various sectors, using the method of steepest descent on the complex contour integral representation (Crothers, 1972, 1976). It is particularly important to recognize the simple topological complexities in moving from one to two transition points. First, strong-coupling asymptotics are required to uniformly account for the approach of the two transition points; and second, for higher values of angular momentum, the bending of the double Stokes line (Barany and Crothers, 1981). In the case of the canonical Nikitin model (1962a,b, 1970), the adiabatic term difference includes a double pole, in the vicinity of which the JWKB approximation is uniformly exact. In this case the comparison equation is solved exactly in terms of Whittaker functions. Strong-coupling asymptotics were derived by Crothers (1978). A full phase-integral treatment, including logarithmically double Stokes lines were given by Bárány and Crothers (1983). This work was extended to nonzero impact parameters (Nesbitt et al., 1996), generalized to complex energies and interactions by O'Rourke and Crothers (1992) and particularized to the Demkov case by O'Rourke (1992). Similarly, the Kummer exponential model developed by O'Rourke and Crothers (1994) was solved exactly in terms of strong-coupling asymptotics of the generalized Kummer functions.

The motivation for this chapter is to review the importance of the role that the Stokes phenomenon plays in nonadiabatic theory. The plan is as follows: In Section II we outline the nature and detail of the Stokes phenomenon with particular reference to the one-transition-point problem and quantum-tunneling calculations. In Section III we then outline how the two- and four-transition-point problems arise in ion–atom scattering theory. In Section IV we examine the comparison equation method and discuss the physical manifestation of the Stokes phenomenon in nonadiabatic collision processes by considering various simple two-state canonical models. Finally, in Section V we present some concluding remarks.

II. THE STOKES PHENOMENON

The Stokes phenomenon (1905) has been well reviewed by Paris and Wood (1995). The best-known example historically concerns the rainbow effect and the homogeneous second-order differential equation

$$\left(\frac{d^2}{dz^2} - z\right)w = 0 \tag{1}$$

a solution of which is the Airy function $Ai(z)$ with asymptotic properties for $|z| \gg 1$ given by

$$Ai(z) \sim \tfrac{1}{2}\pi^{-1/2}z^{-1/4}\exp(-\zeta)\sum_{k=0}^{\infty}(-1)^k c_k \zeta^{-k}, \qquad |\arg(z)| < \pi \tag{2}$$

$$Ai(-z) \sim \pi^{-1/2}z^{-1/4}\left[\sin\left(\zeta + \frac{\pi}{4}\right)\sum_{k=0}^{\infty}(-1)^k c_{2k}\zeta^{-2k}\right.$$
$$\left. - \cos\left(\zeta + \frac{\pi}{4}\right)\sum_{k=0}^{\infty}(-1)^k c_{2k+1}\zeta^{-2k-1}\right], \qquad |\arg(z)| < \frac{2\pi}{3} \tag{3}$$

where

$$c_0 = 1 \tag{4}$$

$$c_k = \frac{\Gamma(3k + \tfrac{1}{2})}{54^k k!\,\Gamma(k + \tfrac{1}{2})}$$

$$= \frac{(2k+1)(2k+3)\cdots(6k-1)}{216^k k!}, \qquad (k = 1, 2, 3, \ldots) \tag{5}$$

and

$$\zeta = \tfrac{2}{3}z^{3/2} \tag{6}$$

Unfortunately this presentation obscures the true nature of the phenomenon, which concerns the abrupt change of the coefficient of the exponential decaying subdominant solution across, in this case, each of the three Stokes lines defined by

$$\text{Im } z^{3/2} = 0 \tag{7}$$

namely, arg $z = 0$, $\pm \tfrac{2}{3}\pi$. These abrupt changes are in fact continuous and may be parametrized in terms of the error function; as $|z|$ increases in magnitude, the more abrupt the changes becomes (Berry, 1989).

However, until Berry's revelation many practitioners have regarded these abrupt changes as discontinuities, as in the Zwaan–Stueckelberg method (Stueckelberg, 1932; Zwaan, 1929), which is based on semiclassical JWKB (Jeffreys–Wentzel–Kramers–Brillouin; also Green–Liouville) phase integrals and their analytic continuation into the complex plane of R, typically the internuclear separation in ion–atom collisions, an important component within the study of atomic and molecular processes.

A.   Application of the Stokes Phenomenon to the JWKB Connection Formulae

1.   The One-Transition-Point Problem

According to Heading (1962), Zwaan (1929) was apparently the first to use the complex plane to investigate transition points [$z = 0$ in Eq. (1)]. Zwaan traced a subdominant JWKB solution from one side of the transition point to the other by tracing it around the point in the complex plane. When the real axis was again reached, the condition was that the solution must be real, allegedly providing sufficient information for the deduction of the appropriate connection formula. Langer (1934) pointed out the weakness of such a treatment. It appeared to Heading (1962) that Furry (1947) was the first author to have treated the idea of the Stokes phenomenon seriously and to have derived the Stokes constants and hence the Jeffreys connection formula for the one-transition-point problem, [Eqs. (1) and (2)] by analyticity arguments. However, as pointed out by Crothers (1971), Stueckelberg (1932) not only preceded Furry (1947) in this respect by 15 years (Pokrovski and Khalatnikov, 1961, by 29 years) but also made an outstanding contribution to the solution of two- and four-transition-point problems in quantum-mechanical scattering problems.

Consider the one-transition-point problem. The general JWKB asymptotic solution, valid for $R \gg R_0$, of the Schrödinger equation

$$\left(\frac{d^2}{dR^2} + v^2(R) \right) y = 0 \tag{8}$$

may be written as

$$y_I(R) \simeq v^{-1/2}(R) \left[A \exp\left(i \int_{R_0}^{R} v(s) \, ds \right) + B \exp\left(-i \int_{R_0}^{R} v(s) \, ds \right) \right] \tag{9}$$

where A and B are arbitrary constants and where the square of the wave number $v^2(R)$ is assumed to have a simple zero at the classical turning point $R = R_0$. The functions $v(R)$ and $v^{1/2}(R)$ are made single-valued by inserting a branch cut from R_0 to ∞ at $\arg(R - R_0) = -2\pi/3$ with $\arg(R - R_0)$ assigned to zero for $R > R_0$ as shown in Fig. 3.1. Clearly, the approximate solution (9) is singular and has a branch point at $R = R_0$, whereas the exact solution (ascending power series à la Frobenius) has not. We wish to know the asymptotic behavior in the classically allowed region $R \ll R_0$, of the exact solution, which is the analytic continuation of the exact solution with asymptotic behavior given by (9) in the classically forbidden region $R \gg R_0$.

The essence of the Zwaan–Stueckelberg method is to continue analytically the asymptotic solution y_I around a simple closed contour in the complex R plane, at sufficient distance from R_0 that the JWKB functions form a valid estimate everywhere on the contour. It is understood throughout that the contour for the line integral $\int_{R_0}^{R} v(s) \, ds$ does not cross the cut. The Stokes lines are defined by

$$\text{Im } i \int_{R_0}^{R} v(s) \, ds \equiv \text{Re} \int_{R_0}^{R} v(s) \, ds = 0 \tag{10}$$

On each of these lines there is one exponentially dominant term and one exponentially subdominant term. It follows that the initial directions of the Stokes lines are given by

$$\arg(R - R_0) = \frac{\pi}{3} + \frac{2n\pi}{3}, \qquad \text{integral } n. \tag{11}$$

Thus there are precisely three Stokes lines on the sheet considered, namely at $\pm \pi/3$ and π. In fact it is easy to show that the π line continues along the negative real axis; on the other hand, the $\pm \pi/3$ lines meander on the way

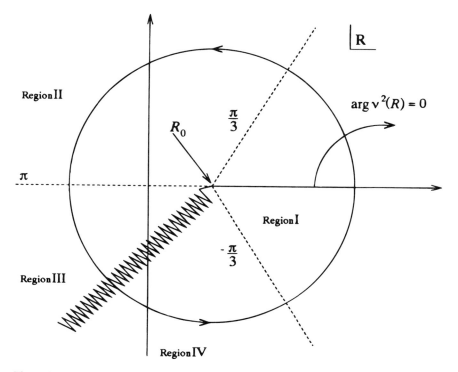

Figure 3.1. The complex R plane for the one-transition-point problem. The dashed Stokes lines emit from the transition point R_0 in directions given by $\arg(R - R_0) = -\pi/3, +\pi/3$, and π on the particular sheet of $v(R)$, which is indicated by a zig-zagging branch cut and the argument assignation of $v^2(R)$.

to infinity. It also follows that $\exp[-i \int_{R_0}^R v(s)\, ds]$ is subdominant, dominant, subdominant, on the $-\pi/3, +\pi/3, +\pi$ Stokes lines, respectively, and conversely for $\exp(+i \int_{R_0}^R v(s)\, ds)$.

Interlacing the Stokes lines are the anti-Stokes lines, including therefore the line $R > R_0$ on which neither is exponentially dominant. As described by Heading (1962), as the Stokes line is crossed the coefficient of the subdominant solution must be changed, effectively discontinuously, in order to emerge on the next anti-Stokes line with the appropriate coefficient required for the next domain in which it is again dominant. This effective discontinuity does not violate continuity of the approximate solution, since it is associated with a term much smaller in magnitude than the error allowed in the dominant term. The change suffered by the coefficient of the subdominant term is known as the Stokes phenomenon. Continuity of the approximate solution does of course require continuity of the coefficient of the dominant term in crossing a Stokes line. Moreover, in the special circumstance that the dominant solution is absent on the Stokes line, the subdominant solution must be unique and continuous. It follows that for $\pi/3 < \arg(R - R_0) < \pi$, having crossed the $\pi/3$ Stokes line from region I into region II (see Fig. 3.1), y_I connects with

$$y_{II} \simeq v^{-1/2} \left[(A + bB) \exp\left(i \int_{R_0}^R v(s) \, ds \right) + B \exp\left(-i \int_{R_0}^R v(s) \, ds \right) \right] \quad (12)$$

which similarly connects, upon crossing the π line into region III, with

$$y_{III} \simeq v^{-1/2} \left[(A + bB) \exp\left(i \int_{R_0}^R v(s) \, ds \right) \right.$$
$$\left. + (b(A + bB) + B) \exp\left(-i \int_{R_0}^R v(s) \, ds \right) \right] \quad (13)$$

We assume the positive (counterclockwise) Stokes constant is the same for all three Stokes lines by rotational invariance. Upon crossing the branch cut into region IV, y_{III} connects with

$$y_{IV} \simeq -iv^{-1/2} \left[(b(A + bB) + B) \exp\left(i \int_{R_0}^R v(s) \, ds \right) \right.$$
$$\left. + (A + bB) \exp\left(-i \int_{R_0}^R v(s) \, ds \right) \right] \quad (14)$$

and finally crossing the $-\pi/3$ Stokes line from region IV back into region I connects with

$$y_I \simeq -iv^{-1/2} \left[(bA + (1 + b^2)B) \exp\left(i \int_{R_0}^R v(s) \, ds \right) \right.$$
$$\left. + \{A + bB + b[bA + (1 + b^2)B]\} \exp\left(-i \int_{R_0}^R v(s) \, ds \right) \right] \quad (15)$$

Equating Eqs. (9) and (15) and invoking analyticity (invariance of y after circulating $R = 0$ once) one deduces that

$$b = i \quad (16)$$

If one requires a purely subdominant solution on the π line so that the solution vanishes exponentially as $R \to -\infty$, then from (13)

$$A + bB = 0 \tag{17}$$

Setting

$$A = \frac{c}{2} \exp\left(-\frac{\pi i}{4}\right) \tag{18}$$

and

$$B = \frac{c}{2} \exp\left(\frac{\pi i}{4}\right) \tag{19}$$

where c is an arbitrary constant, gives the well-known Gans–Jeffreys connection formula

$$\frac{c}{2} |v|^{-1/2} \exp\left(-\int_R^{R_0} |v| \, ds\right) \to c v^{-1/2} \sin\left(\int_{R_0}^R v \, ds + \frac{\pi}{4}\right) \tag{20}$$

on which most quantum-tunneling calculations are based. Equation (20) generalizes (2) and (3) and was derived by Stueckelberg (1932) apart from the $\pi/4$ being misprinted as $\pi/2$. The net result of Berry (1989) is that, if $A + iB \neq 0$, then precisely on the π Stokes line, y is given by

$$
\begin{aligned}
y \simeq v^{-1/2} \Bigg[&(A + iB) \exp\left(i \int_{R_0}^R v(s) \, ds\right) \\
&+ \left(\frac{i}{2}(A + iB) + B\right) \exp\left(-i \int_{R_0}^R v(s) \, ds\right) \Bigg]
\end{aligned}
\tag{21}
$$

that is, precisely one-half of the abrupt change in the subdominant coefficient has been developed.

Setting the subdominant coefficient to zero in (21) gives

$$A = iB = \frac{c'}{2} \exp\left(-\frac{i\pi}{4}\right) \tag{22}$$

where c' is an arbitrary constant, which in turn gives the second connection formula

$$c' |v|^{-1/2} \exp\left(+\int_R^{R_0} |v| \, ds\right) \leftarrow c' v^{-1/2} \cos\left(\int_{R_0}^R v \, ds + \frac{\pi}{4}\right) \tag{23}$$

which, however, is usually considered less reliable in the absence of an independent check such as unitarity; the reason is that a very small error in the classically forbidden exponentially increasing solution must surely result in some nonzero component of the (classically forbidden) exponentially decreasing solution. In any case, internuclear repulsion normally dictates the use of Eq. (20) and not Eq. (23).

III. SEMICLASSICAL THEORY FOR NONADIABATIC TRANSITIONS

A. Semiclassical Formalism

In treating low-energy inelastic atomic collision processes (such as excitation, charge transfer, transfer ionization, etc.), one is faced with a complicated quantum-mechanical many-body system. Through a series of physically reasoned approximations one may, in many cases, reduce the problem to a manageable semiclassical form, namely, a set of equations describing the quantum evolution of the electronic states as the nuclei follow classical trajectories (see, for example, Delos et al., 1972). Using these classical trajectory equations together with semiclassical phase shifts for the nuclear motion, many collision processes may be explained.

Let us consider the general formulation of a typical rearrangement process involving the elementary three-body problem consisting of two ionic cores A and B of masses M_A and M_B, respectively, and an electron e^-. Initially the electron is assumed to be bound to center A and state i. The general problem is to determine the capture of the electron to center B in a bound state f. Then the time-independent, nonrelativistic Schrödinger equation for the three-body system is

$$H\Psi(s, \mathbf{R}) = E\Psi(\mathbf{s}, \mathbf{R}) \tag{24}$$

where

$$H = T + V \tag{25}$$

and T and V are the kinetic and potential energy operators, respectively, E being the total energy.

The coordinate system $(\mathbf{s}, \mathbf{R})$ (Fig. 3.2) is particularly suitable for describing the pseudomolecule formed when all three particles are interacting at short distances, $\mathbf{s}$ is the position vector of the electron with respect to the center of mass of A and B, while $\mathbf{R}$ is the relative position vector of B with

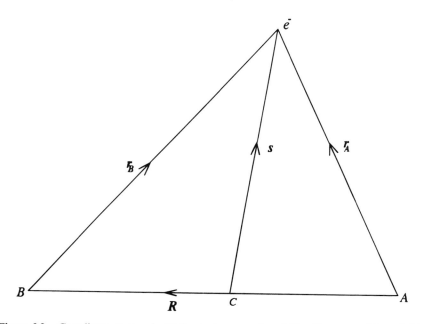

Figure 3.2. Coordinate system (**s**, **R**) for ionic cores A and B, with masses M_A and M_B, respectively, and the electron e^-. C is the center of mass of A and B.

respect to A (internuclear separation). Therefore, the kinetic energy operator T may be expressed as

$$T = -\frac{1}{2m}\nabla_s^2 - \frac{1}{2\mu}\nabla_R^2 \qquad (26)$$

where atomic units have been adopted and where the reduced masses are

$$m = \frac{M_A + M_B}{M_A + M_B + 1} \qquad (27)$$

and

$$\mu = \frac{M_A M_B}{M_A + M_B} \qquad (28)$$

The potential energy of the system is

$$V = V^A(r_A) + V^B(r_B) + W(R) \tag{29}$$

where the potentials are given by

$$V^A(r_A) = -\frac{Z_A}{r_A} \tag{30}$$

$$V^B(r_B) = -\frac{Z_B}{r_B} \tag{31}$$

$$W(R) = \frac{Z_A Z_B}{R} \tag{32}$$

The total energy eigenvalues E is given by

$$E = \varepsilon_1^A + \frac{M_B(M_A + 1)}{2(M_A + M_B + 1)} v_1^2 \tag{33}$$

where v_1 is the impact velocity and the initial target bound state eigenfunction ϕ_1^A is satisfied by

$$\left(-\frac{1}{2\mu_A} \nabla_{r_A}^2 + V^A(r_A) - \varepsilon_1^A\right)\phi_1(\mathbf{r}_A) = 0 \tag{34}$$

where

$$\mu_A = \frac{M_A}{M_A + 1} \tag{35}$$

Ignoring problems concerning rearrangement collisions, such as electron translation factors and assuming excitation of the target, one poses the two-state approximation of Mott and Massey (1965)

$$\Psi = \phi_1(\mathbf{r}_A)F_1(\mathbf{R}) + \phi_2(\mathbf{r}_A)F_2(\mathbf{R}) \tag{36}$$

Applying the Kohn variational principle (Brown et al., 1994) yields the coupled differential equations

$$(\nabla_{\mathbf{R}}^2 + k_1^2)F_1 = U_{12}F_2 + U_{11}F_1 \tag{37}$$

$$(\nabla_{\mathbf{R}}^2 + k_2^2)F_2 = U_{21}F_1 + U_{22}F_2 \tag{38}$$

where

$$k_i = Mv_i, \qquad i = 1, 2 \tag{39}$$

$$M \approx \mu \tag{40}$$

States 1 and 2 are not degenerate, and the interaction potential matrix elements are given by

$$U_{ij}(\mathbf{R}) = 2M \langle \phi_i | (V^B + W) | \phi_j \rangle \tag{41}$$

For simplicity let us restrict attention to two s states. Then following a partial-wave analysis

$$F_j = (k_1 R)^{-1} \sum_{l=0}^{\infty} (2l + 1) G_j^l(R) P_l(\cos \Theta), \qquad j = 1, 2 \tag{42}$$

with Θ the polar angle of $\mathbf{R}$, $\mathbf{v}_1$ as the polar axis, we obtain the two-state wave equations for the radial-wave amplitudes G_j^l

$$\left(\frac{d^2}{dR^2} + K_{jl}^2(R) \right) G_j^l(R) = U_{jk}(R) G_k^l(R), \qquad \{j, k\} = \{1, 2\}, j \neq k \tag{43}$$

where

$$K_{jl}^2 \approx k_j^2(\infty) - U_{jj} - \frac{(l + \frac{1}{2})^2}{R^2} \tag{44}$$

$$\equiv k_j^2(R) - \frac{(l + \frac{1}{2})^2}{R^2} \tag{45}$$

Here we have incorporated the Langer modification (1934) in anticipation since the domain is $R \in [0, +\infty]$, rather than the usual $[-\infty, +\infty]$. We must solve Eq. (43) subject to the following boundary conditions

$$G_1^l(0) = 0 \tag{46}$$

$$G_2^l(0) = 0 \tag{47}$$

$$\lim_{R \to \infty} G_1^l(R) = i^l \sin\left(k_1(\infty)R - \frac{l\pi}{2} \right) + \alpha_l \exp[ik_1(\infty)R] \tag{48}$$

$$\lim_{R \to \infty} G_2^l(R) = \beta_l \exp[ik_2(\infty)R] \tag{49}$$

where α_l and β_l are constants that are related to the partial elastic and inelastic cross-sections Q_l^{11} and Q_l^{12}

$$Q_l^{11} = \frac{4\pi(2l+1)|\alpha_l|^2}{k_1^2(\infty)} \tag{50}$$

$$Q_l^{12} = \frac{4\pi k_2(\infty)(2l+1)|\beta_l|^2}{k_1^3(\infty)} \tag{51}$$

Equations (43) may be solved in one of two ways: The first is to follow the analysis of Bates and Crothers (1970) and derive the generalized impact-parameter equations, and the second is to apply the Zwaan–Stueckelberg phase-integral method developed by Crothers (1971).

B. Generalized Impact-Parameter Equations

By forcing a common turning point on the diabatic JWKB functions given by R_{0l} the outermost zero of

$$R^2 k_1(R) k_2(R) - (l + \tfrac{1}{2})^2 \tag{52}$$

Bates and Crothers (1970) were able to obtain generalized impact-parameter equations for the state amplitudes c_m given by

$$i\frac{dc_1}{dZ} = \frac{U_{12}(R)}{2k_1^{1/2}(R)k_2^{1/2}(R)} c_2 \exp\left(-i \int_0^Z [k_1(R') - k_2(R')]\, dZ'\right) \tag{53}$$

$$i\frac{dc_2}{dZ} = \frac{U_{21}(R)}{2k_1^{1/2}(R)k_2^{1/2}(R)} c_1 \exp\left(+i \int_0^Z [k_1(R') - k_2(R')]\, dZ'\right) \tag{54}$$

where

$$Z = \mp \int_{R_{0l}}^R \left[1 - \frac{(l+\tfrac{1}{2})^2}{(R')^2 k_1(R') k_2(R')}\right]^{-1/2} dR' \tag{55}$$

according as we are on the ingoing or outgoing half of the trajectory and where a prime distinguishes a dummy variable of integration. By analogy with purely classical motion, the impact parameter ρ is given by

$$\rho^2 k_1(\infty) k_2(\infty) = (l + \tfrac{1}{2})^2 \tag{56}$$

that is (Bates and Sprevak, 1970),

$$\rho^2 = \rho_1 \rho_2 \tag{57}$$

We may then deduce the familiar diabatic straight-line impact-parameter equations

$$i \frac{dc_1}{dt} = H_{12} c_2 \exp\left(-i \int_0^t (H_{22} - H_{11})\, dt'\right) \tag{58}$$

$$i \frac{dc_2}{dt} = H_{21} c_1 \exp\left(+i \int_0^t (H_{22} - H_{11})\, dt'\right) \tag{59}$$

where $c_2(-\infty) = 0$ and where the internuclear separation is

$$\mathbf{R} = \mathbf{\rho} + \mathbf{v}t \tag{60}$$

and where

$$H_{12} = \frac{U_{12}(R)}{2k_1(R)k_2(R)} = H_{21} \tag{61}$$

$$H_{22} - H_{11} = [k_1(R) - k_2(R)]v$$

$$\approx \frac{k_1^2(R) - k_2^2(R)}{2M}$$

$$= \frac{k_1^2(\infty) - k_2^2(\infty) + U_{22} - U_{11}}{2M} \tag{62}$$

The diabatic amplitudes $c_m(t)$ are closely related to the phase-integral amplitudes, negative (positive) time corresponding to ingoing (outgoing) waves.

C. Zwaan–Stueckelberg Phase-Integral Treatment

Alternatively, Eqs. (43) may be solved by adopting the Zwaan–Stueckelberg phase-integral treatment developed by Crothers (1971). Eliminating G_2^l from Eq. (43) gives the fourth-order equation

$$G_1^{IV} - 2(\ln U_{12})^l G_1^{III} + \left(K_1^2 + K_2^2 - \frac{U_{12}^{II}}{U_{12}} - 2\{(\ln U_{12})^l\}^2\right)G_1^{II}$$

$$+ \{2U_{12}^l - 2(\ln U_{12})^l K_1^2\}G_1^l$$

$$+ \left[K_1^2 K_2^2 - U_{12}^2 + 2U_{12}^l \left(\frac{K_1^2}{U_{12}}\right)^I - \frac{U_{12}^{II}}{U_{12}} K_1^2\right]G_1 = 0 \tag{63}$$

where l dependence has been suppressed in the notation and where differentiation with respect to R is indicated by a roman numerical superscript. We may now apply the Zwaan–Stueckelberg phase-integral method of Section II.A.1 (for details see Crothers, 1971). The significant point is that the generalized wave numbers are given by

$$\tilde{v}_j^2(R) = \frac{k_1^2 + k_2^2}{2} + (-1)^{j-1}\left[\left(\frac{k_1^2 - k_2^2}{2}\right)^2 + U_{12}^2\right]^{1/2} \tag{64}$$

so that there is a complex adiabatic degeneracy at the zeros of the square root in (64) comprising a pair of complex-conjugate transition points, say R_c and R_c^*, which are joined by a double Stokes line. This is the two-transition-point problem. Since the Stokes line must be traversed by both ingoing and outgoing waves, in reality we have a four-transition-point problem. Just as in Section II.A.1, the coefficients of the generalized JWKB adiabatic phase integrals suffer discontinuities, so that a nonadiabatic transition from state 1 to state 2 is the physical manifestation of the Stokes phenomenon. The inelastic transition amplitude is then found by invoking analyticity, unitarity, and symmetry and is given by

$$2i\sqrt{\frac{k_2(\infty)}{k_1(\infty)}}\,\beta_l = -2i\,\exp(i\eta_1 + i\eta_2)\sqrt{P}\sqrt{1 - P}\,\sin(x + \arg a) \tag{65}$$

where P, the Landau–Zener–Stueckelberg single transition probability for a pseudo or avoided crossing is given by

$$P = e^{-2y} \tag{66}$$

The adiabatic phase shifts η_j are given by

$$\eta_j = \frac{\pi}{4} + \frac{l\pi}{2} - k_j(\infty)R_j + \int_{R_j}^{\infty}[\tilde{v}_j(R) - k_j(\infty)]\,dR \tag{67}$$

where R_j is given by the classical turning point. The quantities x and y are given by

$$x + iy = \int_{R_1}^{R_c}\tilde{v}_1\,dR - \int_{R_2}^{R_c}\tilde{v}_2\,dR \tag{68}$$

There are several significant aspects to this approach. Firstly, the extra phase $\arg a$ in (65) is the argument of the Stokes constant and is indeterminate in the phase-integral method: It may be calibrated using the comparison-equation method or purely numerically. Second, identifying an impact parameter ρ with $(l + \frac{1}{2})/k_1(\infty)$, we may deduce from Eq. (65), averaging over the rapidly varying frequency x, the standard Landau–Zener impact-parameter probability $2P(1 - P)$, which comprises $P(1 - P)$ and $(1 - P)P$, that is, transition followed by no transition, or vice versa, at consecutive crossings (ingoing followed by outgoing). And finally, Eq. (65) does, however, illustrate Stueckelberg oscillations concerning the quantal interference between ingoing and outgoing amplitudes, the frequency being of the order of magnitude given by the inverse impact velocity.

IV. THE COMPARISON EQUATION METHOD

Within a semiclassical JWKB framework, the comparison-equation method has been applied to various one-pole two-transition-point problems. These include the Demkov model (O'Rourke, 1992; Demkov, 1964), the Nikitin model (Nikitin, 1962a,b, 1970; Crothers, 1978; Nesbitt et al., 1996; O'Rourke and Crothers, 1992), the Kummer model (O'Rourke and Crothers, 1994) and the non-crossing parabolic model (Nesbitt et al., 1977). (In truth each of these models have four transition points, but we can split the four-transition-point problem into two independent two-transition-point problems by assuming that there is no direct coupling between ingoing and outgoing waves. Bearing in mind the generalized impact parameter treatment, this is to be expected on physical grounds.)

The quantity of interest to be determined is the transition probability:

$$P = |c_2(+\infty)|^2 \tag{69}$$

in the region of nonadiabatic strong coupling. This can be obtained by solving the straight-line impact parameter equations in Section III.B:

$$i\frac{dc_m}{dt} = H_{mn}c_n \exp\left(-i\int_0^t (H_{nn} - H_{mm})\, dt'\right), \qquad m, n = 1, 2; m \neq n \tag{70}$$

subject to the initial conditions:

$$c_1(-\infty) = 1, \qquad c_2(-\infty) = 0 \tag{71}$$

We proceed by introducing a new reduced time variable (Delos and Thorson, 1972):

$$\tau = \int_0^t H_{12}(t')\, dt' \tag{72}$$

with the supposition that

$$\tau_\infty = \int_0^\infty H_{12} \, dt' < +\infty \tag{73}$$

It is assumed that $H_{12} > 0$ for all t, which implies that τ is a monotonic function and $\tau_\infty > 0$. It is also assumed that H_{12} is an even function of time, which is the case in low-energy atomic collisions where the matrix elements depend on time through the internuclear separation R.

The Stueckelberg variable T is defined as

$$T = \frac{H_{22} - H_{11}}{2H_{12}} \tag{74}$$

Therefore Eq. (70) transforms to

$$i \frac{dc_m^+}{d\tau} = c_n^+ \exp\left((-1)^m 2i \int_0^\tau T(\tau') \, d\tau'\right), \qquad m, n = 1, 2; m \neq n \tag{75}$$

Changing the dependent variable to

$$a_m^+ = c_m^+ \exp\left((-1)^{m-1} i \int_0^\tau T(\tau') \, d\tau'\right), \qquad m = 1, 2 \tag{76}$$

we find that Eq. (75) transforms to

$$i \frac{da_m^+}{d\tau} = (-1)^m T a_m^+ + a_n^+, \qquad m, n = 1, 2; m \neq n \tag{77}$$

which can be uncoupled to yield second-order differential equations

$$\left(\frac{d^2}{d\tau^2} + 1 + (-1)^m i \frac{dT}{d\tau} + T^2\right) a_m^+ = 0, \qquad m = 1, 2 \tag{78}$$

The amplitudes a_m^+ denote that we are considering $t > 0$. Similarly the amplitudes for a_m^- may be found by considering $t < 0$. Continuity of the a_m at $\tau = 0$ now requires the following matching conditions

$$a_2^+(\tau = 0) = a_2^-(\tau = 0) \tag{79}$$

$$\left. \frac{da_2^+}{d\tau} \right|_{\tau = 0} = - \left. \frac{da_2^-}{d\tau} \right|_{\tau = 0} \tag{80}$$

It should be noted that although the continuity in the total wavefunction leads to a matching of the amplitudes and the first-order derivatives at $\tau = 0$, it is easy to show that second-order derivatives are not matched; this is due to the nonanalytic nature of the interaction potentials H_{mn}.

The transition probabilities are derived from the comparison equation method supplemented by strong- and weak-coupling asymptotics. The definition of weak coupling and strong coupling is related to the ratio between λ and γ [given by Eqs. (89) and (95), respectively] but not to the value of the coupling H_{12}, although the two are related to some extent. The case of weak coupling ($\gamma \gg \lambda$, 1) corresponds to a small energy difference and large coupling. Then the transitions mainly occur at medium and large internuclear separations, and there will be two well-separated transition regions if the impact parameter is small compared to the typical transition-point difference. Then one should expect an appearance of Stueckelberg (1932) oscillations in the transition probability due to quantum interference between the two possible classical-event sequences. The case of strong coupling (γ, $\lambda \gg 1$) corresponds to a large energy difference and large coupling. In this case, the transitions mainly occur near the classical turning point, which means that there is only one transition region and hence, the Stueckelberg oscillations may be damped. Alternatively, we may say that the impact velocity is significantly less than 1 a.u.

A.   The Nikitin Model

The interaction potentials for the Nikitin model (Nikitin, 1962a,b, 1970; Crothers, 1978) are

$$H_{22} - H_{11} = \Delta\varepsilon - A \cos \theta \, e^{-\alpha R} \tag{81}$$

$$H_{12} = H_{21} = -\tfrac{1}{2}A \sin \theta \, e^{-\alpha R} \tag{82}$$

where R is the internuclear separation

$$R = \sqrt{\rho^2 + Z^2} \tag{83}$$

$$Z = vt \tag{84}$$

The three independent positive parameters are θ, $\Delta\varepsilon$, and A. It is assumed that A is greater than $\Delta\varepsilon$ and that θ lies in the range $0 < \theta < \pi$. $\Delta\varepsilon$ is the resonance defect, and α is a positive scaling parameter closely related to the single-ionization potentials of the two states. Given α and $\Delta\varepsilon$, A effectively determines the location of the nonadiabatic transition region, while θ determines the diabatic mixing of terms, which includes both pseudocrossings and noncrossings. That is, the Nikitin model comprises the curve-crossing models of Landau (1932), Zener (1932), and Stueckelberg (1932) and the noncrossing models of Rosen–Zener (1932) and Demkov (1964) as particular cases.

Setting the impact parameter to zero for simplicity and applying (74) to Eq. (78), we obtain the comparison equation for the Nikitin model, which has been reduced to the Whittaker differential equation

$$\left(\frac{d^2}{dz^2} - \frac{1}{4} + \frac{\kappa}{z} + \frac{\frac{1}{4} - \mu^2}{z}\right)a_m^+ = 0, \qquad m = 1, 2 \tag{85}$$

where

$$\kappa = \frac{i}{2}\lambda\cos\theta \tag{86}$$

$$z = 2i(\tau_\infty - \tau)\operatorname{cosec}\theta \tag{87}$$

$$T = \frac{\lambda}{2(\tau_\infty - \tau)} - \cot\theta \tag{88}$$

$$\lambda = \frac{\Delta\varepsilon}{\alpha v} \tag{89}$$

$$\mu = \frac{1}{2} + \frac{i}{2}\lambda \tag{90}$$

Therefore the comparators for (85) are Whittaker functions, the general solution being given as

$$a_2^+ = A_1 M_{\kappa,\,\mu*}(z) + A_2 M_{\kappa,\,-\mu*}(z) \tag{91}$$

and similarly

$$a_2^- = A_3 M_{\kappa,\,\mu}(z) + A_4 M_{\kappa,\,-\mu}(z) \tag{92}$$

where A_1, A_2, A_3, and A_4 are constants determined by applying the initial conditions. Therefore

$$A_4 = 0 \tag{93}$$

$$A_3 = \frac{\sin\theta}{2(1 + i\lambda)}\exp\left(\frac{\pi\lambda}{4} + \frac{i}{2}(\gamma\cos\theta - \lambda\ln\gamma)\right) \tag{94}$$

where

$$\gamma = \frac{A}{\alpha v} \tag{95}$$

and where we have used the expansion of the Whittaker function and its derivative at $z = 0$ (Abramowitz and Stegun, 1970) to satisfy the initial conditions. The quadrature in Eq. (76) has been effected so that no unwarranted Coulomb phases remain. The constants A_1 and A_2 are determined by using the matching conditions (79) and (80)

$$A_1 = -\frac{A_3}{2\mu^*} \frac{d}{dz} z^{1+i\lambda} {}_1F_1(\tfrac{1}{2} + \mu - \kappa; 1 + 2\mu; z)$$

$$\times {}_1F_1(\tfrac{1}{2} - \mu^* - \kappa^*; 1 - 2\mu^*; z^*) \tag{96}$$

$$A_2 = \frac{A_3}{2\mu^*} \frac{d}{dz} z^2 {}_1F_1(\tfrac{1}{2} + \mu - \kappa; 1 + 2\mu; z)$$

$$\times {}_1F_1(\tfrac{1}{2} + \mu^* - \kappa^*; 1 + 2\mu^*; z^*) \tag{97}$$

where the derivatives are to be evaluated at the classical turning point $z = i\gamma$. The final-state amplitude for the Nikitin model is then given by

$$c_2^{N}(+\infty) = A_2 \exp\left(-\frac{\pi\lambda}{4} - \frac{i}{2}(\gamma \cos\theta - \lambda \ln\gamma)\right) \tag{98}$$

Therefore, the transition probability for the Nikitin model expressed in terms of confluent hypergeometric functions is

$$p_{12}^{N} = \left\{\left(\frac{\gamma \sin\theta}{1 + \lambda^2}\right) \mathrm{Re}\left[(1 - i\lambda){}_1F_1\left(1 + i\frac{\lambda}{2}(1 - \cos\theta); 2 + i\lambda; i\gamma\right)\right.\right.$$

$$\left.\left.\times \left\{{}_1F_1\left(i\frac{\lambda}{2}(1 - \cos\theta); i\lambda; i\gamma\right)\right\}^*\right]\right\}^2 \tag{99}$$

Equation (99) represents the exact solution to the two-state impact parameter equations (70) ($\rho = 0$) for the Nikitin model. However, even the relatively well-studied ${}_1F_1(a; b; z)$ are not analytically convenient, which suggests the necessity of physically reasoned approximations.

In order to facilitate the derivation of the asymptotic expansions of the transition probability, it is convenient to write (99) in terms of the hypergeometric function of the second kind $U(a; b; z)$ related to $_1F_1(a; b; z)$ by

$$_1F_1(a; b; z) = \Gamma(b)\left(\frac{e^{\pm i\pi a}U(a; b; z)}{\Gamma(b-a)} + \frac{e^{\pm i\pi(a-b)+z}U(b-a; b; -z)}{\Gamma(a)}\right) \quad (100)$$

the upper sign being for $\arg(z) > 0$ and the lower for $\arg(z) \le 0$. Then Eq. (99) becomes

$$p_{12}^N = \left[\left(\frac{\sqrt{1+\cos\theta}\,\gamma^2\,e^{(\pi\lambda/2)\cos\theta}}{\sinh(\pi\lambda)}\right)\right.$$

$$\times \text{Re } \gamma^{i\gamma}\left(i[e^{(\pi\lambda/2)\cos\theta} - e^{(-\pi\lambda/2)\cos\theta}\cosh(\pi\lambda)]XY\right.$$

$$+ \frac{\pi\,e^{(-\pi\lambda/2)+i\gamma}X^2}{\Gamma(1+i(\lambda/2)(1+\cos\theta))\Gamma(-i(\lambda/2)(1-\cos\theta))}$$

$$\left.\left.- \frac{\pi\,e^{(\pi\lambda/2)+i\lambda}Y^2}{\Gamma(1+i(\lambda/2)(1-\cos\theta))\Gamma(-i(\lambda/2)(1+\cos\theta))}\right)\right]^2 \quad (101)$$

where

$$X = U\left(1 + \frac{i}{2}\lambda(1-\cos\theta); 2 + i\lambda; i\gamma\right) \quad (102)$$

$$Y = U\left(1 + \frac{i}{2}\lambda(1+\cos\theta); 2 + i\lambda; -i\gamma\right) \quad (103)$$

Strong-coupling asymptotic first-order approximations for X and Y are obtained by using the method of steepest descent (Crothers, 1978). It should be borne in mind that the phase-integral method, on which the model theory of nonadiabatic collisions is based, requires only that the leading term be retained in each of the asymptotic expansions of X and Y, which will in general occur in the complete asymptotic expansion of any solution to a second-order differential equation.

Therefore, the strong-coupling approximation to the inelastic transition probability at a zero impact parameter is given by

$$p_{12}^N = 4e^{\pi\lambda - 2y}\frac{\sinh(\pi\lambda - y)\sinh(y)}{\sinh^2(\pi y)}\sin^2\Phi \quad (104)$$

where

$$\Phi = x + f\left(\frac{y}{\pi}\right) - f\left(\frac{\eta}{\pi}\right) \tag{105}$$

The function f is the phase of the Stokes constant and is defined by Bayfield et al. (1973) as

$$f(h) = \frac{\pi}{4} + h \ln h - h - \arg \Gamma(1 + ih), \qquad h > 0 \tag{106}$$

The parameters x and y are given by

$$x + iy = \frac{1}{v} \int_0^{Z_c} \sqrt{4H_{12}^2 + (H_{22} - H_{11})^2} \, dZ \tag{107}$$

where x is the usual Zwaan–Stueckelberg phase (Crothers, 1971), which represents the adiabatic action difference integrated between the classical turning point 0 and the complex transition point Z_c, η allows for interference between the primary pair of conjugate transition points and a secondary pair lying further from the real axis

$$\eta = |\pi\lambda - y| \tag{108}$$

and y is a nonadiabatic parameter. We point out that Eq. (107) is equivalent to Eq. (68) in Section III.C. The complex transition point Z_c is given by the zero of the integrand [Eq. (107)] and in this case lies in the first quadrant

$$Z_c = [(R_c)^2 - \rho^2]^{1/2} \tag{109}$$

$$R_c = \alpha^{-1}\left[\ln\left(\frac{\gamma}{\lambda}\right) + i\theta\right] \tag{110}$$

The restriction to a zero impact parameter may now be removed by invoking analytical continuation (Nesbitt et al., 1996; Crothers and Todd, 1980) to give the following transition probability for strong coupling

$$p_{12}^N = 4P_{12}^N(1 - P_{12}^N) \sin^2 \Phi \tag{111}$$

where

$$P_{12}^{N} = e^{-y} \frac{\sinh(\eta)}{\sinh(y + \eta)} \qquad (112)$$

In common with the by now standard Landau–Zener–Stueckelberg formula, P_{12}^{N} represents a single transition probability at the real non-adiabatic transition point, whether on the ingoing or outgoing half of the trajectory, while Φ represents the phase interference between the ingoing and outgoing transition regions.

Equation (108) only accounts exactly for interference in the precise context of the two-state exponential model. But this definition of η is inaccurate for nonzero impact parameters because the expression

$$F(Z) = \sqrt{4H_{12}^{2} + (H_{22} - H_{11})^{2}} \qquad (113)$$

is not now given by a pure exponental model. Therefore, from Nesbitt et al. (1996) we have

$$\eta = \mathrm{Re}\left(\frac{i}{2v} \int_{Z_{c}^{A}}^{Z_{c}} F(Z) \, dZ \right) \qquad (114)$$

where

$$Z_{c}^{A} = [(R_{c}^{A})^{2} - \rho^{2}]^{1/2} \qquad (115)$$

$$R_{c}^{A} = \alpha^{-1}\left[\ln\left(\frac{\gamma}{\lambda}\right) + i(2\pi - \theta) \right] \qquad (116)$$

B.   The Demkov Model

A particular case of the Nikitin model is the noncrossing Demkov model (O'Rourke, 1992; Demkov, 1964), where the mixing parameter is $\theta = \pi/2$. Since the Demkov model is a particular case of the Nikitin model, we can deduce the transition amplitude $c_{2}^{D}(\infty)$ from Eqs. (97) and (98) by setting $\kappa = 0$; that is

$$c_{2}^{D} = \tfrac{1}{2}i\pi \ \mathrm{sech}\left(\frac{\pi\lambda}{2}\right) \frac{d}{dz}\left[zJ_{\mu}(z)J_{\mu*}(z) \right] \qquad (117)$$

where the confluent hypergeometric function has been converted to the Bessel function and where

$$\mu = \frac{1}{2} + \frac{i}{2}\lambda \qquad\qquad (118)$$

$$z = 2i(\tau_\infty - \tau) \qquad\qquad (119)$$

Equation (117) may be solved exactly by developing first-order asymptotic expansions of the Bessel functions in various sectors, using the method of steepest descent on complex contour integral representations (O'Rourke, 1992). Therefore, the transition probability in terms of real energies and interactions is given by

$$p_{12}^D = |c_2^D(\infty)|^2 = \sin^2(x)\,\text{sech}^2(y) \qquad\qquad (120)$$

where x and y are given by Eq. (107), with the transition point lying in the first quadrant.

At this point it is worth stressing that inappropriate application of the JWKB method can lead to unphysical transition probabilities. Vitanov (1993) included second-order terms in his derivation of the generalized Demkov model, which was based on neglecting discontinuities in second-order derivatives. This small second-order correction to the Stueckelberg frequency of oscillation has been interpreted as the argument of a Stokes constant (Crothers and O'Rourke, 1993). Vitanov's transition probability for the Demkov model is given as

$$p_{12}^D = \text{sech}^2\,y\left[\sin x + \frac{y\tau_\infty}{2\pi(\tau_\infty^2 + y^2/\pi^2)^{3/2}}\,\sinh y\right)^2 \qquad\qquad (121)$$

which comes from deriving second-order expressions when applying the method of steepest descent to obtain asymptotic expansions for the Bessel functions. Vitanov (1994) repeats this strategy in the Nikitin model by using the method of steepest descent to derive second-order expressions for the functions X and Y [Eqs. (102) and (103)], which resulted in the following transition probability:

$$p_{12}^N = \frac{4e^{\eta-y}}{\sinh^2(\eta+y)}\left((\sinh\eta\,\sinh y)^{1/2}\sin\Phi\right.$$

$$\left. + \frac{\lambda\tau_\infty}{\zeta^3}\,(e^{-1/2(\eta-y)}\cosh(\eta+y) - e^{1/2(\eta-y)})\right)^2 \qquad\qquad (122)$$

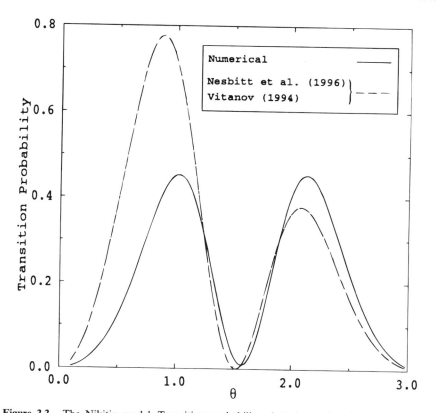

Figure 3.3. The Nikitin model. Transition probability plotted as a function of the mixing parameter θ where $\Delta\varepsilon = 0.074$, $A = 1000$, $\alpha = 1.5$, $0.5 \le \theta \le 3$, $\rho = 0$, and $v = 0$. This is a case of weak coupling. Equations (111) and (122) are shown to be in exact agreement, but for acute θ they overestimate the numerical results, whereas for obtuse θ the results are underestimated. In the case of Eq. (111) (Nesbitt et al., 1996), this defect would appear to be unique for a zero-impact parameter only. The potentials H_{11} and H_{22} cross for acute θ, while they are noncrossing for obtuse θ. The angle $\theta = \pi/2$ corresponds to perturbed symmetric resonance (Crothers, 1971), that is, to potentials that run approximately parallel (Demkov, 1964).

where

$$\zeta = \left(\frac{\pi^2 \tau_\infty^2 \, \lambda^2}{y\eta} + \lambda^2 - \frac{2\lambda\tau_\infty(\eta - y)}{\sqrt{y\eta}} \right)^{\frac{1}{2}} \tag{123}$$

The formulae of Nesbitt et al. (1996) and Vitanov (1994) [Eqs. (111) and (122), respectively] are compared with exact numerical solutions of the two-state equations (70) for the Nikitin and Demkov models (Figs. 3.3–3.6 and 3.7–3.10, respectively).

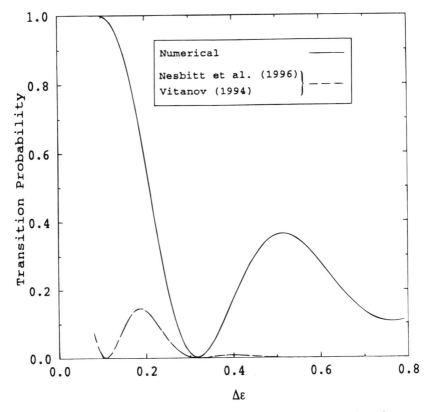

Figure 3.4. The Nikitin model. Transition probability plotted as a function of resonance defect $\Delta\varepsilon$, whre $0.08 \leq \Delta\varepsilon \leq 0.8$, $A = 10150$, $\alpha = 1.5$, $\theta = 2$, $\rho = 0$, and $v = 0.2$. Again we see that the formulae of Nesbitt et al. (1996) and Vitanov (1994) agree but differ greatly from the numerical solutions.

It should be pointed out that there is nothing wrong with Vitanov's derivations of the strong-coupling asymptotic expansions to second order. In fact, for a zero impact parameter both Eqs. (121) and (122) produce excellent results (Figs. 3.7 and 3.8), but for nonzero impact parameters the formulae break down, and there are two main reasons for this. The first is the failure to observe the cusplike discontinuity in the first-order derivative of the diabatic Hamiltonian matrix elements H_{12} and $(H_{22} - H_{11})$ at $t = 0$; thus it follows that only the leading terms are of any significance in determining first- and second-order derivatives. Second, the formulation of Vitanov (1993, 1994) does not incorporate the essential physics of the models such as the bending of the double Stokes line and the movement of

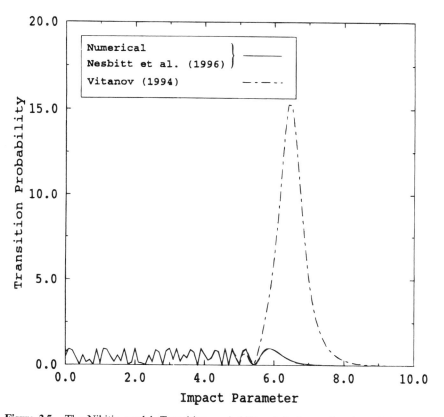

Figure 3.5. The Nikitin model. Transition probability plotted as a function of the impact parameter ρ, where $\Delta\varepsilon = 0.074$, $A = 1000$, $\alpha = 1.5$, $\theta = 1$, $0 \le \rho \le 10$, and $v = 0.2$. Vitanov's formulae [Eq. (122)] are shown to break down for intermediate and large impact parameters producing unphysical results (probabilities greater than unity) even where the coupling is weak. On the other hand, the results of Nesbitt et al. (1996) agree favorably with the numerical solutions.

the unphysical nonanalytical cusp into the remote R plane, especially for intermediate and large impact parameters, and hence leads to unphysical results.

C.   The Kummer Model

The interaction potentials for the Kummer model (O'Rourke and Crothers, 1994) are

$$H_{22} - H_{11} = \Delta\varepsilon - A \cos \theta \, e^{-2\alpha R} \tag{124}$$

$$H_{12} = H_{21} = -\tfrac{1}{2}A \sin \theta \, e^{-\alpha R} \tag{125}$$

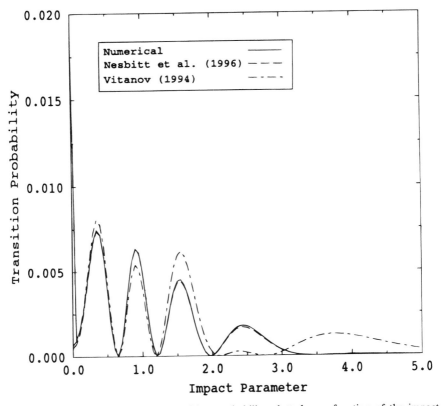

Figure 3.6. The Nikitin model. Transition probability plotted as a function of the impact parameter ρ, where $\Delta\varepsilon = 0.1$, $A = 1$, $\alpha = 0.9$, $\theta = 2$, $0 \le \rho \le 5$, and $v = 0.08$. Again we see that as the impact parameter increases, Vitanov's formulae break down. It is also worth pointing out that the errors occurring in the formulae of Nesbitt et al. (1996) for zero impact parameters (Figs. 3.3 and 3.4) no longer appear for nonzero impact parameters. One could then conclude that the error would have little or no significant effect when calculating cross-sections over a wide range of impact parameters.

The approximation for the Hamiltonian matrix elements for the generalized Kummer model differs from those chosen by Nikitin in the off-diagonal matrix elements, in the exponential time factor only, while the interaction potentials for the diagonal elements are retained. As with the Nikitin model, the Kummer model comprises the curve-crossing models of Landau, Zener, and Stueckelberg and the noncrossing models of Rosen–Zener and Demkov. Again by setting the impact parameter to zero and applying (74) to Eq. (78), we obtain the comparison equation for the Kummer model

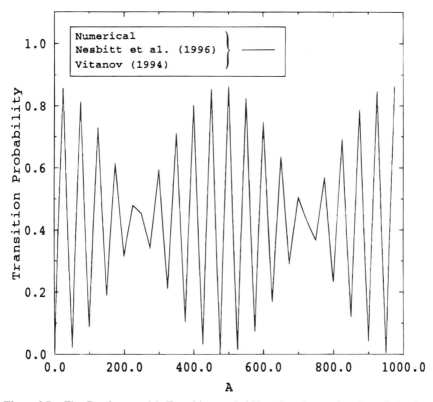

Figure 3.7. The Demkov model. Transition probability plotted as a function of A, where $\Delta\varepsilon = 0.074$, $1 \le A \le 1000$, $\alpha = 1.5$, $\theta = \pi/2$, $\rho = 0$, and $v = 0.2$. This is a case where the coupling becomes much weaker as A increases in magnitude.

given by

$$\left(\frac{d^2}{dz^2} + \frac{1}{2z}\frac{d}{dz} + \frac{1}{4} + \frac{\delta^2 - \frac{1}{2}\beta - (-1)^{m-1}(i/4)}{z} + \frac{\frac{1}{4}\beta^2 - (-1)^{m-1}\frac{1}{4}i\beta}{z^2}\right)a_m^+ = 0,$$

$$m = 1, 2 \quad (126)$$

where

$$2\delta z^{1/2} = (\tau_\infty - \tau) \tag{127}$$

$$\delta^2 = \frac{\gamma \sin^2 \theta}{8 \cos \theta} \tag{128}$$

$$\beta = \lambda/2 \tag{129}$$

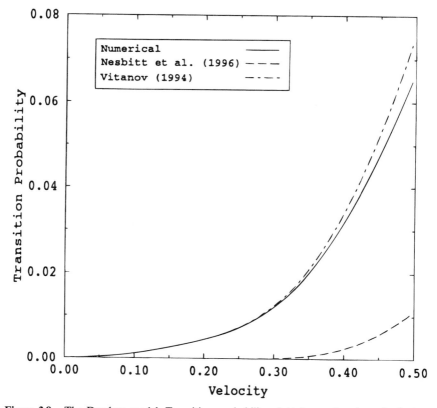

Figure 3.8. The Demkov model. Transition probability plotted as a function of velocity v, where $\Delta\varepsilon = 0.84$, $A = 1$, $\alpha = 0.9$, $\theta = \pi/2$, $\rho = 0$, and $0.01 \leq v \leq 0.05$. This figure represents the case of very strong coupling and shows that by going beyond the leading terms of the confluent hypergeometric functions, Vitanov's results are indeed much better than Nesbitt et al., but only at a zero impact parameter.

Similarly for $t < 0$ one obtains

$$\left(\frac{d^2}{dz^2} + \frac{1}{2z}\frac{d}{dz} + \frac{1}{4} + \frac{\delta^2 - \frac{1}{2}\beta + (-1)^{m-1}(i/4)}{z} + \frac{\frac{1}{4}\beta^2 + (-1)^{m-1}\frac{1}{4}i\beta}{z^2}\right)a_m^- = 0,$$

$$m = 1, 2 \quad (130)$$

The comparators for (126) and (130) are the confluent hypergeometric func-

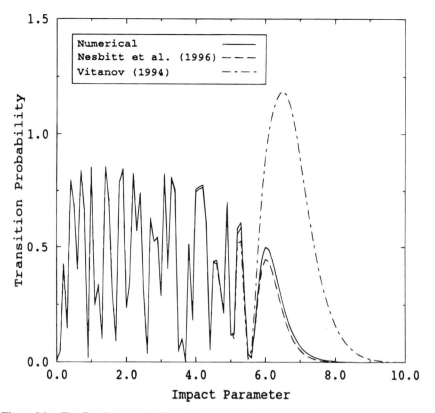

Figure 3.9. The Demkov model. Transition probability plotted as a function of the impact parameter ρ, where $\Delta\varepsilon = 0.074$, $A = 1000$, $\alpha = 1.5$, $\theta = \pi/2$, $0 \leq \rho \leq 10$, and $v = 0.2$. Again Vitanov's formulae break down for intermediate and large impact parameters, whereas the results of Nesbitt et al. agree with the numerical solutions.

tions. Therefore, the general solution for $t < 0$ is

$$a_2^- = B_1 z^{1/2 + (i/2)\beta} \, e^{-(1/2)iz} \, {}_1F_1(1 + i\delta^2; \tfrac{3}{2} + i\beta; iz)$$
$$+ B_2 \, z^{-(i/2)\beta} \, e^{-(1/2)iz} \, {}_1F_1(\tfrac{1}{2} + i\delta^2 - i\beta; \tfrac{1}{2} - i\beta; iz) \qquad (131)$$

The constants B_1 and B_2 are determined by using the first-order equations and initial conditions. Thus

$$B_1 = \frac{\delta}{\beta - \tfrac{1}{2}i} \, \exp\{\tfrac{1}{2}i[\tfrac{1}{2}\gamma \cos \theta - \beta \ln(\tfrac{1}{2}\gamma \cos \theta)]\} \qquad (132)$$

$$B_2 = 0 \qquad (133)$$

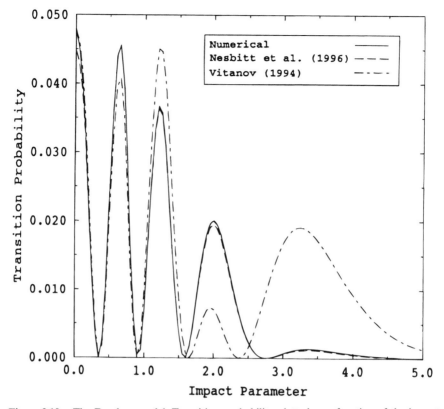

Figure 3.10. The Demkov model. Transition probability plotted as a function of the impact parameter ρ, where $\Delta\varepsilon = 0.1$, $A = 1$, $\alpha = 0.9$, $\theta = \pi/2$, $0 \leq \rho \leq 5$, and $v = 0.08$.

Without loss of generality we take that

$$
\begin{aligned}
a_2^+ = {}& B_3 \, z^{1/2 - (i/2)\beta} \, e^{-(i/2)z} {}_1F_1(\tfrac{1}{2} + i\delta^2 - i\beta; \tfrac{3}{2} - i\beta; \, iz) \\
& + B_4 \, z^{(i/2)\beta} \, e^{-(i/2)z} {}_1F_1(i\delta^2; \tfrac{1}{2} + i\beta; \, iz)
\end{aligned}
\tag{134}
$$

The localized nonadiabatic transition zones merge at the classical turning point $z = \frac{1}{2}\gamma \cos\theta$, and this is the region in which the JWKB connection formulae are applied to link the incoming and outgoing amplitudes in passing from one semiclassical region to another. Therefore, applying the matching conditions (79) and (80) yields

$$B_3 = -\left(\frac{B_1 z^{1/2}}{\frac{1}{2} - i\beta}\right) \frac{d}{dz} z^{1/2 + i\beta} {}_1F_1(1 + i\delta^2; \tfrac{3}{2} + i\beta; iz)$$

$$\times {}_1F_1(\tfrac{1}{2} - i\delta^2 + i\beta; \tfrac{1}{2} + i\beta; -iz) \tag{135}$$

$$B_4 = \left(\frac{B_1 z^{1/2}}{\frac{1}{2} - i\beta}\right) \frac{d}{dz} z \,|\, {}_1F_1(1 + i\delta^2; \tfrac{3}{2} + i\beta; iz)|^2 \tag{136}$$

where the derivatives are to be evaluated at $z = \frac{1}{2}\gamma \cos \theta$. Therefore, the final-state amplitude $c_2^K(\infty)$ for the Kummer model is

$$c_2^K(\infty) = B_4 \exp\left\{-\frac{i}{2}\left[\frac{\gamma}{2}\cos\theta - \beta \ln\left(\frac{\gamma}{2}\cos\theta\right)\right]\right\} \tag{137}$$

Strong-coupling first-order asymptotic expansions for the confluent hypergeometric functions which occur in (135) and (136) have been derived by O'Rourke and Crothers (1994). Therefore, the transition probability is

$$p_{12}^K = |c_2^K(\infty)|^2 \tag{138}$$

$$= 4e^{\pi\beta - 2y} \frac{\cosh(\pi\beta - y)\sinh(y)}{\cosh^2(\pi\beta)} \sin^2(\Omega_1 + \Omega_2) \tag{139}$$

where

$$\Omega_1 = f\left(\frac{y}{\pi}\right) + g(\beta) + \frac{\pi}{4} \tag{140}$$

$$\Omega_2 = x + g\left(\beta - \frac{y}{\pi}\right) - g(\beta) - \frac{\pi}{4} \tag{141}$$

The functions f and g are phases of Stokes constants and are defined as

$$g(h) = -g(-h) = h - h \ln|h| + \arg \Gamma(\tfrac{1}{2} + ih) \tag{142}$$

$$f(h) = \frac{\pi}{4} + h \ln h - h - \arg \Gamma(1 + ih), \qquad h > 0 \tag{143}$$

To ensure conservation of flux in Eq. (142), we define $h = |h|$ for $h > 0$ and $h = e^{i\pi}|h|$ for $h < 0$. Again x and y are defined by Eq. (107), although the transition point Z_c in this case lies closest to the real $Z = vt$ axis in the fourth quadrant. The nonadiabatic phases $(\Omega_1 + \Omega_2)$ represent the phase interference between ingoing and outgoing transition regions.

In the Demkov limit of $\theta = \pi/2$, assuming that β is finite and allowing $\gamma \to \infty$, the transition probability (139) reduces to give

$$p^{\text{K}}_{12} = \text{sech}^2 \, \pi\beta \, \sin^2 x \tag{144}$$

which agrees with the Demkov result upon identifying

$$x \approx \int_{-\infty}^{0} H_{12} \, dt \tag{145}$$

and $\beta = \alpha/\gamma$, where α and γ are given by Demkov (1964).

D. The Noncrossing Parabolic Model

Typical interaction potentials for the noncrossing parabolic model (Nesbitt et al., 1997) are

$$H_{22} - H_{11} = \Delta\varepsilon \tag{146}$$

$$H_{12} = H_{21} = Af(t) \tag{147}$$

where $f(t)$ is normalized to unity and in this case it is chosen to be the Lorentzian function

$$f(t) = \frac{1}{\pi(1 + t^2)} \tag{148}$$

Other possible forms of $f(t)$ include

$$f(t) = \frac{1}{\pi(1 + t^2)^2} \tag{149}$$

or the hyperbolic secant of the Rosen–Zener model (1932)

$$f(t) = \tfrac{1}{2} \, \text{sech}\left(\frac{\pi}{2} t\right) \tag{150}$$

As well as describing certain types of low-energy ion–atom collisions, the noncrossing parabolic model has also been applied to some quantum-optics two-level problems such as the coupling of two levels of an atom by a laser pulse that has a temporal width that is small compared to the natural life-times of the levels. In quantum optics the parameter $\Delta\varepsilon$ is related to the atom–field detuning, and $Af(t)$ is defined as the pulse area. Previous work in this field for an arbitrary pulse area $Af(t)$ has led to analytic formulae by Bambini and Berman (1981) and perturbation calculations (Robinson and Berman, 1983; Molander et al., 1983).

The comparison equation for the noncrossing parabolic model is obtained by transforming the diabatic equations (70) to an adiabatic formulation. According to adiabatic theory (Crothers, 1981), the standard straight-line impact parameter adiabatic equations are given by

$$\frac{da_1}{dt} = \frac{dT/dt}{2(1 + T^2)} a_2 \exp\left(-i \int_0^t \sqrt{4H_{12}^2 + (H_{22} - H_{11})^2}\, dt'\right) \quad (151)$$

$$\frac{da_2}{dt} = \frac{-dT/dt}{2(1 + T^2)} a_1 \exp\left(i \int_0^t \sqrt{4H_{12}^2 + (H_{22} - H_{11})^2}\, dt'\right) \quad (152)$$

where, as before, T is the Stueckelberg variable given by Eq. (74) and

$$a_m(+\infty) = c_m(+\infty) \exp\left(i \int_0^\infty \sqrt{E_m + E_1 - H_{mm} - H_{11}}\, dt'\right), \quad m = 1, 2$$

$$(153)$$

where a_m are the adiabatic amplitudes and E_m are the adiabatic energies given by

$$2E_m = H_{22} + H_{11} + (-1)^m \sqrt{4H_{12}^2 + (H_{22} - H_{11})^2} \quad (154)$$

The strongly coupled adiabatic equations may then be specifically written as

$$\frac{da_1}{dt} = \frac{\Omega t}{(1 + t^2)^2 + \Omega^2} a_2 \exp\left(-i\, \Delta\varepsilon \int_0^t \sqrt{1 + \frac{\Omega^2}{(1 + t'^2)^2}}\, dt'\right) \quad (155)$$

$$\frac{da_2}{dt} = \frac{-\Omega t}{(1 + t^2)^2 + \Omega^2} a_1 \exp\left(i\, \Delta\varepsilon \int_0^t \sqrt{1 + \frac{\Omega^2}{(1 + t'^2)^2}}\, dt'\right) \quad (156)$$

where

$$\Omega = \frac{2\gamma}{\pi\lambda} \quad (157)$$

Within the framework of the Zwaan–Stueckelberg method and following on from the parabolic model analysis (Crothers, 1976) enables one to uncouple the equations to obtain a comparison equation in the form of the second-order Weber differential equation. By embracing the appropriate first-order expansions for the parabolic cylinder functions $D_p(z)$ of large order and magnitude (Crothers, 1972), one then obtains the following final amplitude

$$a_2(+\infty) = -i \sin x \operatorname{sech} y \tag{158}$$

where we have used the method of steepest descent, concentrating on the transition points in the upper half-plane. As before the nonadiabatic parameters x and y are given by Eq. (107).

Careful analysis of the phase integral yields

$$x + iy = \Delta\varepsilon \sqrt{-1 + i\Omega} \sum_{n=0}^{\infty} \frac{(-\frac{1}{2})_n(-\Omega^2)^n}{n!} \, _2F_1(\tfrac{1}{2}; 2n; \tfrac{3}{2}; 1 - i\Omega) \tag{159}$$

$$\approx \frac{A}{2} + i\,\Delta\varepsilon + O(\Omega^2) \tag{160}$$

for small Ω, and where $(-\frac{1}{2})_n$ is a Pochhammer symbol. Therefore the strong-coupling transition probability is

$$p_{12}^{\text{PR}} = |a_2(+\infty)|^2 = \sin^2\left(\frac{A}{2}\right) \operatorname{sech}^2(\Delta\varepsilon) \tag{161}$$

In the limit $|\Delta\varepsilon| \gg 1$

$$p_{12}^{\text{PR}} \sim 4e^{-2\,\Delta\varepsilon} \sin^2\left(\frac{A}{2}\right) \tag{162}$$

As pointed out by Nesbitt et al. (1997), adiabatic perturbation theory can at best only provide weak-coupling approximations:

$$p_{12} \sim \pi^2 \, e^{-2\,\Delta\varepsilon} \sin^2\left(\frac{A}{2}\right) \tag{163}$$

which contains an error of $4/\pi^2$ when compared with Eq. (162), showing the inadequacy of perturbation theory when applied to this model.

Another approximation based on perturbation theory [47] yields

$$p_{12} \sim \frac{\pi^2}{9} \left| \sum_i (\pm)_i e^{s_i} \right|^2 \tag{164}$$

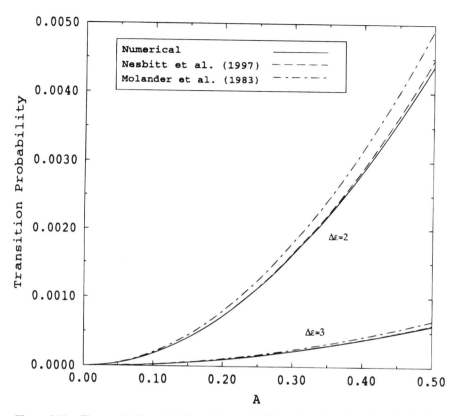

Figure 3.11. The parabolic model. Transition probability plotted as a function of A, where $0 \leq A \leq 0.5$, $\Delta\varepsilon = 2$, and $\Delta\varepsilon = 3$. Numerical results and Eq. (162) are shown to be in exact agreement for $\Delta\varepsilon$.

where

$$s_i = i \int_0^{t_i} \sqrt{4H_{12}^2 + (H_{22} - H_{11})^2} \, dt' \tag{165}$$

and t_i are the complex transition points in the upper half-plane and the $(\pm)_i$ are chosen according to whether

$$4H_{12}(t_i) = \pm i(H_{22} - H_{11}) \tag{166}$$

Equation (164) then reduces to

$$p_{12} \sim \frac{\pi^2}{9}\, 4e^{-2\,\Delta\varepsilon}\, \sin^2\!\left(\frac{A}{2}\right) \tag{167}$$

Equation (162) and the perturbation approximation of Molander et al. (1983) [Eq. (167)] are compared with the exact numerical solutions of the coupled equations. Figures 3.11–3.13 show that the strong-coupled approximation of Nesbitt et al. (1997) [Eq. (162)] is more or less in exact agreement with the numerical solutions, even for small values of $\Delta\varepsilon$ (far outside the specified limit $|\Delta\varepsilon| \gg 1$). For relatively small values of $\Delta\varepsilon$ (within the region of strong coupling) the approximation of Molander et al. (1983) is shown to deviate from the numerical solution. Therefore, one must con-

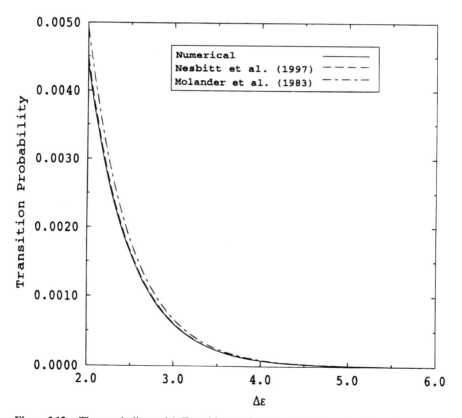

Figure 3.12. The parabolic model. Transition probability plotted as a function of resonance defect $\Delta\varepsilon$, where $2 \le \Delta\varepsilon \le 6$, $A = 0.5$.

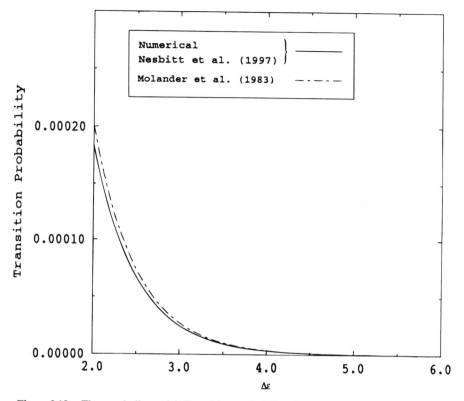

Figure 3.13. The parabolic model. Transition probability plotted as a function of resonance defect $\Delta\varepsilon$, where $2 \leq \Delta\varepsilon \leq 6$, $A = 0.1$. Equation (162) and the numerical solution are in exact agreement.

clude that the Zwaan–Stueckelberg phase-integral method produces the correct strong-coupling transition amplitude when supplemented with the appropriate first-order asymptotic expansions for the parabolic cylinder functions of large order and magnitude.

V. SUMMARY

This chapter has concentrated on the mathematical development of the Stokes phenomenon and its applications within the field of semiclassical atomic and molecular physics. The nature of the phenomenon, which was initiated by Stokes (1905) and developed by Zwaan (1929), has been examined in detail by considering the one-transition-point problem. We then discussed the phase-integral method pioneered by Stueckelberg (1932), who made an outstanding contribution to the solution of two- and four-transition-point problems in quantum-mechanical scattering problems, which led to the development of the Zwaan–Stueckelberg phase-integral method of Crothers (1971).

The main strength of the Zwaan–Stueckelberg phase-integral method is that when supplemented with appropriate asymptotic expansions, it produces strong-coupling approximations that incorporate the essential physics of the problem such as the bending of the double Stokes line, and ensures analyticity, unitarity, and symmetry. Finally, within a Zwaan–Stueckelberg framework we examined various four-transition-point problems, namely, the Nikitin, Demkov, Kummer, and parabolic models, and we pointed out the weakness of adiabatic perturbation theory. The exponential Nikitin model has been applied to a variety of reactions, either in its original form or in a slightly modified version. Of particular interest is the work of Crothers and Todd (1978, 1980) who investigated electron capture of H^+ from Be^+ and by the ions Zn^{2+}, Cd^{2+}, B^{2+}, Mg^{2+} and C^{6+} from $H(1s)$, for low impact velocities $v_0 < 1$ a.u. An important conclusion reached by comparison with the numerical solution of $C^{6+} + H(1s) \rightarrow C^{5+} + H^+$ is that given favorable conditions the model solutions can indeed represent the exact two-state solutions with useful accuracy.

ACKNOWLEDGMENTS

SFCO'R and DSFC acknowledge EPSRC support under grant GR/L 21891. BSN acknowledges financial support from the Department of Education, Northern Ireland (DENI).

REFERENCES

Abramowitz, M., Stegun, I. A. (1970) *Handbook of Mathematical Functions*, Dover: New York.

Bambini, A., Berman, P. R. (1981) *Phys. Rev. A* **23**, 2496–501.

Bárány, A., Crothers, D. S. F. (1981) *Phys. Scr.* **123**, 1096–103.

Bárány, A., Crothers, D. S. F. (1983) *Proc. R. Soc. Lond. A* **385**, 129–43.

Bates, D. R., Crothers, D. S. F. (1970) *Proc. R. Soc. Lond. A* **315**, 465–78.

Bates, D. R., Sprevak, D. (1970) *J. Phys. B* **3**, 1483–91.

Bayfield, J. E., Nikitin, E. E., Reznikov, A. I. (1973) *Chem. Phys. Lett.* **19**, 471–75 (Erratum **21**, 212).

Berry, M. V. (1989) *Proc. R. Soc. Lond. A* **422**, 7–21.

Brown, G. J. N., O'Rourke, S. F. C., Crothers, D. S. F. (1994) *Physica Scripta* **49**, 129–34.

Child, M. S. (1978) *Adv. At. Mol. Phys.* **14**, 225–80.

Crothers, D. S. F. (1971) *Adv. Phys.* **20**, 504–51.

Crothers, D. S. F. (1972) *J. Phys. A* **5**, 1680–88.

Crothers, D. S. F. (1976) *J. Phys. B* **9**, 635–43.

Crothers, D. S. F. (1978) *J. Phys. B* **11**, 1025–37.

Crothers, D. S. F. (1981) *Adv. At. Mol. Phys.* **17**, 55–98.

Crothers, D. S. F., O'Rourke, S. F. C. (1993) *J. Phys. B* **26**, L547–L553.

Crothers, D. S. F., Todd, N. R. (1978) *J. Phys. B* **11**, L663–L668.

Crothers, D. S. F., Todd, N. R. (1980) *J. Phys. B* **13**, 547–63.

Delos, J. B., Thorson, W. R. (1972) *Phys. Rev. A* **6**, 728–45.

Delos, J. B., Thorson, W. R., Knudson, S. K. (1972) *Phys. Rev. A* **6**, 709–20.

Demkov, Y. N. (1964) *Sov. Phys. JETP* **18**, 138–42.

Furry, W. H. (1947) *Phys. Rev.* **71**, 360–71.

Heading, J. (1962) *Phase Integral Methods*, London: Methuen.

Landau, L. D. (1932) *Phys. Z. Sowjetunion* **2**, 46–51.

Langer, R. E. (1934) *Bull. Am. Math. Soc.* **40**(2), 545–82.

McCarroll, R., Crothers, D. S. F. (1994) in *Advances in Atomic, Molecular and Optical Physics*, Academic Press, New York, **32**, 253–78.

Molander, W., Belsely, M., Burnett, K., Farrelly, D. (1983) *J. Chem. Phys.* **79**, 1297–300.

Mott, N. F., Massey, H. S. W. (1965) *The Theory of Atomic Collisions*, 3rd ed., Clarendon Press, Oxford.

Nakamura, H. (1991) *Int. Rev. Phys. Chem.* **10**, 123–88.

Nesbitt, B. S., Crothers, D. S. F., O'Rourke, S. F. C., Berman, P. R. (1997) *Phys. Rev. A.* **56**, 1670–73.

Nesbitt, B. S., O'Rourke, S. F. C., Crothers, D. S. F. (1996) *J. Phys. B* **29**, 2515–28.

Nikitin, E. E. (1962a) *Optics Spectrosc.* **13**, 431–33.

Nikitin, E. E. (1962b) *Discuss. Faraday Soc.* **33**, 14–21.

Nikitin, E. E. (1970) *Adv. Quantum Chem.* **5**, 135–44.

O'Rourke, S. F. C. (1992) *Phys. Scripta* **45**, 292–301.

O'Rourke, S. F. C., Crothers, D. S. F. (1992) *Proc. R. Soc. Lond. A* **438**, 1–22.

O'Rourke, S. F. C., Crothers, D. S. F. (1994) *J. Phys. B* **27**, 2497–509.

Paris, R. B., Wood, A. D. 1995) *IMA Bulletin* **31**, 21–28.

Pokrovskii, V. L., Khalatnikov, I. M. (1961) *Sov. Phys. JETP* **13**, 1207–10.

Robinson, E. J., Berman, P. R. (1983) *Phys. Rev. A* **27**, 1022–29.

Rosen, N., Zener, C. (1932) *Phys. Rev.* **40**, 502–7.

Smith, F. T. (1969) *Lect. Theor. Phys.* **11c**, 95.

Stokes, G. G. (1905) Reprints of papers in *Mathematical and Physical Papers* by the late Sir George Gabriel Stokes, Cambridge University Press, Vols. IV, V.

Stueckelberg, E. C. G. (1932) *Helv. Phys. Acta.* **5**, 369–422.

Vitanov, N. V. (1993) *J. Phys. B* **26**, L53–L60.

Vitanov, N. V. (1994) *J. Phys. B* **27**, 1791–805.

Zener, C. (1932) *Proc. R. Soc. Lond. A* **137**, 696–702.

Zwaan, A. (1929) Thesis, Utrecht.

FINITE INTEGRAL REPRESENTATION OF CHARACTERISTIC TIMES OF ORIENTATIONAL RELAXATION PROCESSES: APPLICATION TO THE UNIFORM BIAS FORCE EFFECT IN RELAXATION IN BISTABLE POTENTIALS

WILLIAM COFFEY

School of Engineering, Department of Electronic & Electrical Engineering, Trinity College, Dublin 2, Ireland

CONTENTS

Advances in Chemical Physics, Volume 103, Edited by I. Prigogine and Stuart A. Rice.
ISBN 0-471-24752-9 © 1998 John Wiley & Sons, Inc.

I. INTRODUCTION

A. Purpose of the Review

The purpose of this chapter is to give a comprehensive account of the methods of calculation of characteristic times of orientational relaxation processes that are governed by the Langevin or Fokker–Planck (Risken, 1989) equations and that have been extensively studied by my collaborators, Professor Derrick Crothers, Professor Yuri Kalmykov, Dr. John Waldron, and I. In particular, an elementary account of the calculation of mean first passage times and correlation times is given and can be conceived of as an extension of the material contained in the recently published book, *The Langevin Equation*, by Coffey, Kalmykov, and Waldron (1996a). Particular attention is paid to the calculation of the Kramers escape rate for rotational Brownian motion in non-axially-symmetric potentials with a view towards establishing the range of friction in which the various formulae for the escape rate are valid. A problem which is of current interest in fine-particle magnetic relaxation. The calculation of the Kramers escape rate for the non axially-symmetric magnetic problem will also be set out in its proper context in the statistical mechanical literature as the determination of the escape rate for a multidimensional system as a function of the damping, pioneered by Brinkman (1956), Landauer and Swanson (1961), Langer (1969), and Matkowsky et al (1984), see Hänggi et al. (1990) for a review. In addition, the recently discovered (Coffey et al., 1995a) uniform bias field effect in a bistable potential will be described in detail. As this is the most novel feature of the work from a physical point of view, I shall begin the review by introducing the effect.

B. Brownian Motion in Bistable Potentials

The Brownian motion in bistable potentials is a crucial factor in the thermally activated switching behavior of systems that may exist in either of two stable states. We mention a chemical reaction where two distinct chemical species are separated by a (internal) potential barrier; the reaction being modeled by thermally activated diffusion over the barrier (Risken, 1989; Coffey et al., 1996a). Another example is a fine single-domain ferromagnetic particle with uniaxial anisotropy, so possessing two stable magnetic states (Coffey et al., 1995a, 1996a; Garanin, 1996). Orientation of the magnetization of such a particle may undergo a reversal (i.e., cross the anisotropy internal potential barrier) due to thermal agitation.

In both problems the time to cross the potential barrier (mean first passage time) in the high-barrier (low-temperature) limit may be obtained from the inverse of the smallest nonvanishing eigenvalue λ_1 of the Sturm–Liouville equation associated with the appropriate FPE. Thus the barrier crossing process in the high-barrier limit is described by a *single eigenmode* of the FPE.

Yet another quantity describing the relaxation process is the correlation time T, which is defined as the area under the curve of the normalized autocorrelation function (ACF) $C(t)$ of the appropriate dynamic quantity (Risken, 1989; Garanin, 1996). The correlation time, which is a global characterization of the relaxation process involved, contains contributions from all the eigenvalues of the FPE and is related to the escape rate over the potential barrier. In the magnetic problem T is the correlation time of the normalized equilibrium ACF of the magnetization of the particle (Garanin, 1996; Coffey et al., 1995a). In the chemical reaction problem T is the correlation time of the normalized ACF of the reaction coordinate (position) (Perico et al., 1993; Kalmykov et al., 1996).

The asymptotic behavior of T is often similar to λ_1^{-1}, showing that the relaxation process is dominated by the "long-lived" barrier crossing mode. Marked differences occur, however, in the magnetic problem if a strong external uniform bias field is applied along the anisotropy axis. Here T will diverge exponentially from λ_1^{-1} in the high-anisotropy (low-temperature) limit (Garanin, 1996; Coffey et al., 1995a). Moreover, such behavior occurs at a critical value h_c of the ratio h (i.e., bias field parameter/anisotropy barrier height parameter), *far less* than the value needed to destroy the bistable nature of the potential. Thus in the low-temperature limit the relaxation process is no longer dominated by the slow-decay mode associated with the barrier crossing at values of h in excess of the critical value. The phenomenon (anti-Arrhenius behavior) was first noted for the magnetic problem in Coffey et al. (1995a) and later explained by Garanin (1996). He showed that it is a natural consequence of the depletion of the shallower of the two potential wells by the uniform field. Thus at low temperatures the fast modes in the deeper well may come to dominate the relaxation.

In order to comprehensively discuss the depletion effect caused by the

bias field, it will be necessary to have precise definitions of the various characteristic times associated with the relaxation process. In particular, it is
necessary to have precise definitions of the integral relaxation time (in linear
response the correlation time) and the mean first passage time, which in the
high-barrier limit is proportional (Coffey et al., 1996a) to the inverse of the
smallest nonvanishing eigenvalue essentially (the Kramers escape rate) λ_1^{-1}
of the governing FPE. This is best accomplished by obtaining wherever
possible finite integral representations of both the integral relaxation time
and the mean first passage time. The ratio of the two times is then a
measure of the depletion effect. We shall commence by reviewing the calculation of the integral relaxation time (Coffey and Crothers, 1996), then we
shall proceed to the discussion of the mean first passage time, demonstrating how it may be expressed in integral form for one-dimensional or quasi-
one-dimensional problems. We shall demonstrate the uniform bias field
effect for the two disparate examples of relaxation in a biased 2–4 potential
and orientational relaxation in a biased uniaxial anisotropy potential. Our
discussion will mainly be couched in the language of magnetic relaxation;
the corresponding quantities for dielectric relaxation (Martin et al., 1971;
Coffey et al., 1994a) may be obtained by appropriate substitution.

II. THE INTEGRAL RELAXATION TIME

A. The Correlation Time

A single-domain ferromagnetic particle with uniaxial anisotropy is characterized by an internal magnetic potential that has two stable stationary
points with a potential barrier between them. If the particle is sufficiently
fine, the direction of the magnetization may undergo a rotation due to
thermal agitation, surmounting the barrier, as first described by Néel (Fig.
4.1; Coffey et al., 1996a).

The calculation of the relaxation behavior of an assembly of such particles is usually accomplished (Risken, 1989; Coffey et al., 1996a; Garanin,
1996) by assuming that the relaxation of the magnetization is dominated by
a single relaxation mode, namely, that associated with the time of reversal
of the magnetization over the energy barrier between two stable orientational states. This means that in the set of eigenvalues $\{\lambda_k\}$ and corresponding amplitudes $\{A_k\}$ of the Sturm–Liouville equation (to which the
Fokker–Planck equation underlying the process may be converted), $\lambda_1 \ll$
λ_k, $k \geq 2$, and $A_1 \gg A_k$, since then the decay functions $A_k \exp(-\lambda_k t/2\tau_N)$,
$k \geq 2$, are small compared to $A_1 \exp(-\lambda_1 t/2\tau_N)$ except in the very early
stages of an approach to equilibrium. The diffusional relaxation time τ_N is

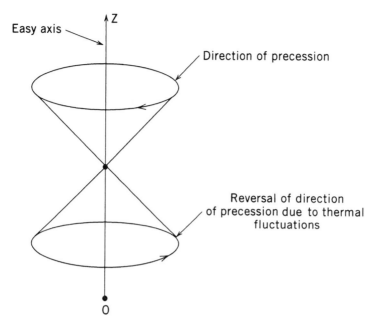

Figure 4.1. Schematic diagram for Néel relaxation. If OZ is the easy axis of magnetization, the magnetization may reverse its direction of precession about OZ due to thermal fluctuations.

defined as (Coffey et al., 1996a)

$$\tau_N = \frac{v}{2\eta kT}\left(\frac{1}{\gamma^2} + \eta^2 M_s^2\right) \tag{1}$$

where γ is the gyromagnetic ratio, M_s is the saturation magnetization, k is the Boltzmann constant, T is the absolute temperature, v is the volume of the particle, and η is the phenomenological damping constant from Gilbert's equation, namely (Coffey et al., 1996a)

$$\dot{\mathbf{M}} = \gamma \mathbf{M} \times (\mathbf{H}_{tot} - \eta \dot{\mathbf{M}}] \tag{2}$$

In Eq. (2) $\mathbf{M}$ denotes the magnetization, the overdot denotes d/dt, and

$$\mathbf{H}_{tot} = \mathbf{h}_r - \partial V/\partial \mathbf{M} \tag{3}$$

where $\mathbf{h}_r$ is the random white-noise field arising from thermal agitation and vV is the barrier potential (Gibbs free energy) including that of the internal crystalline anisotropy and the applied external field $\mathbf{H}$.

In view of the above considerations, the early studies (Coffey et al., 1996a; Brown, 1963) of the relaxation process were confined to the calculation of the smallest nonvanishing eigenvalue of the Sturm–Liouville equation, making the assumption that the process is dominated by a single relaxation mode with the time constant

$$\tau \simeq \frac{2\tau_N}{\lambda_1} \qquad (4)$$

Recently Garanin et al. (1990) and earlier Moro and Nordio (1985) (in the context of the analogous problem in chemical physics) have introduced the concept of the "integral relaxation time," which in the present context is proportional to the area under the curve that describes the relaxation of the magnetization after an abrupt change of a magnetic field that had been applied along the anisotropy axis. Furthermore in linear response (that is for an infinitesimal change in the magnetic field) Garanin et al. (1990) were able to obtain a general expression for the integral relaxation time of a single-domain ferromagnetic particle for all values of the barrier height parameter. Here the integral relaxation time is identical to the correlation time of linear response theory, since the decay of the magnetization is now proportional to the magnetization autocorrelation function (Coffey and Crothers, 1996). This formulation has been used extensively by Coffey et al. (1996a). According to Garanin et al. (1990), Coffey et al. (1996a), and Coffey and Crothers (1996), the integral relaxation time presents a more complete picture of the relaxation process rather than the approximation (Coffey et al., 1996a) of a "long-lived" (Garanin et al., 1990) exponential decay mode, which underlies Eq. (4) as the contribution of all the decay modes are included in the integral relaxation time. This is of particular importance in the problem of the response following an infinitesimal change in a strong bias field applied along the anisotropy axis, where, as discovered by Coffey et al. (1995a), at a value of the bias field far less than the critical value required to maintain the two-well structure of the potential, there is an abrupt departure of the decay time of the longest-lived (Néel) mode from the correlation time. Another advantage of the integral relaxation time is that it is possible to obtain exact analytic solutions for it in a number of problems (Coffey et al., 1996a). Indeed this property has very recently allowed Garanin (1996) to give a physical explanation of the constant magnetic field effect described by Coffey et al. (1995a).

The analysis given by Garanin et al. (1990) and Moro and Nordio (1985) was carried out by converting the Fokker–Planck equation underlying the problem to a Sturm–Liouville equation by using in the case of Moro and Nordio (1985) the Laplace transform and linear response theory and in Garanin et al. (1990) assuming a solution where the time variation of the bias field is exp $i\omega t$ and confining the response to small impressed fields. The solution of the resulting Sturm–Liouville equation in the zero-frequency limit then yields the integral relaxation time. In the special case of an applied field parallel to the anisotropy axis so that the problem is axially symmetric, a first integral of the Sturm–Liouville equation may be immediately written down so that the complete solution is easily given by quadratures (Risken, 1989), so yielding an *exact* expression (in integral form) for the correlation time.

A different approach to the problem has been made in the series of papers by Coffey et al. (1996a). Instead of representing the solution (of the Fokker–Planck equation) as a Sturm–Liouville problem, they have used a Floquet approach (Coffey et al., 1995a; Coffey and Crothers, 1996), whereby the solution is expanded in a Fourier series, in the manner often used for the solution of Hill's equation. Thus the calculation of the decay of the magnetization invariably reduces to the solution of a set of differential-recurrence relations. On taking the Laplace transform this set may be solved numerically using the standard methods of linear algebra. In a few special cases (Coffey et al., 1996a) we cite in particular (Brown, 1963) the simple uniaxial potential of the crystalline anisotropy (after a weak collinear dc field has been removed):

$$vV(\vartheta) = Kv(1 - z^2) = Kv \sin^2 \vartheta \qquad (5)$$

(where K is the anisotropy constant and ϑ is the polar angle specifying the direction of the magnetization), the set of algebraic recurrence relations (which now reduce to a three-term one) may be solved analytically in the zero-frequency limit so that the correlation time may be calculated. In particular, for the potential of Eq. (5) the correlation time may be expressed as a sum of products of Kummer functions (Coffey et al., 1996a), which in turn may be expressed in integral form [Eqs. (40) and (54) of Coffey and Crothers, 1996]. The equivalence of the two approaches has been demonstrated in Coffey and Crothers (1996), and for convenience the result will be reproduced below.

B. Calculation of the Integral Relaxation Time

In general the Fokker–Planck equation for the probability density $W(\vartheta, \varphi, t)$ of orientations of the magnetization vector **M** on the unit sphere is

(Coffey et al., 1996a; Brown, 1963) (φ is the azimuthal angle)

$$2\tau_N \frac{\partial W}{\partial t} = \Lambda^2 W + \frac{\beta}{\sin \vartheta} \frac{\partial}{\partial \vartheta} \left(\sin \vartheta \frac{\partial V}{\partial \vartheta} W - \frac{1}{a} \frac{\partial V}{\partial \phi} W \right)$$
$$+ \frac{\beta}{\sin \vartheta} \frac{\partial}{\partial \phi} \left(\frac{1}{\sin \vartheta} \frac{\partial V}{\partial \phi} W + \frac{1}{a} \frac{\partial V}{\partial \vartheta} W \right) \tag{6}$$

where $\beta = v/kT$, $a = \eta \gamma M_s$ is a dimensionless damping parameter, $V(\vartheta, \phi)$ is the Gibbs free energy density, and Λ^2 denotes the angular part of the Laplacian

$$\Lambda^2 = \frac{1}{\sin \vartheta} \frac{\partial}{\partial \vartheta} \left(\sin \vartheta \frac{\partial}{\partial \vartheta} \right) + \frac{1}{\sin^2 \vartheta} \frac{\partial^2}{\partial \phi^2}$$

Equation (6) has the generic form

$$\frac{\partial W}{\partial t} = LW \tag{7}$$

where L denotes the Fokker–Planck operator in Eq. (6). The potential energy $vV(\vartheta, \varphi)$ arises from the crystalline anisotropy potential and the action of an external uniform magnetic field $\mathbf{H}$ characterized by the external field parameter

$$\xi = v \frac{M_s H}{kT} \tag{8}$$

If we now suppose that ξ is decreased abruptly by a small amount ξ_1 such that $\xi_1 \ll 1$ so that the ensuing response is linear in ξ_1 the solution of Eq. (7) at any time after the perturbation has been made will be of the form (Coffey and Crothers, 1996)

$$W_t = W^{(0)} + W^{(1)} + W^{(2)} + \cdots \tag{9}$$

where $W^{(0)}$ is the new equilibrium distribution (attained after the relaxation from the former equilibrium distribution W_0, which had prevailed up to the time $t = 0$ before the abrupt change in the field), $W^{(1)}$ is the portion of the response linear in ξ_1, $W^{(2)}$ that quadratic in ξ_1, and so on. Since

$$LW^{(0)} = 0 \tag{10}$$

we shall have

$$\frac{\partial W^{(1)}}{\partial t} = LW^{(1)} \tag{11}$$

The formal solution of Eq. (11) is

$$W^{(1)} = (\exp Lt)W_0^1 \tag{12}$$

where W_0^1 denotes the value of $W^{(1)}$ at $t = 0$, that is, before ξ_1 has been removed. Hence the time-dependent orientational distribution function is, in linear response,

$$W_t = W^{(0)} + (\exp Lt)W_0^{(1)} \tag{13}$$

The ratio of potential energy to thermal energy is for uniaxial anisotropy

$$\frac{vV}{kT} = -\sigma(\mathbf{u} \cdot \mathbf{n})^2 - \xi(\mathbf{u} \cdot \mathbf{h}) \tag{14}$$

where

$$\sigma = \frac{Kv}{kT} \tag{15}$$

is the barrier height parameter, $\mathbf{u}$, $\mathbf{h}$, and $\mathbf{n}$ are unit vectors in the direction of $\mathbf{M}$, the field axis $\mathbf{H}$, and the anisotropy axis is denoted by $\mathbf{n}$. If we alter ξ by the small amount ξ_1 in order to apply the perturbation, we have

$$W_0 \equiv W_0(\vartheta, \varphi, 0) = \frac{e^{-\beta V}[1 + \xi_1(\mathbf{u} \cdot \mathbf{h})]}{\int_\Omega e^{-\beta V}[1 + \xi_1(\mathbf{u} \cdot \mathbf{h})] \, d\Omega} \tag{16}$$

where $d\Omega$ denotes the solid angle element

$$d\Omega = \sin\vartheta \, d\vartheta \, d\phi$$

Thus ignoring terms $O(\xi_1^2)$, we have the linear approximation

$$W_0 \equiv W(\vartheta, \phi, 0) = W^{(0)}(\vartheta, \phi)\{1 + \xi_1[(\mathbf{u} \cdot \mathbf{h}) - \langle \mathbf{u} \cdot \mathbf{h}\rangle_0]\} \tag{17}$$

where the symbol $\langle\ \rangle_0$ denotes the equilibrium ensemble average over $W^{(0)}$. Thus Eq. (13) becomes, the zero on the angular braces indicating that the average is to be performed in the absence of the perturbation ξ_1,

$$W_t = W^{(0)}(\vartheta, \phi) + \xi_1(\exp Lt)W^{(0)}(\vartheta, \phi)[(\mathbf{u} \cdot \mathbf{h}) - \langle \mathbf{u} \cdot \mathbf{h}\rangle_0] \qquad (18)$$

The mean change in the magnetic moment in the direction of $\mathbf{h}$ following the perturbation is then

$$\langle \Delta m \rangle_{\xi_1}(t) = M_s[\langle \mathbf{u} \cdot \mathbf{h}\rangle - \langle \mathbf{u} \cdot \mathbf{h}\rangle_0]_{\xi_1}$$

$$= \xi_1 M_s \int_\Omega (\mathbf{u} \cdot \mathbf{h} - \langle \mathbf{u} \cdot \mathbf{h}\rangle_0) \, e^{Lt}W^{(0)}(\mathbf{u} \cdot \mathbf{h} - \langle \mathbf{u} \cdot \mathbf{h}\rangle_0) \, d\Omega$$

$$= \xi_1 M_s \langle (\mathbf{u} \cdot \mathbf{h} - \langle \mathbf{u} \cdot \mathbf{h}\rangle_0)(0)(\mathbf{u} \cdot \mathbf{h} - \langle \mathbf{u} \cdot \mathbf{h}\rangle_0)(t)\rangle_0$$

$$= \frac{H_1}{kT} \langle \Delta\mathbf{m}(0) \cdot \Delta\mathbf{m}(t)\rangle_0$$

$$= \frac{H_1}{kT} [\langle \mathbf{m}(0) \cdot \mathbf{m}(t)\rangle_0 - \langle m\rangle_0^2] = \frac{H_1}{kT} b_m(t) \qquad (19)$$

and we have noted that $\langle m\rangle_0 = \langle m(0)\rangle_0 = \langle m(\infty)\rangle_0$. Thus the time-dependent magnetic moment due to the small change in the external field has been expressed in terms of the *equibilium correlation function* $b_m(t)$. The *integral relaxation time* T is (Coffey et al., 1996a) the area under the slope of the normalized decay function, so that

$$T = \frac{\int_0^\infty \langle \Delta m \rangle_{\xi_1}(t) \, dt}{\langle \Delta m \rangle_{\xi_1}(0)} \qquad (20)$$

Equation (20) may be written in terms of the zero-frequency limit of its Laplace transform as

$$T = \lim_{s \to 0} \left\{ \int_0^\infty \frac{\langle \Delta m \rangle_{\xi_1}(t) \, e^{-st} \, dt}{\langle \Delta m \rangle_{\xi_1}(0)} \right\}$$

$$= \lim_{s \to 0} \frac{\langle \Delta\tilde{m} \rangle_{\xi_1}(s)}{\langle \Delta m \rangle_{\xi}(0)} = \frac{\langle \Delta\tilde{m} \rangle_{\xi_1}(0)}{\langle \Delta m \rangle_{\xi_1}(0)} \qquad (21)$$

Equation (21) is a general expression for the integral relaxation time and does not rely on the assumption of linear response per se. It may be related, however, to the response in the presence of an alternating field only when linearity of the response is assumed, in which case Eq. (21) may be related to the magnetization correlation function $b_m(t)$ by means of Eq. (19) as follows. We have $\mathcal{L}$ denoting the Laplace transformation,

$$T = \lim_{s \to 0} \mathcal{L}\{\langle \mathbf{m}(0) \cdot \mathbf{m}(t) \rangle_0 - \langle m \rangle_0^2\}/\{\langle m^2 \rangle_0 - \langle m \rangle_0^2\} = \frac{\tilde{b}_m(0)}{b_m(0)} = \tilde{c}_m(0) \quad (22)$$

or in view of Eq. (19)

$$T = \lim_{s \to 0} \left\{ \int_\Omega (\mathbf{u} \cdot \mathbf{h} - \langle \mathbf{u} \cdot \mathbf{h} \rangle_0)(s - L)^{-1} W^{(0)}[\mathbf{u} \cdot \mathbf{h} - \langle \mathbf{u} \cdot \mathbf{h} \rangle_0] \, d\Omega \right\}$$

$$\times \left\{ \int_\Omega W^{(0)}[(\mathbf{u} \cdot \mathbf{h}) - \langle \mathbf{u} \cdot \mathbf{h} \rangle_0]^2 \, d\Omega \right\}^{-1} \quad (23)$$

$$= \int_\Omega \frac{[(\mathbf{u} \cdot \mathbf{h}) - \langle \mathbf{u} \cdot \mathbf{h} \rangle_0](-L)^{-1} W^{(0)}[\mathbf{u} \cdot \mathbf{h} - \langle \mathbf{u} \cdot \mathbf{h} \rangle_0]}{\langle [\mathbf{u} \cdot \mathbf{h} - \langle \mathbf{u} \cdot \mathbf{h} \rangle_0]^2 \rangle_0} \, d\Omega \quad (24)$$

The *formal* solution for T, Eq. (24), although derived by referring to the uniaxial potential of Eq. (14), is a general formula that applies *regardless* of the precise form of the potential $vV(\vartheta, \phi)$ as long as the perturbation ξ_1 is applied parallel to the field axis. The task of calculating the integral relaxation time from Eq. (24) always reduces to the determination of

$$(-L)^{-1} \frac{e^{-\beta V}}{Z} [\mathbf{u} \cdot \mathbf{h} - \langle \mathbf{u} \cdot \mathbf{h} \rangle_0] \quad (25)$$

where the partition function is

$$Z = \int_{-1}^{+1} e^{-\beta V(z)} \, dz$$

Analytic solutions of this problem may be determined in only a few specialized cases, in particular those where a first integral (Coffey and Crothers, 1996) may be determined when the problem is couched in Sturm–Liouville form or when the zero-frequency limit of the continued-fraction solution (Coffey et al., 1996a) in the Floquet representation may be recognized as a known function.

C. Integrable Cases of Eq. (25) for the Correlation Time

Equation (25) and so T may be calculated exactly for the case of axial symmetry. Here Eq. (6) with $z = \cos \vartheta$ becomes

$$2\tau_N \frac{\partial}{\partial t} W(z, t) = \frac{\partial}{\partial z} \left[(1 - z^2) \left(\frac{\partial W(z, t)}{\partial z} + \beta V'(z) W(z, t) \right) \right] \quad (26)$$

and taking the Laplace transform

$$2\tau_N[s\tilde{W}(z, s) - W(z, 0)] = \frac{\partial}{\partial z}\left[(1 - z^2)\left(\frac{\partial\tilde{W}(z, s)}{\partial z} + \beta V'(z)\tilde{W}(z, s)\right)\right] \quad (27)$$

and in the limit $s \to 0$ using the final value theorem of Laplace transformation

$$2\tau_N[W(z, \infty) - W(z, 0)] = \frac{\partial}{\partial z}\left[(1 - z^2)\left(\frac{\partial\tilde{W}(z, 0)}{\partial z} + \beta V'(z)\tilde{W}(z, 0)\right)\right]. \quad (28)$$

The unknown functions W on the left-hand side may now be determined explicitly allowing us to integrate Eq. (28). We have

$$2\tau_N[W(z, \infty) - W(z, 0)] = 2\tau_N\left(\frac{e^{-\beta V}}{Z} - \frac{e^{-\beta V}[1 + \xi_1 z]}{Z[1 + \xi_1\langle z\rangle_0]}\right)$$

$$= -2\xi_1\tau_N\frac{e^{-\beta V}}{Z}(z - \langle z\rangle_0) + O(\xi_1^2) \quad (29)$$

thus Eq. (28) becomes in the linear approximation in ξ_1 (all sub- and superscripts and arguments are dropped from W, as the meaning of W is now obvious)

$$-2\tau_N\xi_1\frac{e^{-\beta V}}{Z}(z - \langle z\rangle_0) = \frac{d}{dz}\left[(1 - z^2)\left(\frac{d\tilde{W}}{dz} + \beta V'(z)\tilde{W}\right)\right]$$

$$= L\tilde{W}(z, 0) \quad (30)$$

The particular integral of Eq. (30) and so Eq. (25) is best calculated by introducing the variable

$$F(z) = e^{\beta V(z)}\tilde{W}(z, 0) \quad (31)$$

so that Eq. (30) now becomes

$$-2\tau_N\xi_1\frac{e^{-\beta V}}{Z}(z - \langle z\rangle_0) = \frac{d}{dz}\left((1 - z^2)e^{-\beta V}\frac{dF(z)}{dz}\right) \quad (32)$$

which, mindful that the interval of solution is $[-1, 1]$, becomes

$$(1 - z^2)e^{-\beta V}\frac{dF(z)}{dz} + c_1 = \frac{-2\tau_N\xi_1}{Z}\int_{-1}^z e^{-\beta V(z_1)}(z_1 - \langle z_1\rangle_0)\,dz_1$$

with, since the probability current must vanish on the boundaries,

$$c_1 = (1 - z^2) \, e^{-\beta V} \left. \frac{dF(z)}{dz} \right|_{z=-1} = 0$$

Note this condition also implies that $F'(z)$ is *finite* on the boundaries, that is, at the poles of the sphere. Thus

$$\frac{dF(z)}{dz} = -\frac{2\tau_N \xi_1}{1 - z^2} \frac{e^{\beta V(z)}}{Z} \int_{-1}^{z} e^{-\beta V(z_1)} (z_1 - \langle z_1 \rangle_0) \, dz_1 \tag{33}$$

and so integrating once more:

$$F(z) = -2\tau_N \xi_1 \int_{-1}^{z} \frac{dz_2}{(1 - z_2^2)} \frac{e^{\beta V}}{Z} \int_{-1}^{z_2} e^{-\beta V(z_1)} (z_1 - \langle z_1 \rangle_0) \, dz_1 + c_2 \tag{34}$$

where

$$c_2 = F(z) \Big|_{z=-1} \tag{35}$$

Thus Eq. (25) reduces in this instance to

$$-\frac{e^{-\beta V(z)}}{2\tau_N \xi_1} F(z) = e^{-\beta V(z)} \int_{-1}^{z} \frac{dz_2}{(1 - z_2^2)} \frac{e^{\beta V}}{Z} \int_{-1}^{z_2} e^{-\beta V(z_1)} (z_1 - \langle z_1 \rangle_0) \, dz$$

$$-\frac{c_2}{\xi_1} e^{-\beta V(z)} \tag{36}$$

and so the exact integral relaxation time (correlation time) is from Eq. (24)

$$\frac{T}{2\tau_N} = -\int_{-1}^{+1} dz \, e^{-\beta V(z)} (z - \langle z \rangle_0) \int_{-1}^{z} \frac{dz_2}{(1 - z_2^2)} \frac{e^{\beta V}}{Z}$$

$$\times \int_{-1}^{z_2} e^{-\beta V(z_1)} (z_1 - \langle z_1 \rangle_0) \, dz_1 / [\langle (z - \langle z \rangle_0)^2 \rangle_0] \tag{37}$$

since the second term in Eq. (36) vanishes in the integral in Eq. (37). We further note on integration by parts that Eq. (37) may be written as

$$\frac{T}{2\tau_N} = \int_{-1}^{+1} \frac{dz}{1 - z^2} \frac{e^{\beta V(z)}}{Z} \frac{\left\{ \int_{-1}^{z} e^{-\beta V(z_1)} (z_1 - \langle z_1 \rangle_0) \, dz_1 \right\}^2}{\langle z^2 \rangle_0 - \langle z \rangle_0^2} \tag{38}$$

Equation (38) is the exact solution for the integral relaxation time for the axially symmetric Fokker–Planck operator of Eq. (26). It was essentially derived in the chemical physics context using a different formulation by Moro and Nordio (1985) and was later rederived in the context of magnetic relaxation by Garanin et al. (1990), who approached the problem by calculating the response due to a weak alternating field. A similar formula is also implicit in the work of Szabo (1980). We shall now evaluate Eq. (38) for the particular case of a weak dc uniform field reduced to zero at time $t = 0$ so that (Coffey et al., 1994b)

$$\beta V(\vartheta) = \sigma \sin^2 \vartheta = \sigma(1 - z^2) \tag{39}$$

and we shall prove that the solution yielded by Eq. (38) is identical in all respects to the integral form of the solution found by summing the series of products of Kummer functions yielded by the continued-fraction solution in Coffey et al. (1994b).

D.   Comparison of Integral Relaxation Time and Kummer Function Solutions

In order to evaluate T using Eq. (38) for the potential of Eq. (39) and to show that it is identical to Eq. (55) of Coffey et al. (1994b), namely, $M(a, b, z)$ denotes Kummer's function; the subscript $\parallel$ denotes the longitudinal correlation time,

$$\frac{T_\parallel}{\tau_N} = M(1, \tfrac{5}{2}, \sigma) - \frac{3e^\sigma \sigma^{-2}}{4M(\tfrac{3}{2}, \tfrac{5}{2}, \sigma)} \int_0^{\pi/2} d\theta \, \frac{\sqrt{1 + \cos \theta}}{\cos \theta}$$

$$\times \left[\frac{1}{2} (e^{\sigma \cos \theta} + e^{-\sigma \cos \theta}) + 2 - \frac{3}{2} \left(\frac{e^{\sigma \cos \theta} - e^{-\sigma \cos \theta}}{\sigma \cos \theta} \right) \right] \tag{40}$$

We note that Eq. (38) reduces for the potential of Eq. (39) to

$$\frac{T}{2\tau_N} = \int_{-1}^{+1} \frac{dz}{1 - z^2} \frac{e^{\sigma(1 - z^2)}}{Z} \frac{1}{4\sigma^2} \frac{(1 - e^{-\sigma(1 - z^2)})^2}{\langle z^2 \rangle_0}$$

$$= \int_{-1}^{+1} \frac{dz}{1 - z^2} \frac{e^{-\sigma(1 - z^2)}}{Z} \frac{1}{4\sigma^2} \frac{(1 - e^{\sigma(1 - z^2)})^2}{\langle z^2 \rangle_0} \tag{41}$$

that is

$$\frac{T}{2\tau_N} = \frac{1}{4\sigma^2} \frac{\langle (1 - z^2)^{-1} (1 - e^{\sigma(1 - z^2)})^2 \rangle_0}{\langle z^2 \rangle_0} \tag{42}$$

or

$$\frac{T}{\tau_N} = \frac{3e^\sigma}{\sigma^2} \left(\int_0^1 \frac{dz}{1 - z^2} \{\cosh[\sigma(z^2 - 1)] - 1\} \right) \Big/ M(\tfrac{3}{2}, \tfrac{5}{2}, \sigma) \qquad (43)$$

where we have noted that (Coffey et al., 1994b)

$$\int_0^1 e^{\sigma z^2} \, dz = M(\tfrac{1}{2}, \tfrac{3}{2}, \sigma) = \tfrac{1}{2} Z e^\sigma \qquad (44)$$

since the partition function $Z = \int_{-1}^1 e^{\sigma(z^2 - 1)} \, dz$ and

$$\frac{dM(a, b, z)}{dz} = \frac{a}{b} M(a + 1, b + 1, z) \qquad (45)$$

We shall now demonstrate that Eqs. (40) and (43) are identical. Our task is to prove that (taking a common denominator)

$$4\sigma^2 M(1, \tfrac{5}{2}, \sigma) M(\tfrac{3}{2}, \tfrac{5}{2}, \sigma) - 3e^\sigma \int_0^{\pi/2} d\theta$$

$$\times \left[\frac{\sqrt{1 + \cos\theta}}{\cos\theta} \left(\cosh(\sigma\cos\theta) - 3\frac{\sinh(\sigma\cos\theta)}{\sigma\cos\theta} + 2 \right) \right]$$

$$= 12e^\sigma \int_0^1 \frac{dz}{1 - z^2} \{\cosh[\sigma(1 - z^2)] - 1\} \qquad (46)$$

We first remark that the leading term on the left-hand side may be written in integral form as follows. We have

$$M(1, \tfrac{5}{2}, \sigma) M(\tfrac{3}{2}, \tfrac{5}{2}, \sigma) = e^\sigma M(1, \tfrac{5}{2}, \sigma) M(1, \tfrac{5}{2}, -\sigma) \qquad (47)$$

also (Coffey and Crothers, 1996; Coffey et al., 1994b),

$$M(1, \tfrac{5}{2}, \sigma) M(1, \tfrac{5}{2}, -\sigma)$$

$$= \frac{[\Gamma(\tfrac{5}{2})]^2 \sigma^{-3/2}}{\Gamma(1)\Gamma(\tfrac{3}{2})} \int_{-\infty}^\infty dt \, e^{t/2} \operatorname{sech} t \, I_{3/2}(\sigma \operatorname{sech} t)$$

$$= 2(\tfrac{3}{2})^2 \tfrac{1}{2} \sqrt{\pi} \, \sigma^{-3/2} \int_0^\infty \operatorname{sech} t \, I_{3/2}(\sigma \operatorname{sech} t) \cosh\frac{t}{2} \, dt$$

$$= -\tfrac{9}{4} \sqrt{\pi} \, \sigma^{-3/2} \int_0^{\pi/2} d\theta \, \frac{dt}{d\theta} \sin\theta \, I_{3/2}(\sigma \sin\theta) \cosh\frac{t}{2} \qquad (48)$$

where $I_v(x)$ is the modified Bessel function of the first kind, sech $t = \sin \theta$.
   Whence we have (Coffey and Crothers, 1996)

$$M(1, \tfrac{5}{2}, \sigma)M(1, \tfrac{5}{2}, -\sigma) = \frac{9}{4\sigma^2} \int_0^{\pi/2} d\theta \left[\frac{\sqrt{1 + \sin \theta}}{\sin \theta} \right.$$

$$\left. \times \left(\cosh(\sigma \sin \theta) - \frac{\sinh(\sigma \sin \theta)}{\sigma \sin \theta} \right) \right] \quad (49)$$

On substituting Eq. (49) into Eq. (46) and using Eq. (47), we find after some algebra that the left-hand side of Eq. (46) is (cancelling the common factor e^σ on both sides)

$$9 \int_0^{\pi/2} d\theta \frac{\sqrt{1 + \sin \theta}}{\sin \theta} \left(\cosh(\sigma \sin \theta) - \frac{\sinh(\sigma \sin \theta)}{\sigma \sin \theta} \right)$$

$$- 3 \int_0^{\pi/2} d\theta \frac{\sqrt{1 + \cos \theta}}{\cos \theta} \left(\cosh(\sigma \cos \theta) - 3 \frac{\sinh(\sigma \cos \theta)}{\sigma \cos \theta} + 2 \right)$$

which duly reduces to

$$6 \int_0^{\pi/2} d\theta \frac{\sqrt{1 + \cos \theta}}{\cos \theta} [\cosh(\sigma \cos \theta) - 1] \quad (50)$$

On writing $(1 - x^2) = \cos \theta$, Eq. (50) becomes

$$12 \int_0^1 \frac{dx}{1 - x^2} \{\cosh[\sigma(1 - x^2)] - 1\} \quad (51)$$

so proving that Eqs. (40) and (43) are identical.

E. Conclusions of the Foregoing Section

We have given a general expression viz Eq. (24) for the integral relaxation time T based on linear response theory so that the integral relaxation time is identical to the correlation time of the autocorrelation function of the change in the magnetization. The result agrees in all respects with that of Garanin et al. (1990) and Moro and Nordio (1985). T may be evaluated explicitly when the inverse of the Fokker–Planck operator L may be evaluated. In practice it appears that this may be accomplished analytically only for problems with axial symmetry when the Sturm–Liouville equation may be integrated exactly, leading to Eq. (38), which now constitutes the general solution of the problem.

We have evaluated Eq. (38) for the simple uniaxial potential of the crystalline anisotropy, and we have demonstrated that the result is identical to the (Floquet) method based on differential–recurrence relations of Coffey and Crothers (1996) and Coffey et al. (1994b). The Sturm–Liouville method has the great merit, in the uniaxial case, that it is relatively simple to calculate T without the complicated mathematical manipulations associated with the calculation of T from the differential–recurrence relations of Coffey et al.[13] (1994b). In addition, it is possible to exactly evaluate T analytically for a small change in a strong uniform bias field applied along the anisotropy axis. An importance consequence of this is that by evaluating the integral for values of the parameter

$$h = \frac{\xi}{2\sigma}$$

in the range of h, 0.1–0.2 Garanin (1996) has been able to give a clear physical explanation of the numerical results of Coffey et al. (1995a). He has accomplished this by demonstrating that in the above range of h the relaxation switches from being dominated by the behavior of the smallest non-vanishing eigenvalue i.e. the Kramers escape rate (Coffey et al., 1996a) to being dominated by the behavior in the wells of the potential. Such an effect appears to be a general feature of relaxation in a bistable potential in the presence of a uniform field. Explicit evaluation of T from the five-term differential–recurrence relation associated with the problem would be very difficult due to the task of identifying the set of hypergeometric functions associated with the solution of these recurrence relations in the zero-frequency limit. Thus the Sturm–Liouville equation constitutes a powerful method of finding analytic solutions in the case of axial symmetry. A drawback of this method however is that if it is generally very difficult to evaluate the complex susceptibility $\chi(\omega)$ and to extend the solution for T to non-axially-symmetric problems, as in both cases, it is impossible to integrate the equations by quadratures in the manner that leads to Eq. (38). The method based on the solution of differential–recurrence relations used by Coffey et al. (1994b, 1996a) and Coffey et al. (1997b) has the advantage that the solution for $\chi(\omega)$ may be often obtained exactly either in continued-fraction (Coffey et al., 1994b, 1996) or matrix-continued-fraction form (Coffey et al., 1996a,b) and in cases where this is not readily obvious (Geoghegan et al., 1997); as in many non-axially-symmetric problems, by computerized matrix inversion. Accurate solution of non-axially-symmetric problems is of particular importance in the constant-magnetic-field effect considered by Coffey et al. (1995a) and Garanin (1996) because application of the bias field at an oblique angle (Coffey et al., 1996a; Geoghegan et al.,

1997) will in general cause this effect to manifest itself (Coffey et al., 1995b, Coffey et al., 1998 in press) at smaller h values than those when the field is collinear with the anisotropy axis.

We remark that our formula (Eq. 38) for the correlation time may also be derived using the method of Szabo (1980) for the calculation of correlation times of autocorrelation functions of the Legendre polynomials, which is based on a generalization of the theory of first passage times (Risken, 1989). Furthermore, an equation similar to Eq. (38) but confined to one-dimensional translational Brownian motion has been given by Risken (1989) in the second edition of his book.

We also remark that the integral relaxation time may be calculated exactly for the step on–step off nonlinear response, as demonstrated recently by Kalmykov et al. (1997). Just as for axial symmetry, one may still integrate the Sturm–Liouville equation by quadratures. We shall now review the mean first passage time calculation and give a detailed review of the circumstance in which that quantity may be calculated exactly in integral form.

III. CALCULATION OF MEAN FIRST PASSAGE TIMES

A. Introduction

First passage time (Risken, 1989) is the time at which a random variable $\xi(t)$ first leaves a given domain. Clearly the first passage time is also a random variable, as it varies from realization to realization (Coffey et al., 1996a), and we may speak of a distribution function of first passage times so that one may compute a mean first passage time (Risken, 1989). The mean first passage time may be computed either by solving the Fokker–Planck equation in the manner described by Risken (Chapter 8, 1989) or by constructing a partial differential equation that yields the MFPT directly from the adjoint of the FPE (forward Kolmogorov equation) for further details see also Hänggi et al. (1990) where a historical account of the problem is also given. However, just as the calculation of the correlation time, the MFPT may at present only be calculated by quadratures for one-dimensional or quasi-one-dimensional potentials, otherwise the problem may be only solved in the high-barrier limit by regarding the MFPT as the inverse of the Kramers escape rate. This problem has been extensively discussed in the recent review in this series by Geoghegan et al. (1997), and will be further discussed in Section V of this chapter. We shall again discuss the MFPT in the context of superparamagnetic relaxation, as this is perhaps the most striking example of its application. We recall that a sufficiently fine ferromagnetic particle with uniaxial anisotropy comprises a single magnetic domain that possesses two stable magnetic states (Coffey et al., 1996a). Such a body has one distinguished axis that is the easiest axis of magnetization, as, for example (Coffey et al., 1996a), a single crystal of cobalt with its hexagonal axis. The orientation of the magnetization of such a particle may undergo a *reversal* (i.e., cross the internal anistropy potential barrier), due to thermal agitation (Arrhenius process), a mechanism that was first recognized by Néel in 1949. In such a mechanism of relaxation the inertia of the particle plays no role; thus the relaxation is governed by a Fokker–Planck equation (FPE) in the space of orientations of the magnetization only.

The time to cross the potential barrier, that is the reversal time τ (mean first passage time, exit time) has hitherto (Risken, 1989; Coffey et al., 1996a; Brown, 1963) usually been obtained from the inverse of the smallest non-vanishing eigenvalue λ_1 of the Sturm–Liouville equation associated with the appropriate FPE. Thus it has been assumed that the relaxation process across the barrier is described by a *single eigenmode* (that associated with λ_1) of the FPE, which, in turn, means that in the high-barrier limit the transition state theory of Kramers (Risken, 1989; Coffey et al., 1996a) may be used to obtain an approximate expression for the exit time. On the other hand, in the limit of vanishing potential barriers the Sturm–Liouville equation reduces to Legendre's equation (Brown, 1963) so that $\lambda_1 = 2$.

The formulation of the problem in this manner, however, has the disadvantage that it is impossible to give an exact analytical formula for the exit time, which is valid for all barrier heights. Particular difficulties (Brown, 1963) arise for moderate barrier heights where neither the low- nor high-barrier approximations may be safely applied. This limitation is particularly marked in relaxation in the presence of a strong constant bias field (Coffey et al., 1995a). Here one often needs to use fields close to the switching field, at which the magnetization will reverse, in order to study the superparamagnetic behavior. Such large fields reduce the effective barrier height so much that very high values of the anisotropy constant are required in order that the Kramers method should be applicable. The direct matrix inversion method of calculation of λ_1 also becomes impractical in this instance because of the very large matrices required for large values of the anisotropy and external field parameters (Coffey et al., 1994b, 1995b, 1996a, 1997b). Furthermore, it is *in principle* impossible to give an exact analytical formula for λ_1. In addition, λ_1^{-1} is not always an accurate approximation to the mean first passage time, which is the *precise* definition of τ (Gardiner, 1985) as the contributions of the other eigenmodes may be significant for moderate effective barrier heights (Coffey et al., 1995a, 1996a)— less than about 1.5 (Coffey et al., 1995b). All the foregoing defects may, however, be overcome by treating the problem using its proper definition as a mean first passage time problem (Gardiner, 1985; Coffey, 1996; Klik and

Gunther, 1990a), which allows one to give an *exact finite integral represen-tation* of the superparamagnetic relaxation time (which may be evaluated using special functions) rather than as the calculation of λ_1^{-1} (a very detailed discussion of mean first passage times and their relation to λ_1^{-1} is given by Gardiner 1985, in particular, Chapters 5 and 9) as we shall now demonstrate. The corresponding problems in chemical physics may again be treated simply by replacing magnetic quantities by their appropriate analogues. The treatment proceeds by generalizing the work of Klein (1952), who studied mean first passage times for the rotational Brownian motion mainly in the absence of a potential.

We again consider Eq. (6) for axially symmetric problems where $V = V(\vartheta)$ and $W(\vartheta, \phi, t) = W(\vartheta, t)$, the gyromagnetic terms in a^{-1} will drop out (Coffey et al., 1996a) and Eq. (6) becomes a partial differential equation in one spatial variable only viz. Eq. (26). We note also the adjoint operator

$$
L_{FP}^\dagger = (\Lambda')^2 - \beta \left[(1 - z'^2) \frac{\partial V}{\partial z'} + \frac{1}{a} \frac{\partial V}{\partial \phi'} \right] \frac{\partial}{\partial z'}
$$

$$
+ \beta \left[\frac{1}{a} \frac{\partial V}{\partial z'} - \frac{1}{1 - z'^2} \frac{\partial V}{\partial \phi'} \right] \frac{\partial}{\partial \phi'} \tag{52}
$$

$$
z' = \cos \vartheta' \tag{53}
$$

We reiterate that with $a^{-1} = 0$, Eq. (6) is similar to the Smoluchowski equa-tion for the probability distribution function W of the orientations of a polar molecule in nematic liquid crystals in the mean-field approximation (Coffey et al., 1996a). In what follows we shall again be mainly concerned with axial symmetry as then the FPE and its adjoint equation may in certain limits be integrated by quadratures. Thus we shall confine our dis-cussion for the greater part to $V(\vartheta)$ with minima at $\vartheta = 0$, π and a maximum at $\vartheta = \vartheta_m$.

B. Exact Expression for the Mean First Passage Time for Axial Symmetry

The MFPT, $\tau(\vartheta')$, for a magnetic moment starting at a definite orientation ϑ' to leave either of the domains $(0 \le \vartheta' \le \vartheta_m)$, $(\vartheta_m \le \vartheta' \le \pi)$ may be calcu-lated by forming, just as described for one-dimensional translational Brownian motion by Risken (1989) the equation

$$
L_{FP}^\dagger \tau(\vartheta') = \frac{1}{2\tau_N} \frac{e^{\beta V(\vartheta')}}{\sin \vartheta'} \frac{\partial}{\partial \vartheta'} \left(\sin \vartheta' \, e^{-\beta V} \frac{\partial \tau(\vartheta')}{\partial \vartheta'} \right) = -1 \tag{54}
$$

where $L_{FP}^\dagger$ is the adjoint Fokker–Planck operator (backward Kolmogorov operator, Risken (1989)). We consider the first of the two domains. The above equation is now integrated subject to the boundary condition that

$$\frac{\partial \tau(\vartheta')}{\partial \vartheta'} \tag{55}$$

is finite everywhere in $0 \le \vartheta' < \vartheta_m$ and so is finite at the reflecting boundary $\vartheta' = 0$, yielding

$$\frac{\partial \tau}{\partial \vartheta'} = -\frac{2\tau_N \, e^{\beta V(\vartheta')}}{\sin \vartheta'} \int_{\vartheta_1=0}^{\vartheta'} \sin \vartheta_1 \, e^{-\beta V(\vartheta_1)} \, d\vartheta_1 \tag{56}$$

for barrier crossings from the region $0 \le \vartheta' < \vartheta_m$. Equation (56) is now integrated over the range $[\vartheta', \vartheta_m]$ employing the second (obvious) boundary condition

$$\tau(\vartheta_m) = 0$$

Hence in the region $0 \le \vartheta' \le \vartheta_m$, τ is given by

$$\tau(\vartheta') = 2\tau_N \int_{\vartheta'}^{\vartheta_m} \frac{d\vartheta \, e^{\beta V(\vartheta)}}{\sin \vartheta} \int_{\vartheta_1=0}^{\vartheta} \sin \vartheta_1 \, e^{-\beta V(\vartheta_1)} \, d\vartheta_1 \tag{57}$$

where we note that ϑ starts at ϑ_m and moves *backwards* to ϑ'. For transitions from the region $\vartheta_m \le \vartheta \le \pi$, the lower limit, namely, zero in the inner integral in Eq. (57), is replaced by π. Thus setting $\vartheta' = 0$ in the integral specified for the region $(0 \le \vartheta' \le \vartheta_m)$ and $\vartheta' = \pi$ for $(\vartheta_m \le \vartheta' \le \pi)$, we obtain exact formulae for $\tau(0)$, $\tau(\pi)$ describing the mean first passage times from the minima at 0, π, respectively, namely;

$$\tau(0) = 2\tau_N \int_0^{\vartheta_m} \frac{d\vartheta \, e^{\beta V(\vartheta)}}{\sin \vartheta} \int_{\vartheta_1=0}^{\vartheta} \sin \vartheta_1 \, e^{-\beta V(\vartheta_1)} \, d\vartheta_1, \qquad 0 \le \vartheta' \le \vartheta_m \tag{58}$$

$$\tau(\pi) = 2\tau_N \int_\pi^{\vartheta_m} \frac{d\vartheta \, e^{\beta V(\vartheta)}}{\sin \vartheta} \int_{\vartheta_1=\pi}^{\vartheta} \sin \vartheta_1 \, e^{-\beta V(\vartheta_1)} \, d\vartheta_1, \qquad \vartheta_m \le \vartheta' \le \pi \tag{59}$$

We remark that Eqs. (58) and (59) (which are *conditional* mean first passage times since there is a *delta function* distribution of the entry points) when inverted coincide *precisely* with Eq. (4.36) of Brown (1963) for the rates for transitions from 0 to ϑ_m and π to ϑ_m, namely

$$v_{ij} = \frac{h'}{\beta I_m I_i}, \qquad i = 1, j = 2 \quad \text{or} \quad i = 2, j = 1 \tag{60}$$

where I_m and I_i are defined by Eqs. (4.27) and (4.32) of his 1963 paper, that is before any steepest descents approximations have been made so that I_i is *not decoupled* from I_m. Thus both the *conditional* mean first passage time and the Kramers (without the steepest descents approximation) approach yield the *same exact analytic* result. The problem when formulated in this way may now be exactly solved for any barrier height as we shall demonstrate.

C.   The Mean First Passage Time in Terms of Kummer (Confluent Hypergeometric) Functions

We have for transitions from the minimum at $\vartheta' = 0$ on setting $z_1 = \cos \vartheta_1$, $z = \cos \vartheta$

$$\tau(1) = 2\tau_N \int_1^{\cos \vartheta_m} \frac{dz}{1 - z^2} e^{\beta V(z)} \int_1^z dz_1 \, e^{-\beta V(z_1)}, \qquad 1 \le z \le \cos \vartheta_m \tag{61}$$

where we note that the potential for a uniform field parallel to the anisotropy axis is

$$\beta V(z) = \sigma(1 - z^2) - 2h\sigma z, \qquad h = \frac{\xi}{2\sigma} \tag{62}$$

In order to calculate the mean first passage time from Eq. (61), it is now necessary to determine the maximum angle ϑ_m (exit angle) from the potential in Eq. (62). Stationary points exist for this potential for $(d/d\vartheta)[\beta V(\vartheta)] = 0$, that is,

$$\beta V'(\vartheta) \equiv 2\sigma \sin \vartheta (\cos \vartheta + h) \tag{63}$$

The nature of these points is determined by

$$\frac{d^2}{d\vartheta^2} [\beta V(\vartheta)]$$

that is

$$\beta V''(\vartheta) = 2\sigma(\cos^2 \vartheta - \sin^2 \vartheta + h \cos \vartheta) \tag{64}$$

It is apparent from Eq. (64) that the stationary points for $\vartheta = 0, \pi$ are minima and $\vartheta = \cos^{-1}(-h)$ is a maximum so that

$$\cos \vartheta_m = -h \tag{65}$$

Hence in terms of z, τ for transitions from the point domain $z = 1$ is

$$\tau(1) = 2\tau_N \int_1^{-h} \frac{dz}{1-z^2} e^{\sigma(1-z^2)-2h\sigma z} \int_1^z dz_1 \, e^{-\sigma(1-z_1^2)+2h\sigma z_1} \tag{66}$$

Now on completing the square in the exponential in the inner integral

$$\int_1^z e^{-\sigma(1-z_1^2)+2h\sigma z_1} \, dz_1 = e^{-\sigma(1+h^2)} \int_1^z e^{\sigma(z_1+h)^2} \, dz_1 \tag{67}$$

On making the substitution $t = \sqrt{\sigma}(z_1 + h)$, this becomes

$$\int_1^z e^{-\beta V(z_1)} \, dz_1 = \frac{e^{-\sigma(1+h^2)}}{\sqrt{\sigma}} \int_{t_1}^{t_2} e^{t^2} \, dt, \qquad t_1 = \sqrt{\sigma}(h+1),$$

$$t_2 = \sqrt{\sigma}(z+h) \tag{68}$$

The integral above can now be expressed in terms of the error functions Abramowitz and Stegun, 1964) of an imaginary argument because

$$\int_0^u e^{-t^2} \, dt = \frac{\sqrt{\pi}}{2} \operatorname{erf}(u) \tag{69}$$

whence

$$\int_0^u e^{t^2} \, dt = \frac{\sqrt{\pi}}{2i} \operatorname{erf}(iu) \tag{70}$$

and so

$$\int_{t_1}^{t_2} e^{t^2} \, dt = \frac{\sqrt{\pi}}{2} \left[\operatorname{erf}(i\sqrt{\sigma}(z+h)) - \operatorname{erf}(i\sqrt{\sigma}(h+1)) \right] \tag{71}$$

so that

$$\int_1^z e^{\beta V(z_1)} \, dz_1 = \frac{e^{-\sigma(1+h^2)}}{2i} \sqrt{\frac{\pi}{\sigma}} \{\mathrm{erf}(i\sqrt{\sigma}(z+h)) - \mathrm{erf}(i\sqrt{\sigma}(h-1))\} \quad (72)$$

Now from Eq. (7.1.21) of Abramowitz and Stegun (1964):

$$\mathrm{erf}(u) = \frac{2u}{\sqrt{\pi}} \, M(\tfrac{1}{2}, \tfrac{3}{2}, -u^2) \quad (73)$$

and in terms of the error function of imaginary argument we have

$$\mathrm{erf}(iu) = \frac{2iu}{\sqrt{\pi}} \, M(\tfrac{1}{2}, \tfrac{3}{2}, u^2) \quad (74)$$

In Eqs. (73) and (74) M denotes the confluent hypergeometric (Kummer) function $_1F_1(\alpha; \gamma, z)$ defined by (Abramowitz and Stegun, 1964)

$$M(\alpha, \gamma, z) = \,_1F_1(\alpha; \gamma, z) = \sum_{n=0}^{\infty} \frac{(\alpha)_n z^n}{(\gamma)_n n!}, \qquad (\alpha)_n = \frac{\Gamma(\alpha+n)}{\Gamma(\alpha)} \quad (75)$$

Hence

$$\int_1^z dz_1 \, e^{-\sigma(1-z_1^2)+2h\sigma z_1} = \frac{e^{-\sigma(1+h^2)}}{2i} \sqrt{\frac{\pi}{\sigma}} \{\mathrm{erf}[i\sqrt{\sigma}(z+h)] - \mathrm{erf}[i\sqrt{\sigma}(h+1)]\}$$

$$= e^{-\sigma(1+h^2)}\{(z+h)M(\tfrac{1}{2}, \tfrac{3}{2}, \sigma(z+h)^2)$$
$$- (h+1)M(\tfrac{1}{2}, \tfrac{3}{2}, \sigma(h+1)^2)\} \quad (76)$$

Thus $\tau(1)$ in terms of Kummer's functions becomes

$$\tau(1) = 2\tau_N \int_1^{-h} \frac{dz}{1-z^2} e^{-\sigma(z+h)^2} \left\{ \begin{matrix} (z+h)M(\tfrac{1}{2}, \tfrac{3}{2}, \sigma(z+h)^2) \\ -(1+h)M(\tfrac{1}{2}, \tfrac{3}{2}, \sigma(h+1)^2) \end{matrix} \right\},$$

$$1 \geq z > -h \quad (77)$$

which is regular at $z = 1$ and where we remark that here the effective barrier height is

$$\sigma(1+h)^2$$

that is, the effective barrier is high. In like manner we have for transitions from the domain $z = -1$,

$$\tau(-1) = 2\tau_N \int_{-1}^{-h} \frac{dz}{1-z^2} e^{-\sigma(z+h)^2} \left\{ \begin{array}{l} (z+h)M(\frac{1}{2}, \frac{3}{2}, \sigma(z+h)^2) \\ +(1-h)M(\frac{1}{2}, \frac{3}{2}, \sigma(1-h)^2) \end{array} \right\},$$

$$-1 \leq z < -h \quad (78)$$

which is regular at $z = -1$. Here the effective barrier height is (Pfeiffer, 1990)

$$\sigma(1-h)^2$$

that is, the effective barrier is low. The overall mean first passage time is then

$$T_m = \frac{\tau(1)\tau(-1)}{\tau(1) + \tau(-1)} \quad (79)$$

which may be written in the more convenient form

$$\frac{T_m}{2\tau_N} = \frac{I(1)I(-1)}{I(1) + I(-1)} \quad (80)$$

where $I(1)$ and $I(-1)$ denote Eqs. (77) and (78) normalized by $2\tau_N$. We remark that if $\tau(1) \gg \tau(-1)$ or vice versa, we shall have

$$T_m \cong \tau(-1), \qquad T_m \cong \tau(+1).$$

D. High- and Low-Barrier Limits of the Mean First Passage Time

Equations (77)–(80) constitute the *exact* solution for the conditional mean first passage time and may be numerically integrated for any value of the anisotropy parameter σ for all values of the external field parameter $h < 1$. In the limit of low and high (effective) barrier heights, however, it is possible to give approximate formulae for T_m. We have when σ and h tend to zero

$$\tau(1) = 2\tau_N \int_1^0 \frac{dz}{1-z^2} \int_1^z dz_1 \quad (81)$$

which immediately yields by elementary integration, the result of Klein (1952) for rotation through a right angle

$$\tau(1) = 2\tau_N \ln 2 \tag{82}$$

Thus the normalized mean first passage time in the very-low-barrier limit is given by

$$\frac{T_m}{2\tau_N} = \tfrac{1}{2} \ln 2 = 0.347 \tag{83}$$

which should be compared with the result gained by considering T_m as given by the inverse of the smallest nonvanishing eigenvalue of the Fokker–Planck equation, namely

$$\frac{T_m}{2\tau_N} = \frac{1}{\lambda_1} = 0.5$$

since $\lambda_1 = 2$ in the zero-potential limit as the Sturm–Liouville equation then reduces to the Legendre equation. Equation (83) may be corrected for small values of σ and h by using the series expansion Eq. (75) of the Kummer functions (Abramowitz and Stegun, 1964). Analytic expressions for T_m in the high-barrier limit may be given as follows: The exact solution for $\tau(0)$ is

$$\tau(0) = 2\tau_N \int_0^{\vartheta_m} \frac{d\vartheta\ e^{\beta V(\vartheta)}}{\sin \vartheta} \int_0^{\vartheta} \sin \vartheta_1\ e^{-\beta V(\vartheta_1)}\ d\vartheta_1 \tag{84}$$

Employing Kramers argument (Risken, 1989; Coffey et al., 1996a) in the manner of Klein (1952) the integral is now evaluated in the limit of very high potential barriers. Since almost all the particles are situated near the minimum at $\vartheta = 0$ (because we have imposed a delta function initial distribution), then ϑ_1 is a very small angle. The inner integral in Eq. (84) may then be approximated by (Coffey et al., 1996a; Brown, 1963)

$$\int_0^{\vartheta} \sin \vartheta_1\ e^{-\beta V(\vartheta_1)}\ d\vartheta_1 \cong \int_0^{\infty} \vartheta_1\ e^{-\beta[V(0) + \vartheta_1^2(V''(0)/2)]}\ d\vartheta_1 \tag{85}$$

The integral on the right in Eq. (85) has been extended to infinity without significant error since all the particles are almost at the origin. Because ϑ_1 is very small, the Taylor series in $V(\vartheta_1)$ can be approximated by its first three terms $[V'(0) = 0]$. This is effectively a *steepest descents* argument (Coffey et al., 1994, 1996a). Hence the integral on the right-hand side in Eq. (85) becomes

$$e^{-\beta V(0)} \int_0^\infty \vartheta_1 \, e^{-\beta \vartheta_1^2 (V''(0)/2)} \, d\vartheta_1 = \frac{1}{\beta V''(0)} \, e^{-\beta V(0)} \tag{86}$$

and thus the two integrals in our expression for $\tau(0)$ effectively decouple (Risken, 1989) from each other.

We now evaluate the outer integral in Eq. (84) following the reasoning of Klein (1952) for the translational double-well potential problem. We have near ϑ_m

$$V(\vartheta) \cong V(\vartheta_m) - \frac{(\vartheta - \vartheta_m)^2}{2} \, V''(\vartheta_m) \tag{87}$$

and hence for the outer integral

$$\int_0^{\vartheta_m} \frac{d\vartheta \, e^{\beta V(\vartheta)}}{\sin \vartheta} \cong \frac{e^{\beta V(\vartheta_m)}}{\sin \vartheta_m} \int_{-\infty}^{\infty} e^{-\beta [(\vartheta - \vartheta_m)^2/2] V''(\vartheta_m)} \, d\vartheta \tag{88}$$

The range of integration in Eq. (88) may be extended from $-\infty$ to ∞ since the integral has its main contribution from values $\vartheta_m + \varepsilon$, $\vartheta_m - \varepsilon$ and almost no contribution from outside these values. Now

$$\frac{1}{\sigma \sqrt{2\pi}} \int_{-\infty}^{\infty} e^{-(x-\mu)^2/2\sigma^2} \, dx = 1 \tag{89}$$

and with

$$\mu = \vartheta_m, \qquad \frac{1}{\sigma^2} = \beta V''(\vartheta_m) \tag{90}$$

we have

$$\int_0^{\vartheta_m} \frac{d\vartheta \, e^{\beta V(\vartheta)}}{\sin \vartheta} \cong \frac{\sqrt{2}}{\sqrt{\beta V''(\vartheta_m)}} \frac{e^{\beta V(\vartheta_m)}}{\sin \vartheta_m} \tag{91}$$

Hence in the *high-barrier limit* the mean first passage time $\tau(0)$ for transitions from the point domain (0) is

$$\tau(0) \cong 2\tau_N \frac{1}{\beta V''(0)} \frac{\sqrt{2\pi}}{\sqrt{\beta V''(\vartheta_m)}} \frac{e^{\beta[V(\vartheta_m) - V(0)]}}{\sin \vartheta_m} \tag{92}$$

In like manner τ for transitions from the point domain π to ϑ_m can be established so that

$$\frac{\tau(\pi)}{2\tau_N} \cong \frac{1}{\beta V''(\pi)} \frac{\sqrt{2\pi}}{\sqrt{\beta V''(\vartheta_m)}} \frac{e^{\beta[V(\vartheta_m) - V(\pi)]}}{\sin \vartheta_m} \tag{93}$$

Thus with $\tau(0)$ and $\tau(\pi)$ defined in Eqs. (92) and (93) we can find the reciprocal p_1 of the longest time constant and hence λ_1, which is given in the high-barrier limit by

$$\lambda_1 \cong 2\tau_N(v_{12} + v_{21}) \tag{94}$$

where

$$v_{12} = \frac{1}{\tau(0)}, \qquad v_{21} = \frac{1}{\tau(\pi)} \tag{95}$$

and so Brown's (1963) approximate formula for the reciprocal p_1 of the longest time constant in the high barrier limit is retrieved viz.

$$p_1 = v_{12} + v_{21} \tag{96}$$

where

$$v_{ij} = \frac{1}{2\tau_N} \beta V_i'' \left(-\frac{\beta V''(\vartheta_m)}{2\pi} \right)^{1/2} \sin \vartheta_m \, e^{-\beta[V(\vartheta_m) - V_i]},$$

$$i = 1, j = 2 \quad \text{or} \quad i = 2, j = 1 \tag{97}$$

In Eq. (97) $V_1 = V''(0)$, $V_2 = V''(\pi)$ are the values at the minima and $V(\vartheta_m)$ is a maximum. Thus for the anisotropy potential

$$\beta V(\vartheta) = \sigma \sin^2 \vartheta - 2h\sigma \cos \vartheta, \qquad h = \frac{\xi}{2\sigma} \tag{98}$$

we can find v_{12}, v_{21} by taking the first and second derivatives of Eq. (98), yielding

$$\beta V'(\vartheta) = 2\sigma \sin \vartheta (\cos \vartheta + h)$$
$$\beta V''(\vartheta) = 2\sigma(2 \cos^2 \vartheta + h \cos \vartheta - 1) \tag{99}$$

In the expressions for v_{12}, v_{21} we note that a maximum exists for $\vartheta_m = \cos^{-1}(-h)$ and hence using the definition for λ_1 in Eq. (94), this becomes

$$\lambda_1 \cong 2\pi^{-1/2}\sigma^{3/2}(1 - h^2)\{(1 + h) \exp[-\sigma(1 + h)^2]$$
$$+ (1 - h) \exp[-\sigma(1 - h)^2]\}$$
$$= \lambda_k \tag{100}$$

so regaining the Brown (1963) formula for λ_1 in the high-barrier limit. Equation (100) also yields

$$\frac{\tau(0)}{2\tau_N} \cong \frac{\sqrt{\pi}}{2}\sigma^{-3/2} \frac{e^{\sigma(1+h)^2}}{(1 + h)^2(1 - h)} \tag{101}$$

which is the asymptotic normalized mean first passage time for transitions over the high barrier $\sigma(1 + h)^2$ and

$$\frac{\tau(\pi)}{2\tau_N} \cong \frac{\sqrt{\pi}}{2} \frac{\sigma^{-3/2} e^{\sigma(1-h)^2}}{(1 + h)(1 - h)^2} \tag{102}$$

which is the asymptotic normalized mean first passage time for transitions over the *low* barrier $\sigma(1 - h)^2$. Thus for large h and σ the mean first passage time will be mainly determined by Eq. (102) provided

$$\sigma \gg \frac{1}{(1 - h)^2} \tag{103}$$

We shall designate values of λ_1 obtained from the asymptotic Eq. (100) by the symbol λ_k.

E. Numerical Calculation of the Mean First Passage Time from Eq. (80) and Comparison with Previous Results

The smallest nonvanishing eigenvalue λ_1 as calculated by numerical solution of the differential–recurrence relations generated by the Fokker–Planck or the Langevin equation underlying the problem, cf. Eq. (17) of Coffey et

TABLE I

The smallest nonvanishing eigenvalue λ_1 for various values of the barrier height (σ) and field (h) parameters as computed by numerical solution of the differential–recurrence relation generated by the Fokker–Planck equation; cf. Coffey et al. *Phys. Rev. B* **51**, 15947 (1995), Eq. (17); note that the value of σ is restricted because of the very large matrix sizes needed for large values of σ. In this reference, λ_1 is labeled as $2\lambda_1$. (This table comprises part of Table 1 of that reference.)

	λ_1			
h	$\sigma = 2$	$\sigma = 5$	$\sigma = 10$	$\sigma = 20$
0.1	0.832	0.179	0.0086	7.79×10^{-6}
0.2	0.906	0.313	0.0383	0.000 198
0.5	1.41	1.4	0.883	0.206
0.8	2.32	3.67	4.52	4.78
1	3.12	5.82	8.78	12.9

al. (1995a) is shown in Table I for values of the anisotropy parameter varying from 2 to 20 and h values varying from 0.1 to 1. The range of σ just as in the earlier approaches of Aharoni (1964, 1969) and Coffey et al. (1995a) has to be restricted. Such a constraint arises here because as discussed by Geoghegan et al. (1997) and Coffey et al. (1995b, 1997b), calculation of the smallest nonvanishing eigenvalue of the system (transition) matrix of problems of this type is a straightforward numerical problem when the value of the effective barrier height parameter is relatively small, since comparatively small matrices are needed to achieve convergence. However, the loss of precision in floating-point calculations makes it difficult to achieve convergence for large h when the anisotropy parameter is very large and the corresponding matrix size has to be drastically increased in order to achieve convergence (the problem becomes particularly acute for non-axially-symmetric Fokker–Planck equations; Geoghegan et al., 1997, Kennedy, 1997). However, the numerical method could be improved by reformulating the problem in matrix continued-fraction form (Coffey et al., 1996a,b). In contrast, numerical integration of the hypergeometric functions poses no such convergence problems for large σ, and so calculations for large anisotropy and large field parameters are simpler to code and run much more efficiently. This is the main advantage of the formulation of the calculation of the mean first passage time in finite integral form. Thus the results obtained from the exact mean first passage time solution Eqs. (77)–(80) in Table II are given for values of σ up to 200. (I am indebted to Dr John Waldron for these numerical calculations.) The results from the asymptotic approximation [Eq. (100) of this paper] of Brown (1963), namely, λ_k, as reformulated by Aharoni (1964) are shown in Table III. In

TABLE II

λ_1 as inverse of the normalized mean first passage time $2\tau_N T_m^{-1}$ from the *exact solution*, Eq. (80), for various values of the barrier height (σ) and field (h) parameters. Note that it is now possible to attain very high values of the anisotropy parameters, unlike in the eigenvalue method.

h	$2\tau_N T_m^{-1}$							
	$\sigma = 2$	$\sigma = 5$	$\sigma = 10$	$\sigma = 20$	$\sigma = 60$	$\sigma = 100$	$\sigma = 140$	$\sigma = 200$
0.1	0.724	0.168	0.00825	7.58×10^{-6}	3.53×10^{-19}	6.52×10^{-33}	9.22×10^{-47}	1.24×10^{-67}
0.2	0.81	0.278	0.035	0.000187	8.02×10^{-15}	1.34×10^{-25}	1.7×10^{-36}	6.16×10^{-53}
0.5	1.54	1.16	0.688	0.17	0.0000511	5.25×10^{-9}	4.04×10^{-13}	2.15×10^{-19}
0.6	2.06	1.75	1.31	0.633	0.00708	0.0000275	7.85×10^{-8}	9.33×10^{-12}
0.8	4.64	4.48	4.22	3.72	2	0.88	0.323	0.0558
0.9	9.7	9.61	9.48	9.21	8.16	7.15	6.2	4.9
0.99	99.7	99.7	99.7	99.7	99.6	99.5	99.4	99.2

TABLE III

$\lambda_1 = \lambda_k$ computed from the *asymptotic* formula Eq. (100) deduced from the exact solution Eq. (80) for the mean first passage time T_m by applying steepest descents for various values of the barrier height (σ) and field (h) parameters. The formula provides a very good approximation to $2\tau_N T_m^{-1}$ of Table II when the effective barrier height $\sigma(1-h)^2$ is sufficiently large.

					λ_k			
h	$\sigma = 2$	$\sigma = 5$	$\sigma = 10$	$\sigma = 20$	$\sigma = 60$	$\sigma = 100$	$\sigma = 140$	$\sigma = 200$
0.1	0.872	0.228	0.00987	8.29×10^{-6}	3.65×10^{-19}	6.68×10^{-33}	9.39×10^{-47}	1.25×10^{-67}
0.2	0.888	0.406	0.0456	0.000214	8.47×10^{-15}	1.39×10^{-25}	1.75×10^{-36}	6.3×10^{-53}
0.5	0.766	1.36	1.1	0.255	0.0000602	5.88×10^{-9}	4.42×10^{-13}	2.31×10^{-19}
0.6	0.613	1.45	1.84	1.05	0.00909	0.0000325	8.95×10^{-8}	1.03×10^{-11}
0.8	0.215	0.744	1.72	3.27	3.43	1.49	0.498	0.0771
0.9	0.0603	0.228	0.613	1.57	5.47	7.89	8.76	8.21
0.99	0.000681	0.00251	0.00709	0.02	0.104	0.222	0.367	6.23

TABLE IV

The product of the exact smallest nonvanishing eigenvalue λ_1 and $(2\tau_N)^{-1}T_m$ Eq. (80) for values of σ and h given in Table I. It is apparent that λ_1 provides an adequate approximation to $2\tau_N T_m^{-1}$ for relatively high effective barriers (cf. $h = 0.1, 0.2$); however, the accuracy reduces as the effective barrier height decreases (h increasing), reflecting the contribution of relaxation modes to T_m other than associated with λ_1.

h	$(2\tau_N)^{-1} T_m \lambda_1$			
	$\sigma = 2$	$\sigma = 5$	$\sigma = 10$	$\sigma = 20$
0.1	1.149	1.065	1.042	1.027
0.2	1.118	1.12	1.094	1.058
0.5	0.915	1.206	1.394	1.211
0.8	0.5	0.819	1.07	1.284

Tables IV and V, in order to test the validity of the various approximations to T_m, we show, respectively, the product of the exact smallest nonvanishing eigenvalue λ_1 (obtained by the numerical solution of the FPE, Coffey et al., 1995a) and the normalized mean first passage time $(2\tau_N)^{-1}T_m$, and the product of the asymptotic solution for λ_1 namely λ_k [rendered by Eq. (100)] and $(2\tau_N)^{-1}T_m$. It is apparent from Table IV that the approximation

$$T_m \cong \frac{2\tau_N}{\lambda_1} \qquad (104)$$

is a reasonable one over a wide range of σ and h parameters, becoming very accurate (cf. $h = 0.1, 0.2$) where the effective height

$$\sigma(1 - h)^2 \qquad (105)$$

of the lowest barrier is still very large for large σ so that the smallest eigenvalue dominates the relaxation process in accordance with the reasoning of Chapter 5, Section 5.4.1 of Gardiner (1985). However, the accuracy of the approximation, Eq. (100), steadily decreases as the height of the lower barrier decreases (increasing h for fixed σ), in which situation [i.e., $\sigma \sim O(1 - h)^{-2}$] it is preferable to use the exact solution Eq. (80). Likewise it is apparent from Table V that the asymptotic solution Eq. (100) enables one to accurately estimate T_m over a wide range of σ and h values. The approximation starts to fail as is to be expected when $\sigma \cong (1 - h)^{-2}$ when the deviation from the exact integral solution is more than the deviation predicted by the exact λ_1. As the effective barrier height is increased, the asymptotic solution provides a more and more accurate approximation to the exact solution. Unfortunately, for large h values very large anisotropy

TABLE V

The product of the asymptotic solution [deduced from the exact solution Eq. (80) for the mean first passage time T_m by applying steepest descents] for the smallest nonvanishing eigenvalue $\lambda_1 = \lambda_k$ Eq. (100) and exact solution for $(2\tau_N)^{-1}T_m$, Eq. (80). It is again apparent that λ_k provides an adequate approximation to $2\tau_N T_m^{-1}$ for relatively high effective barriers; however, the accuracy again reduces as the effective barrier height decreases (h increasing), indicating that it is again preferable to use the exact integral Eq. (80) in this instance.

				$(2\tau_N)^{-1}\lambda_k T_m$				
h	$\sigma = 2$	$\sigma = 5$	$\sigma = 10$	$\sigma = 20$	$\sigma = 60$	$\sigma = 100$	$\sigma = 140$	$\sigma = 200$
0.1	1.204	1.357	1.196	1.094	1.034	1.025	1.019	1.008
0.2	1.096	1.46	1.302	1.144	1.056	1.037	1.029	1.022
0.5	0.49	1.17	1.599	1.5	1.178	1.12	1.094	1.074
0.6	0.297	0.828	1.404	1.658	1.283	1.181	1.14	1.103
0.8	0.046	0.166	0.407	0.879	1.715	1.693	1.541	1.381
0.9	6.21×10^{-3}	0.023	0.064	0.170	0.670	1.103	1.412	1.67

values are required in order that the condition for the validity of the asymptotic solution should be fulfilled. Thus the exact solution is again useful, as there is no such restriction of the σ values.

F. Discussions of Results and Conclusions

We have demonstrated how the time of reversal of the magnetization in the presence of axially symmetric potentials may be exactly calculated for all potential barrier heights provided the calculation is formulated as a mean first passage time problem rather that as the calculation of the inverse of the smallest nonvanishing eigenvalue of the Fokker–Planck equation. The resulting *finite* integral representation of the solution leads to substantial mathematical and conceptual simplifications [as conjectured by Klein (1952), who suggested that the calculation of the Kramers escape rate for the translational Brownian motion of a particle in a double-well potential is best posed as a mean first passage time problem as substantiated by the arguments given by Gardiner (1985)]. One of the main simplifications resulting from the present treatment is that it is no longer necessary to resort to matching (Aharoni, 1996) the low- and high-barrier solutions in the intermediate-barrier-height regime (where neither limit of the solution accurately obtains) using curve-fitting procedures, in order to produce an approximate formula for the reversal time that is valid for all barriers. Such a procedure must always be followed when the definition of T_m as λ_1^{-1} is used; moreover it is very difficult to match the solutions when $h \neq 0$. In addition, the finite integral representation lends itself to easy calculation of asymptotic corrections to the Kramers formulae, Eqs. (101) and (102) (in the manner described for the correlation time in Coffey et al., (1994b) as well as allowing T_m to be easily calculated numerically for very large values of the anisotropy and external field parameters (cf. Table II). Yet another advantage of the mean first passage time formulation is that now (cf. Section II) exact finite integral representations exist for *two* characteristic times associated with the relaxation process. These are the *integral relaxation time* (area under the curve of the normalized decay of the magnetization following the removal of a previously steady magnetic field, which in the linear response approximation to the perturbed response reduces to the correlation time T of the magnetization autocorrelation function) and the *mean first passage time* T_m (Néel time). It has been shown numerically by Coffey et al. (1995a) (by computing the eigenvalues of the Fokker–Planck equation and the corresponding amplitudes) and analytically by Garanin (1996) (who used the exact finite integral representation of the correlation time originally derived by Garanin et al., 1990) that for sufficiently large h and σ the two times T and T_m may differ exponentially if h exceeds a critical value h_c while remaining substantially the same if $h < h_c$. The critical h_c for the onset of this behavior is far less that the critical value $h_s = 1$ needed to destroy the two-minima structure of the potential. As we shall establish in Section IV, such behavior is due (Garanin, 1996) to the depletion of the population of the shallower of the two potential wells by the action of the applied field (characterized by the parameter h) and indicates a substantial contribution for $h > h_c$ of the fast decay modes in the deeper well to the relaxation process. This behavior suggests that experiments should be set up so that both characteristic times T_m and T can be measured in order that precise information (as evidenced by a substantial difference between T_m and T about the contribution of decay modes other than the barrier crossing one to the overall relaxation process may be gained.

We were able to obtain our exact solution for the Néel time because the form of the adjoint Fokker–Planck operator for axial symmetry allows one to integrate the differential equation, Eq. (54), for the mean first passage times by quadratures. In other words we have an *effective* one-dimensional problem. We illustrated our procedure by considering a potential that has minima at 0 and π. Exactly the same reasoning is true in respect of a potential that has minima at ϑ_A, ϑ_B as arises in formulating (Geoghegan et al., 1997; Coffey et al., 1995b) an axially symmetric approximation for the purpose of calculating the Néel time for certain non-axially-symmetric potentials. The most important direction for future work to take, however, is to discover a method of exactly integrating the adjoint Fokker–Planck equation in both polar and azimuthal variables (ϑ, φ) so that the present approach may be extended to non-axially-symmetric potentials such as the cubic anisotropy potential. It is usually very difficult (Aharoni, 1996) to numerically calculate λ_1 for such potentials due to the small effective barrier heights, so that a finite integral representation of the problem now becomes more or less essential in order to ascertain the range of applicability of Kramers transition-state theory asymptotic approximations. Such approximations have been obtained in the case of two variables ϑ, ϕ (Klik and Gunther, 1990; Smith and de Rozario, 1976; Brown, 1979) by extending the Kramers intermediate to high damping treatment (Chandrasekhar, 1943; Kramers, 1940) of the escape rate problem to curvilinear coordinates with the role of the friction coefficient being placed essentially by the dimensionless damping parameter a. The calculations, which are very lengthy, have been comprehensively reviewed by Geoghegan et al. (1997) in Vol. 100 of this series and will be considerably expanded upon in Section V below. The general thrust of the calculations, although they are rendered much more complicated by virtue of the curvilinear coordinates, is similar to the calculation of Kramers, as indeed is the asymptotic form of the solution. The asymptotic formula so obtained when further applied to a uniform bias field applied at an oblique angle to the easy axis of magnetization (Coffey et al., 1995b, 1998 in press; Pfeiffer, 1990) again renders a close approximation to λ_1^{-1} and hence T_m for large effective potential barriers. We shall now consider in detail (Coffey et al., 1997a) the effect of a uniform bias force on a bistable system. In order to emphasize the generality of our approach we (Coffey et al., 1997a) illustrate the effect for the two disparate examples of relaxation in a biased 2–4 potential and orientational relaxation in a biased uniaxial anisotropy potential given below.

IV. DEPLETING EFFECT OF A UNIFORM BIAS FIELD

A. Relaxational Dynamics of a Brownian Particle in a Biased Double-well Potential

The one-dimensional noninertial translational Brownian motion of a particle in a double-well potential $V(x)$ is governed by the Langevin equation (Coffey et al., 1996a)

$$\zeta \frac{d}{dt} x(t) + \frac{\partial}{\partial x} V = f(t) \tag{106}$$

where $x(t)$ specifies the position of the particle at time t, ζ is the friction coefficient, and $f(t)$ is a white noise force so that

$$\overline{f(t)} = 0$$
$$\overline{f(t)f(t')} = 2kT\zeta \, \delta(t - t') \tag{107}$$

the overbar means "the statistical average of" over an ensemble of particles starting at time t with the same initial position $x(t) = x$.

The noninertial Fokker–Planck (Smoluchowski) equation for the probability distribution function W of the position is (Risken, 1989)

$$\zeta \frac{\partial}{\partial t} W = \frac{\partial}{\partial x} \left(W \frac{\partial}{\partial x} V \right) + kT \frac{\partial^2}{\partial x^2} W \tag{108}$$

where k is the Boltzmann constant and T is the absolute temperature.

The quantities of interest are the positional ACF

$$C(t) = \langle x(0)x(t) \rangle_0 - \langle x(0) \rangle_0^2 \tag{109}$$

where the symbol $\langle \ \rangle_0$ designates the equilibrium ensemble average, and the correlation time

$$T = \frac{\int_0^\infty C(t)\, dt}{C(0)} \tag{110}$$

A general equation for the correlation time T for a system governed by Eq. (106) has been given by Risken [1989, Eq. (S9.14)] viz. [this is simply the translational version of Eq. (38)]

$$T = \beta\zeta \, \frac{\int_{-\infty}^\infty e^{\beta V(x)}(\int_{-\infty}^x (x' - \langle x \rangle_0)\, e^{-\beta V(x')}\, dx')^2\, dx}{Z[\langle x^2 \rangle_0 - \langle x \rangle_0^2]} \tag{111}$$

where

$$Z = \int_{-\infty}^\infty e^{-\beta V(x)}\, dx, \qquad \langle x \rangle_0 = \frac{1}{Z} \int_{-\infty}^\infty x\, e^{-\beta V(x)}\, dx,$$

$$\langle x^2 \rangle_0 = \frac{1}{Z} \int_{-\infty}^\infty x^2\, e^{-\beta V(x)}\, dx \tag{112}$$

and $\beta = (kT)^{-1}$ with in this instance T the absolute temperature.

For the purpose of our discussion we consider a biased double-well potential

$$\beta V(x) = \sigma U(x) - \xi x \tag{113}$$

with a maximum at x_B and minima at x_A and x_C and without loss of generality we let $V(x_C) > V(x_A)$ (see Fig 4.2).

At low temperatures ($\sigma,\ \xi \gg 1$) $C(t)$ may be approximated (Garanin, 1996) by two exponentials

$$C(t) \approx \Delta_{\text{well}}\, e^{-t/\tau_{\text{well}}} + \Delta_B\, e^{-t\lambda_1} \tag{114}$$

where τ_{well} is the relaxation time in the deep well and so has a weak temperature dependence, λ_1 is the smallest eigenvalue of the Sturm–Liouville equation associated with the FPE and so has Arrhenius (exponential) tem-

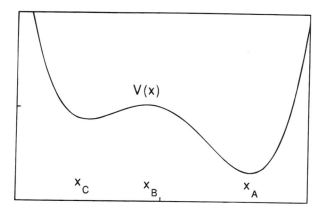

Figure 4.2. Double-well potential with local minima at x_A and x_C and a local maximum at x_B.

perature dependence, and correspondingly $\Delta_{\text{well}} \gg \Delta_B$. These two exponentials correspond to *intrawell* and *overbarrier* relaxation processes. According to Eq. (110) the correlation time is given by, on using Eq. (114),

$$T \approx \frac{\tau_{\text{well}} \Delta_{\text{well}} + \Delta_B \lambda_1^{-1}}{\Delta_{\text{well}} + \Delta_B} \approx \tau_{\text{well}} + \Delta_B \lambda_1^{-1}/\Delta_{\text{well}} \tag{115}$$

Recognizing (Garanin, 1996) that the partition function Z is the sum of the contributions due to the two potential wells and applying the steepest descent method to the evaluation of Z, just as in Section III, we obtain

$$\Delta_{\text{well}} \approx \frac{1 - 2/\pi}{\sigma U''(x_A)} \tag{116}$$

$$\Delta_B \approx \frac{e^{-\sigma[U(x_C) - U(x_A)] - \xi(x_A - x_C)}}{\sigma U''(x_A)} \sqrt{\frac{U''(x_A)}{U''(x_C)}}$$

$$\times \left(\frac{4}{\pi} - 1 + \frac{4}{\pi} \sqrt{\frac{U''(x_A)}{U''(x_C)}} + \frac{U''(x_A)}{U''(x_C)} \right) \tag{117}$$

According to the Kramers escape rate theory (Coffey et al., 1996a), the inverse of the smallest nonvanishing eigenvalue has exponential dependence on the barrier height (Risken, 1989; Chandrasekhar, 1943; Kramers, 1940)

$$\lambda_1^{-1} \sim e^{\sigma[U(x_B) - U(x_C)] + \xi(x_C - x_B)} \tag{118}$$

Thus

$$\Delta_B \, \lambda_1^{-1}/\Delta_{\text{well}} \sim e^{\sigma[U(x_B) + U(x_A) - 2U(x_C)] + \xi(2x_C - x_B - x_A)} \tag{119}$$

Now the argument of the exponential, viz.

$$\sigma[U(x_B) + U(x_A) - 2U(x_C)] + \xi(2x_C - x_B - x_A) \tag{120}$$

may *change its sign* at some critical value of ξ_c. This is the crucial fact underpinning all that follows. Thus if $\xi < \xi_c$ the quantity $\Delta_B \lambda_1^{-1}/\Delta_{\text{well}}$ *increases exponentially* as the temperature decreases and so *determines completely* the temperature dependence of the correlation time T. If $\xi > \xi_c$, on the other hand, the quantity $\Delta_B \lambda_1^{-1}/\Delta_{\text{well}}$ decreases exponentially as the temperature decreases; thus T *no longer has Arrhenius behavior.* At this critical value (Coffey et al., 1995a) of the bias parameter the relaxation switches from being dominated by the behavior of the smallest non-vanishing eigenvalue λ_1 (inverse of the MFPT) to being dominated by the fast relaxation processes in the deep well of the potential because of the depletion of the upper (shallow) well at low temperatures (Garanin, 1996). We shall now demonstrate that this behavior occurs at values of ξ such that the double-well structure of the potential still persists.

B.   Relaxational Dynamics in a Biased 2–4 Potential

As a definite example consider the double-well potential (Coffey et al., 1997a)

$$\beta V(x) = -\xi x - \sigma x^2 + \frac{\sigma}{4} x^4$$

$$= -\frac{\sigma}{4}(hx + 4x^2 - x^4) \tag{121}$$

where x is a dimensionless coordinate and $h = 4\xi/\sigma$ is the bias parameter.

At $\xi = 0$ the potential Eq. (121) is symmetrical and has a barrier at $x = 0$ where the potential has a maximum with height relative to the minimum equal to σ (in dimensionless units). The relaxational dynamics in this case are well known (see, e.g., Perico et al., 1993; Kalmykov et al., 1996) and may be described as follows. The correlation time for all barrier height parameters is given (Perico et al., 1993) by Eq. (111) with $\langle x \rangle_0 = 0$. In the low temperature (high-barrier) limit the ACF may be approximated by (Perico et al., 1993; Kalmykov et al., 1996)

$$C(t)/C(0) \approx \Delta_{\text{well}} \, e^{-t/\tau_{\text{well}}} + \Delta_B \, e^{-t\lambda_1} \tag{122}$$

where

$$\tau_{well} = \beta\zeta(\langle x^2\rangle_0 - \langle x\rangle^2_{well}) \tag{123}$$

$$\lambda_1^{-1} \approx \frac{\pi\beta\zeta\langle x^2\rangle_0\, e^\sigma}{4\sqrt{2}\,\sigma} \tag{124}$$

$$\Delta_{well} \approx \langle x\rangle^2_{well}, \qquad \Delta_B \approx \langle x^2\rangle_0 - \langle x\rangle^2_{well} \tag{125}$$

and $\langle x\rangle_{well}$ is the average of x in one of the wells (e.g., over the range $0 \leq x < \infty$). Thus T has the pronounced Arrhenius behavior

$$T \approx \lambda_1^{-1}\, \frac{\Delta_B}{\Delta_{well}} \sim e^\sigma$$

For $\xi \neq 0$ the potential becomes asymmetrical (see Fig. 4.3). Moreover, the double-well structure of the potential disappears at

$$h_s = 16\sqrt{2/27} \approx 4.354\,65 \tag{126}$$

Now the correlation time is also given by the exact Eq. (111). The results of the exact calculation of T are shown in Fig. 4.4. It is apparent that T

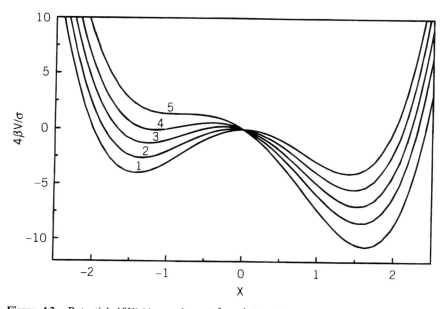

Figure 4.3. Potential $4\beta V(x)/\sigma = -hx - 4x^2 + x^4$ from Eq. (121) as a function of x for various values of the bias parameter $h = 4\xi/\sigma$: 0 (curve 1), 1 (curve 2), 2 (curve 3), 3 (curve 4), and $16\sqrt{2/27}$ (curve 5).

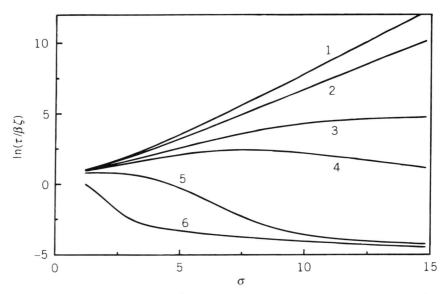

Figure 4.4. $\ln(T/\beta\zeta)$ calculated from Eqs. (111) and (121) as a function of σ for various values of the bias parameter $h = 4\xi/\sigma$: 0 (curve 1), 0.5 (curve 2), 0.95 (curve 3), 1.2 (curve 4), 2 (curve 5), and 4.5 (curve 6).

ceases to have Arrhenius behavior at values of h that are considerably smaller than the critical parameter h_s of Eq. (126). Furthermore, one can see in Fig. 4.4 that for a *given shape of the potential* ($h = $ const) the behavior of T may dramatically alter from exponentially increasing to exponentially decreasing (see curve 4). Applying the approximate treatment of Section A by means of Eq. (119), we see that

$$\lambda_1^{-1} \frac{\Delta_B}{\Delta_{\text{well}}} \sim \exp\left[\sigma\left(1 - \frac{3h}{2\sqrt{2}}\right)\right] \tag{127}$$

alters its behavior from exponential increase to exponential decrease at

$$h_c \approx 2\sqrt{2}/3 \approx 0.943 \tag{128}$$

C.   Orientational Relaxation in a Biased Uniaxial Anisotropy Potential

The FPE for the probability distribution function W of the orientations of magnetization of an assembly of single-domain ferromagnetic particles is

(Sections II, III)

$$2\tau_0 \frac{\partial}{\partial t} W = \frac{\beta}{\sin \vartheta} \frac{\partial}{\partial \vartheta} \left(\sin \vartheta W \frac{\partial}{\partial \vartheta} V \right) + \frac{1}{\sin \vartheta} \frac{\partial}{\partial \vartheta} \left(\sin \vartheta \frac{\partial}{\partial \vartheta} W \right) \quad (129)$$

or with $z = \cos \vartheta$

$$2\tau_0 \frac{\partial}{\partial t} W = \frac{\partial}{\partial z} \left[(1 - z^2) \left(\frac{\partial}{\partial z} W + \beta W \frac{\partial}{\partial z} V \right) \right] \quad (130)$$

where τ_0 is a characteristic (diffusion) relaxation time. We again note that the FPE (129) is similar to the noninertial Smoluchowski equation for the probability distribution function W of the orientations of a polar molecule in nematic liquid crystals in the mean-field approximation (Martin et al., 1971).

In applications of linear response theory to dielectric and magnetic relaxation the quantity of interest is the dipole moment equilibrium ACF (Coffey et al., 1996a) (as in Section II)

$$C(t) = \langle \cos \vartheta(0) \cos \vartheta(t) \rangle_0 - \langle \cos \vartheta(0) \rangle_0^2 \quad (131)$$

which describes the stepoff linear response of polarization and magnetization. The correlation time

$$T = \frac{\int_0^\infty [\langle \cos \vartheta(0) \cos \vartheta(t) \rangle_0 - \langle \cos \vartheta(0) \rangle_0^2] \, dt}{\langle \cos^2 \vartheta(0) \rangle_0 - \langle \cos \vartheta(0) \rangle_0^2} \quad (132)$$

for an arbitrary potential $V(z)$ is (Garanin, 1996; Coffey and Crothers, 1996) as proved in Section II

$$T = 2\tau_0 \frac{\int_{-1}^1 e^{\beta V(z)} (\int_{-1}^z (z' - \langle z \rangle_0) e^{-\beta V(z')} \, dz')^2 \frac{dz}{1 - z^2}}{[\langle z^2 \rangle_0 - \langle z \rangle_0^2] Z} \quad (133)$$

where

$$Z = \int_{-1}^1 e^{-\beta V(z)} \, dz, \qquad \langle z \rangle_0 = \frac{1}{Z} \int_{-1}^1 z \, e^{-\beta V(z)} \, dz$$

$$\langle z^2 \rangle_0 = \frac{1}{Z} \int_{-1}^1 z^2 \, e^{-\beta V(z)} \, dz \quad (134)$$

As an example consider the simplest uniaxial potential

$$\beta V(\cos \vartheta) = -\sigma \cos^2 \vartheta - \xi \cos \vartheta$$

$$= \beta V(z) = -\sigma(z^2 + 2hz) \tag{135}$$

where σ and ξ are the dimensionless anisotropy and external field parameters, respectively, and $h = \xi/2\sigma$.

At $\xi = 0$ the potential Eq. (135) is symmetrical and has a barrier at $\vartheta = \pi/2$, where the potential has a maximum where the height relative to the minima at $\vartheta = 0$ and π is equal to σ. For $\xi \neq 0$ the potential becomes asymmetrical (see Fig. 4.5) and the double well structure disappears at

$$h_s = 1 \tag{136}$$

The relaxational dynamics have been investigated in detail in (Coffey et al., 1995a; Garanin, 1996) and may be described as follows. The correlation time T is given by the exact Eq. (133). The results of the exact calculation of T are shown in Fig. 4.6. It is apparent that T ceases to display Arrhenius

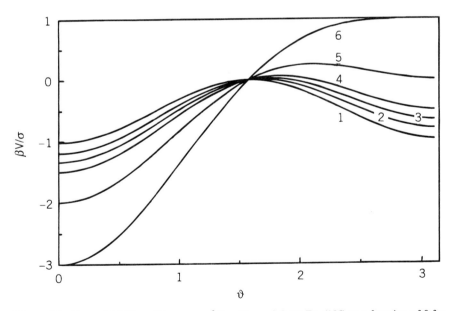

Figure 4.5. Potential $\beta V(\cos \vartheta)/\sigma = -\cos^2 \vartheta - 2h \cos \vartheta$ from Eq. (135) as a function of ϑ for various values of the bias parameter $h = \xi/2\sigma$: 0 (curve 1), 0.1 (curve 2), 0.17 (curve 3), 0.25 (curve 4), 0.5 (curve 5), 1 (curve 6).

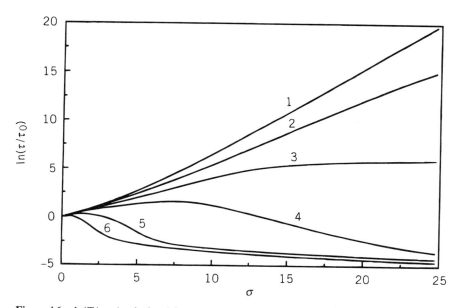

Figure 4.6. $\ln(T/-\tau_0)$ calculated from Eq. (133) as a function of σ for different values of the bias parameter $h = \xi/2\sigma$: 0 (curve 1), 0.1 (curve 2), 0.17 (curve 3), 0.25 (curve 4), 0.5 (curve 5), and 1 (curve 6).

behavior at values of h that are considerably smaller than h_s given by Eq. (133), that is, when the double-well structure of the potential is still present.

In the low-temperature limit the ACF may be approximated by (Garanin, 1996)

$$C(t) \approx \Delta_{\text{well}} \, e^{-t/\tau_{\text{well}}} + \Delta_B \, e^{-t\lambda_1/2} \tag{137}$$

where

$$\tau_{\text{well}} \approx \frac{\tau_0}{2\sigma + \xi} \tag{138}$$

and as in Section III

$$\lambda_1^{-1} \cong \frac{1}{2}\{\tau_0^{-1}\pi^{-1/2}\sigma^{3/2}(1 - h^2)[(1 + h) \, e^{-\sigma(1+h)^2} + (1 - h) \, e^{-\sigma(1-h)^2}]\}^{-1}$$

$$\approx \frac{\tau_0}{2} \, \pi^{1/2}\sigma^{-3/2}(1 + 3h) \, e^{\sigma(1-2h)} + O(h^2) \tag{139}$$

$$\Delta_B/\Delta_{\text{well}} \approx 16\sigma^2 \, e^{-4\sigma h} + O(h^2) \tag{140}$$

Thus

$$\lambda_1^{-1} \frac{\Delta_B}{\Delta_{\text{well}}} \approx 8\tau_0(1 + 3h)(\pi\sigma)^{1/2}\, e^{\sigma(1-6h)} \tag{141}$$

Whence

$$\lambda_1^{-1} \frac{\Delta_B}{\Delta_{\text{well}}} \sim e^{\sigma(1-6h)} \tag{142}$$

alters its behavior from exponential increase to exponential decrease at (Garanin, 1996)

$$h_c \approx 1/6 \approx 0.166\,67 \tag{143}$$

which is again considerably smaller than h_s from Eq. (136). This is in agreement with the originally reported critical value of $h_c \approx 0.17$ (Coffey et al., 1995a).

The above results may be used for the evaluation of the linear response of diverse physical systems. In particular, they can be applied with a small modification to the calculation of the linear dielectric response of nematic liquid crystals and systems of polar and polarizable molecules in a strong dc bias field and to the corresponding magnetic response of an assembly of single-domain ferromagnetic particles. In all cases the longitudinal relaxation (dielectric and magnetic, respectively) of these systems is governed by the FPE with the uniaxial potential given by Eq. (135) and appropriate interpretation of the parameters τ_0, ξ, and σ (for details see Martin et al., 1971; Coffey et al., 1994a,b; 1996a, 1997a; Brown, 1963; Garanin et al., 1990; Moro and Nordio, 1985; Szabo, 1980). Finally, we remark that (on account of linear response theory) the bias field effect is of particular significance for the linear alternating current (ac) response of particles in a bistable potential. It appears that even for a very small change in h near the critical value of h_c that there will be a marked change (Coffey et al. 1996b) in the ac response as the relaxation switches from being dominated by the low-frequency barrier crossing mode to the high-frequency modes associated with the intrawell relaxation (e.g., Coffey et al., 1995a). This may be considered as a type of phase transition similar to the nematic/isotropic phase transition in liquid crystals induced by an external uniform electric field.

V. MEAN FIRST PASSAGE TIMES FOR MULTIVARIABLE SYSTEMS: APPLICATION TO ORIENTATIONAL RELAXATION

A. Introduction

In Section III.F we have alluded to the calculation of mean first passage times of multivariable systems and the fundamental difficulty inherent in such problems arising from the question of how one may integrate the adjoint Fokker–Planck equation where more than one variable e.g. position and momentum governs the problem. (The root of the difficulty consists in reducing the multivariable problem to an equivalent one-dimensional problem. Some progress towards this has, however, been made recently by Crothers, 1996.) Thus at present the multivariable MFPT problem is best treated by regarding the MFPT as the inverse of the Kramers escape rate. The multivariable problem has recently assumed a new importance on account of the experiments of Wernsdorfer et al. (1997a, b). One of their goals is to verify the Néel–Brown conception of magnetization reversal by studying the behavior of the magnetization of a single high-anisotropy particle at very low temperatures. This requires an accurate asymptotic expression for the inverse of the smallest nonvanishing eigenvalue (rate constant) as we have emphasised (Coffey et al., 1995a) that the relevant quantity in the study of the magnetization reversal is the mean first passage time rather than the correlation time. A secondary goal is to test the hypothesis of reversal of the magnetization by quantum tunneling through the anisotropy potential barrier originally advanced by Bean and Livingston in 1959 in their famous review of superparamagnetism. This hypothesis, if it can be verified would be a prime example of the problem of macroscopic quantum tunnelling (MQT) (Bean and Livingston, 1959; Dormann et al., 1997; Benderskii et al., 1994; Kim and Hwang, 1997). In order to examine MQT in a superparamagnet, it is essential to critically analyze the conditions (values of the friction) under which the Kramers escape rate theory (Kramers, 1940) (as adapted to the present problem) is applicable as in any non-axially-symmetric problem the value of the friction will determine which of the various formulae of Kramers (1940) is valid. Such a difficulty does not arise for axially symmetric problems which are governed by one variable such as we have previously discussed for reasons we shall explain in detail below; for example, Eq. (97) will be valid for all values of the dimensionless friction coefficient a. In order to clarify the discussion of the magnetic problem in the non-axially-symmetric case, it is necessary to briefly review the Kramers (1940) calculation of the escape rate for various values of the friction. We remark that although in the Kramers calculation (which is set in the context of a mechanical system) the various formulae have their seat in the relative values of the friction and inertia of the system while in the magnetic problem the inertia plays no role, nevertheless there exists a strong analogy between the various cases of the Kramers calculation for a mechanical system and the superparamagnetic calculation. This analogy was highlighted by Klik and Gunther (1990a,b) and moreover exploited by them in order to find a solution for extremely small values of the dimensionless damping constant. The original investigators of the non-axially-symmetric problem (Smith and de Rozario, 1976; Brown, 1979), while fully appreciating the analogy with the Kramers calculation appeared to have confined their calculations to the intermediate to high-damping case only, as is so in the recent review (Geoghegan et al., 1997). The disadvantage of this is that it is not then possible to specify the range of values of the dimensionless damping factor for which a particular Kramers formula will be valid.

The starting point of the Kramers investigation of the problem of the escape of particles with one degree of freedom over potential barriers due to the shuttling action of the Brownian movement arising from the heat bath is essentially the Langevin equation

$$m\ddot{x} + \zeta\dot{x} + \frac{\partial V(x)}{\partial x} = \lambda(t) \tag{144}$$

where $x(t)$ (the displacement of the Brownian particle of mass m) is a random variable and $\lambda(t)$ is a white noise driving force arising from the bath obeying the equation

$$\overline{\lambda(t_1)\lambda(t_2)} = 2kT\zeta\ \delta(t_1 - t_2) \tag{145}$$

$$\overline{\lambda(t)} = 0 \tag{146}$$

where the overbar means the statistical average over an ensemble of particles starting at time t with initial position x and initial velocity $\dot{x}$. $\lambda(t)$ must also obey (Scaife, 1989) Isserliss' theorem for the mean of a number of observations, namely

$$\overline{\lambda_1\lambda_2 \cdots \lambda_{2n}} = \overline{\lambda(t_1)\lambda(t_2)} \cdots \overline{\lambda(t_{2n})} = \sum \prod_{k_i < k_j} \overline{\lambda(t_k)\lambda(t_j)} \tag{147}$$

The sum is over all distinct products of expectation value pairs, each of which is formed by selecting n pairs of subscripts from $2n$ subscripts, also

$$\overline{\lambda_1\lambda_2 \cdots \lambda_{2n+1}} = \overline{\lambda(t_1)\lambda(t_2)} \cdots \overline{\lambda(t_{2n+1})} = 0 \tag{149}$$

The quantity $\zeta\dot{x}$ in Eq. (144) is the 'systematic' force again arising from the bath. One may then show that the transition probability in phase space (x, p)

$$\rho(x, p \mid x(t), p(t), t) \equiv \rho(x, p, t)$$

satisfies a Fokker–Planck equation (Kramers, 1940) in phase space (this particular equation is now known in the literature as the Klein–Kramers equation; Risken, 1989; Chandrasekhar, 1943)

$$\frac{\partial\rho}{\partial t} + m^{-1}p\frac{\partial\rho}{\partial x} - \frac{\partial}{\partial p}\left[\rho\frac{\partial V}{\partial x} + \frac{\zeta}{m}\left(\rho p + mkT\frac{\partial\rho}{\partial p}\right)\right] = 0 \qquad (150)$$

where the momentum $p = m\dot{x}$. Thus the evolution of ρ is in general governed by two variables (x, p) in phase space. This is the origin of the multivariable escape rate problem.

B. The Range of Validity of the Kramers Intermediate to High Damping Formula

The potential $V(x)$ in accordance with the original notation of Kramers has a minimum at A; a maximum at C; the zero of the potential is chosen to be at the barrier top located at $x = 0$ with the initial condition that there were no particles at the outside of the barrier at $t = 0$; Kramers (1940) demanded the probability that a particle originally caught in the potential hole at A will escape to another potential hole at B, crossing the barrier at C. The depth V_1 of the potential well is large compared to the thermal energy kT so that (Mel'nikov and Meshkov, 1986) a Brownian particle trapped in a deep potential well will reside there for an exponentially long time, exceeding all the other relaxation times so as to allow a sensible definition of the decaying state and the introduction of the concept of a lifetime. (See Fig. 4.7.) Equation (150) must be solved subject to the boundary condition $\rho(x, p, t) \to 0$, $x \to \infty$; it has a solution of the form

$$\rho(x, p, t) = F(x, p)\, e^{-t/\tau} \qquad (151)$$

where τ is the longest relaxation time of the system, the relaxation being due to escapes of the particles over the potential barrier. We have

$$\tau^{-1}F - m^{-1}p\frac{\partial F}{\partial x} + \frac{\partial}{\partial p}\left[F\frac{\partial V}{\partial x} + \eta\left(pF + mkT\frac{\partial F}{\partial p}\right)\right] = 0 \qquad (152)$$

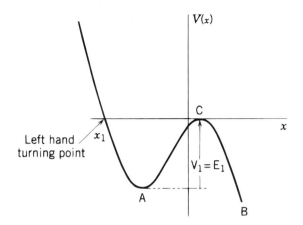

Figure 4.7. Schematic representation of a single well potential as used in the discussion of Section V. x is the reaction coordinate, the values of x at A and B denoted by x_A and x_B denote the reactant and product states respectively. The maximum of $V(x)$ at x_C separating these states corresponds to the transition state or activated complex. All the remaining degrees of freedom of both reaction and solvent molecules constitute a heat bath at a temperature T whose total effect on the reacting particle is represented by the random and systematic damping forces in the Langevin equation. The point C is taken as the point of zero potential energy in this translational Brownian motion problem so that $V(x_C) = 0$, $V(x_A) = V_1$, $E_C = V_1$.

where

$$\eta = \zeta/m \tag{153}$$

and where $F(x, p)$ must approach equilibrium distribution as we move away from the barrier at x_c. For an inverted oscillator potential

$$V(x) = -\tfrac{1}{2}m\omega_c^2 x^2 \tag{154}$$

$$\omega_c = \sqrt{\frac{-V''(x_c)}{m}} \tag{155}$$

Kramers (1940) was able to obtain an *exact* solution of Eq. (152) in integral form, which satisfies the boundary condition that $F(x, p)$ must tend to the *equilibrium* distribution (Maxwell–Boltzmann) far from the barrier *provided that η is large enough to satisfy the following condition on the displacements*

$$\frac{p}{m\omega_c^2}\left[\left(\omega_c^2 + \frac{\eta^2}{4}\right)^{1/2} - \frac{\eta}{2}\right] - x \gg \left(\frac{\eta kT}{m\omega_e^2}\right)^{1/2}\left[\left(\omega_c^2 + \frac{\eta^2}{4}\right)^{1/2} - \frac{\eta}{2}\right]^{1/2} \tag{156}$$

If this condition is not satisfied, then the boundary condition above is violated. On further assuming that the mean contribution (compare the method of steepest descents of III) to the normalized density (Mel'nikov and Meshkov, 1986)

$$\int_{-\infty}^{0} dx \int_{-\infty}^{\infty} \rho(x, p, t) \, dp = N(t) \tag{157}$$

comes from the region near the bottom of the well where the potential can be represented by the harmonic oscillator potential

$$V(x) = -V_1 + \tfrac{1}{2} m\omega_A^2 (x - x_A)^2 \tag{158}$$

where

$$\omega_A = \sqrt{m^{-1} V''(x_A)} \tag{159}$$

is the angular frequency at the bottom of the well, Kramers 1940; Mel'nikov and Meshkov, 1986; Hänggi et al., 1990, Kramers finally obtained (his Eq. 25, note that ω, ω' in Eq. 25 represent frequencies (Hz))

$$\tau^{-1} = \kappa = \frac{\omega_A}{2\pi} \left[\left(1 + \frac{\eta^2}{4\omega_c^2} \right)^{1/2} - \frac{\eta}{2\omega_c} \right] e^{-\beta V_1} \tag{161}$$

which on account of the restriction of Eq. (156) on the value of η is now called the Kramers intermediate to high damping (IHD) formula. The restriction on the value of η of Eq. (156) has been radically simplified by Mel'nikov and Meshkov (1986), who effectively state it is as for large to moderate dampings (part of their reasoning is reproduced below)

$$\eta \geq \omega_c \tag{162}$$

we must have

$$-x \gg \left(\frac{kT}{m\omega_c^2} \right)^{1/2} \tag{163}$$

meaning that $F(x, p)$ deviates from equilibrium in a rather *narrow* region of x where

$$|V(x)| \approx kT \ll V_1 \tag{164}$$

In the underdamped regime $\eta \ll \omega_c$; on the other hand,

$$-x \gg \left(\frac{\omega_c}{\eta}\right)^{1/2}\left(\frac{kT}{m\omega_c^2}\right)^{1/2} \tag{165}$$

and $F(x, p)$ deviates from equilibrium in a much wider region so that as $\eta \to 0$ the inverted oscillator approximation for the potential $V(x)$ breaks down. Arguing that the width of a potential well is of order of magnitude

$$\left(\frac{V_1}{m\omega_c^2}\right)^{1/2} \tag{166}$$

and substituting this value for x into Eq. (165), Eq. (165) becomes

$$\frac{V_1}{kT} \gg \frac{\omega_c}{\eta} \tag{167}$$

so that with

$$\alpha = \frac{\eta}{\omega_c}$$

Eq. (167) becomes

$$\alpha \beta V_1 \gg 1 \tag{168}$$

with, of course, $\beta V_1 \gg 1$ which is the condition of validity of the IHD formula Eq. (161) in the low-damping limit. Equation (161) may be written in the more compact form

$$\tau^{-1} = \kappa = \frac{\omega_A \Omega_c}{2\pi\omega_c} e^{-\beta V_1} \tag{169}$$

with

$$\Omega_c = \omega_c\left[\left(1 + \frac{\alpha^2}{4}\right)^{1/2} - \frac{\alpha}{2}\right] \tag{170}$$

and with a region of validity

$$\frac{\alpha V_1}{kT} \gg 1 \tag{171}$$

We remark that in the very high damping (noninertial) limit, where the inertial parameter

$$\gamma = \frac{kTm}{\zeta^2} \ll 1 \tag{172}$$

we have, on writing

$$\Omega_c = \frac{\omega_c \alpha}{2} \left[\left(1 + \frac{4}{\alpha^2} \right)^{1/2} - 1 \right] \tag{173}$$

and using the binomial theorem

$$\tau^{-1} = \frac{\omega_A}{2\pi\alpha} e^{-\beta V_1} = \frac{\omega_A \omega_c}{2\pi\eta} e^{-\beta V_1} \tag{174}$$

This is the result we would have obtained using the Smoluchowski equation for the distribution function in configuration space, which assumes that equilibrium of the momentum has been attained (Kramers, 1940) so that the origin of Eq. (174) is the *strong damping* of the momentum. In order to complete our discussion of Eq. (161), we write down from that equation the transition-state theory result, namely,

$$\tau^{-1} = \frac{\omega_A}{2\pi} e^{-\beta V_1} \tag{175}$$

with again

$$\omega_c \gg \eta \gg \frac{\omega_c kT}{V_1}$$

and once again

$$\alpha\beta V_1 \gg 1. \tag{176}$$

The problem arising from Eq. (175) for small α values is that all frictional dependence of the prefactor vanishes and that $\alpha\beta V_1 \gg 1$ still must apply.

We remark as well that the transition to small α values such that

$$\alpha\beta V_1 \ll 1$$

is not uniform. Hence we must use a different approach as Eq. (161) will always overestimate (cf. Eq. 175) the true escape rate (and so underestimate the true relaxation time) for small friction. For a detailed discussion the reader is referred to p. 273 of Hänggi et al. (1990) and to the discussion centring on Fig. 5 in the Kramers paper (1940).

C. The Kramers Low Damping Formula

It is apparent that Eqs. (161) and (169) are invalid Mel'nikov and Meshkov (1986), if V_1 being the barrier height

$$\alpha\beta V_1 \ll 1, \qquad \beta V_1 \gg 1$$

that is, for small α, there is significant depopulation of particles below the barrier top. Kramers (1940) considered the motion of the Brownian particle in the extremely underdamped regime as that of an almost conservative system of oscillatory type (so that the trajectories in the phase plane in the domain of the stable equilibrium approximate to ellipses in the absence of stochastic forces) with a very small loss of energy that is a very small diffusion in energy space and by writing the FPE (Eq. (152)) in terms of angle-action variables (Goldstein, 1980, Praestgaard and van Kampen, 1981, Zwanzig, 1959) where the action is a slow variable and the angle is a fast variable, obtained in the limit (Eq. (4.48) of Hänggi et al., 1990)

$$\alpha\beta V_1 \ll 1, \qquad \beta V_1 \gg 1$$

$$\kappa = \frac{\eta\beta\omega_A I_c(E_1)\, e^{-\beta E_1}}{2\pi} \tag{177}$$

The action (the line integral is over the cycle of the motion)

$$I(E) = \oint p\, dq = \oint \sqrt{2m(E - V)}\, dq \tag{178}$$

(p denoting momentum, q position) is the area inside a curve of constant energy, and the action is related to the energy E by

$$\frac{\partial E}{\partial I} = \frac{\omega(I)}{2\pi} \tag{179}$$

where $\omega(I)$ is the angular frequency at the action I. The quantity

$$I_c = I_c(E_1) = \oint_{E = E_c} p_c(x)\, dx$$

is the action corresponding to the (constant) energy trajectory Figs. (4.8, 4.9, 4.10), (the contour integral is taken along this trajectory)

$$E = \frac{p^2}{2m} + V(x) = E_1 = E_c = V(x_c) - V(x_A) = V_1$$

whose equation is

$$p = p_c(x) = [2m\,(E_c - V(x))]^{1/2},$$

through the *maximum* at C (for a detailed review of this calculation see Hänggi et al., 1990). On noting that (Kramers 1940; Hänggi et al., 1990)

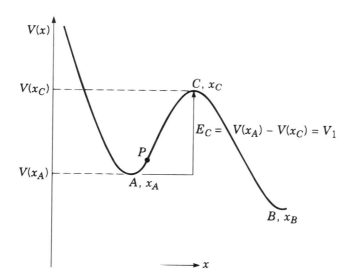

Figure 4.8. Sketch of the potential well in which the particle is confined. Point A is a point of stable equilibrium and Point C is a point of unstable equilibrium. If the noise $\lambda(t)$ in the Langevin equation, Eq. (144), is not zero the trajectories leave the domain of stable equilibrium D in a finite, though possibly large, time. If the thermal energy kT is small a trajectory starting in D, spends a long time $\tau(A)$, fluctuating about x_A, however due to the random noise $\lambda(t)$ the trajectory (phase point) eventually reaches Γ and either leaves or returns to D with equal probabilities (Matkowsky et al., 1984).

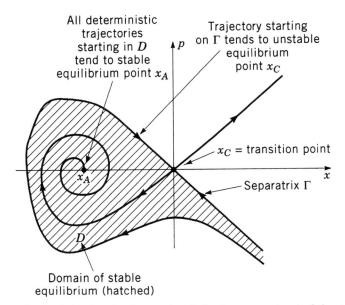

Figure 4.9. Sketch of the domain of attraction D, in phase space (x, p) of the stable equilibrium. The boundary of D is the separatrix Γ which separates D from the remainder of phase space. All deterministic trajectories starting in D tend to the stable equilibrium point x_A, Matkowsky et al. (1984).

$I_c(E_1)$ is of order of magnitude

$$\frac{2\pi E_1}{\omega_A} \tag{180}$$

because the action of a harmonic oscillator of energy E and angular frequency Ω is (Goldstein, 1980)

$$I = \frac{2\pi E}{\Omega}, \tag{181}$$

We have for the system shown in Fig. 4.7, Eq. (28) of Kramers (1940), namely

$$\tau^{-1} = \kappa \approx \eta \beta V_1 \, e^{-\beta V_1}. \tag{182}$$

Here (Landau and Lifshitz, 1976) the phase space is a two dimensional space (plane) with coordinates (p, q) and the phase path or energy trajectory of a system executing a periodic motion is a closed curve in the plane, in this instance the curve is an ellipse with semi-axes $\sqrt{2mE}$, $\sqrt{2E/m\Omega^2}$, the equation of the phase path or trajectory is $H(q, p) = E$ where H is the

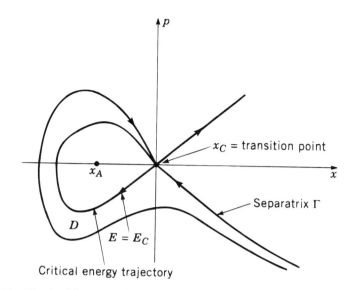

Critical energy trajectory

Figure 4.10. Sketch of the separatrix Γ and the energy curve $E = E_C$ (Matkowsky et al., 1984) in phase space. The mean time to escape the well centred on x_A is (Matkowsky et al., 1984) the sum of the mean time $\tau_1(A)$ to reach the critical energy level E_C from the bottom of the well and the mean time to proceed from E_C to the separatrix Γ and then escape the well. This time is twice the MFPT $\tau_2(E_C)$ from E_C to Γ since trajectories that reach Γ are equally likely to leave or return to $E = E_C$. The inclusion of τ_2 is the crucial factor in the construction of the uniform expansion of the first passage time of Matkowsky et al. (1984) which yields a formula for the escape rate which is valid for all values of the friction η. The Kramers low damping formula, Eq. (183) below is effectively derived by ignoring the contribution of $\tau_2(E_C)$ to the overall MFPT as is also done in the calculation of Section VG below which leads to the magnetic version, Eq. (218), of the low damping formula. The contribution of τ_2 is neglected in this chapter as our purpose here is merely to delineate a region of validity for the various Kramers formulae in the magnetic problem.

Hamiltonian of the system. Equation (181) also follows immediately from Eq. (179) since ω is constant in this case.

We remark that Eq. (182) may also be written since $\omega_c \approx \omega_A$

$$\tau^{-1} = \kappa \approx \alpha \omega_A \beta V_1 \, e^{-\beta V_1} \tag{183}$$

in our notation.

This is the formula (Hänggi et al., 1990) which must be used for small α. In the crossover region $\alpha \beta V_1 \approx 1$ from Eq. (161) to (183), no simple formula for τ^{-1} applies. Mel'nikov and Meshkov (1986) have however applied the Wiener–Hopf method to obtain a formula (their Eq. (6.3)) which is valid for all values of η subject to certain limitations which are discussed in detail in Section VI of Hänggi et al., (1990). In conclusion of this section, we remark (noting that the treatment of Section VB involves essentially order of magnitude estimates only) that the condition of applicability of the IHD formula, Eq. (161) has following the discussion given in p. 299 of the Kramers paper (Kramers, 1940) being put more succinctly by Hänggi et al., (1990) as

$$\eta I(E_1) > kT.$$

Such a condition means (Hänggi et al., 1990) that the IHD formula may be safely applied only if in one round trip of the particles between source and sink the energy dissipated is greater than the thermal energy. If we apply the harmonic oscillator approximation (Mel'nikov, Meshkov, 1986) for $I(E_1)$, the above condition becomes

$$\frac{\eta \beta 2\pi V_1}{\omega_c} > 1$$

or

$$2\pi \alpha \beta V_1 > 1$$

or if (essentially following the discussion of Kramers (1940) in p. 299 we redefine our α as

$$\alpha_\kappa = 2\pi \alpha = \frac{2\pi \eta}{\omega_c} = \frac{\eta}{f_c}$$

where f_c is the frequency of oscillation at the maximum we have

$$\alpha_\kappa \beta V_1 > 1, \qquad \alpha_\kappa \beta V_1 < 1$$

as the conditions under which, either the IHD or LD formula applies, a condition which is in better accord with the original discussion of Kramers.
The low damping formula Eq. (183) then becomes

$$\tau^{-1} = \kappa = \frac{\alpha_\kappa}{2\pi} \omega_A \beta V_1 \, e^{-\beta V_1}. \tag{183a}$$

This equation when combined with the transition state theory result (which is the very low damping limit of the IHD result) Eq. (175) namely

$$\kappa_{TS} = \frac{\omega_A}{2\pi} e^{-\beta V_1}$$

provides a very simple method of establishing the criteria

$$\alpha_\kappa \beta V_1 > 1, \qquad \alpha_\kappa \beta V_1 < 1, \qquad \beta V_1 \gg 1$$

for the use of the IHD or LD formula and also establishes that α_κ is the appropriate damping parameter, because following the discussion of Kramers (1940) p. 229, we expect in the crossover region that the ratio

$$\frac{\kappa}{\kappa_{TS}} = \alpha_\kappa \beta V_1 \approx 1$$

so establishing without any detailed calculation that $\alpha_\kappa \beta V_1 > 1$ pertains to the IHD regime while $\alpha_\kappa \beta V_1 < 1$ pertains to the LD one, and that the parameter α_κ is the appropriate quantity for the characterization of the various damping regimes. The method of identification of the appropriate damping factor by calculation of the ratio of the LD escape rate to the TS result or very low damping limit of the IHD formula, which is essentially due to Kramers (1940) is very useful for the identification of the appropriate damping parameter in other relaxation problems where the theory is used such as the Néel relaxation where the correct damping parameter is not immediately obvious, as the TS result is very easy to calculate.

We remark in passing that the Kramers low damping formula Eq. (183) is in effect the asymptotic solution of the FPE when that equation is written in angle-action variables and an average is taken over the fast angular (phase) variable; the exact solution of the averaged FPE when the energy is a slow variable is found by quadratures in a manner similar to that described for the MFPT problem of Section III.B. For details of the application to the problem at hand see Hänggi et al. (1990). The explicit conversion of the FPE to an equation in angle-action variables is accomplished using the transformation formulae given in Section IV.9 of Risken, 1989. For a definite example, see the transformation leading to Eq. (42) of Praestgaard and van Kampen (1981). The averaging procedure is carried out implicitly in Kramers (1940) by averaging the FPE in the variables (x, p) over a ring of thickness dI. We reiterate, Kramers (1940), that small viscosity or low damping means that the Brownian forces cause only a small variation of the energy during a period of oscillation. (The reader should always recall that Kramers considers the case where the particle performs a motion of oscillatory type (so that the particle oscillates in a well until it reaches the transition point) in the absence of the Brownian forces, (see his Fig. 4 showing the trajectories in phase space.)

A detailed discussion of the transformation of the FPE to angle-action variables is also given by Zwanzig (1959).

We shall now demonstrate the importance of the above results in the context of superparamagnetism starting with the axially symmetric case.

D. Brown's Axially Symmetric Formula for the Escape Rate

Brown's (1963) axially symmetric formula for the escape rate is from Eq. (92), (for a potential with minima at 0 and π and considering escape from the minimum at 0)

$$\kappa = \frac{1}{2\tau_N} \beta V''(0) \frac{\sqrt{\beta V''(\vartheta_m)}}{\sqrt{2\pi}} \sin \vartheta_m \, e^{-\beta[V(\vartheta_m) - V(0)]} \qquad (184)$$

with

$$\frac{1}{2\tau_N} = \frac{\eta k T}{v} \left(\frac{1}{\gamma^2} + \eta^2 M_s^2\right)^{-1}$$

$$= \frac{a\gamma(kT/v)}{(1 + a^2)M_s} = \frac{a}{1 + a^2} \frac{\gamma}{\beta M_s} \qquad (185)$$

where we have written

$$\beta = \frac{v}{kT} \qquad (186)$$

Equation (184), taking due account of the corrections due to the different geometry, superficially appears to bear a marked resemblance to the Kramers very high damping formula, Eq. (174). Such a comparison is, however, highly misleading, as first pointed out by Klik and Gunther (1990a,b), as the physical origins of Eqs. (174) and (184) are entirely different; Eq. (174) arises from the *strong* damping of the momentum, which effectively reduces the mean first passage time problem to a one-dimensional problem. On the other hand, the mean first passage time problem, which has as its asymptotic solution Eq. (184), reduces to an effective one-dimensional problem by *symmetry* so Eq. (184), *unlike* Eq. (174), is valid for *all values* of the damping factor a, being proportional to a for small a and a^{-1} for large a. We may summarize our conclusions as follows. Equation (184) has its origin in a *one-dimensional* Fokker–Planck equation (FPE), which arises from *symmetry*, while Eq. (174) arises from a Smoluchowski equation (SE), which is derived by making a *strong-damping anzatz* on the momentum p in the FPE in the *two variables* (p, x). In order to effect a true comparison between the Kramers 1940 theory of escape rates for translational Brownian motion of a particle with one degree of freedom and the magnetic relaxation problem, it is necessary to consider the non-axially-symmetric magnetic problem; only then will the analogy between the two problems become obvious. The analogy between the problems has the merit of yielding, just as in the translational case, the range of values of the dimensionless friction constant a for which a particular superparamagnetic relaxation time formula is valid, a point that was not emphasized in the earliest investigations (Smith and de Rozario, 1976; Brown, 1979) of the non-axially-symmetric magnetic problem in the 1970s nor in the very recent review (Geoghegan et al., 1997), and that has only been comparatively recently (1990a,b) addressed in detail by Klik and Gunther.

E. Magnetic Version of the Kramers IHD Formula

Equation (5.60) of Geoghegan et al. 1997 and Eq. (3) of Klik and Gunther 1990b are the magnetic version of the Kramers IHD formula, Eq. (161). Both these equations have the form (referring to well 1 of the bistable potential)

$$\kappa = \frac{1}{2\pi} \frac{\omega_A}{\omega_c} \Omega_c \, e^{-\beta(V_0 - V_1)} \tag{187}$$

where the quantities ω_A, ω_c, Ω_c have the same meaning as in Eq. (161) and will be precisely defined below, Geoghegan et al. (1997), V_0 is the saddle point of the potential energy, V_1 the minimum energy. The saddle point energy is taken as V_0 in order to concurr with Geoghegan et al. (1997) so that the barrier height is $V_0 - V_1$. In order to derive Eq. (187), Klik and Gunther (1990a,b) use as variables

$$p = \cos \vartheta \tag{188}$$

and the azimuth ϕ; the Hamiltonian of the system is then the Gibbs free energy

$$vH(p, \phi) = vV(\vartheta, \phi) \tag{189}$$

so that the FPE, Eq. (6), for the density of magnetic moment orientations W on surface of the unit sphere becomes

$$\beta = \frac{v}{kT},$$

$$\frac{\partial W}{\partial t} = \frac{\partial}{\partial p}(1 - p^2)\frac{\partial W}{\partial p} + \frac{1}{1 - p^2}\frac{\partial^2 W}{\partial \phi^2}$$
$$+ \beta \frac{\partial}{\partial p}\left[\left((1 - p^2)H_p + \frac{H_\phi}{a}\right)W\right] + \beta \frac{\partial}{\partial \phi}\left[\left(-\frac{H_p}{a} + \frac{H_\phi}{1 - p^2}\right)W\right] \tag{190}$$

where H_p and H_ϕ are the partial derivatives of the effective Hamiltonian $H(p, \phi)$, whence they deduce, without giving details of their calculations, that in Eq. (187) the well and saddle angular frequencies are

$$\omega_A^2 = \frac{\gamma^2}{M_s^2} H_{pp}^{(A)} H_{\phi\phi}^{(A)} \tag{191}$$

$$\omega_c^2 = -\frac{\gamma^2}{M_s^2} H_{pp}^{(c)} H_{\phi\phi}^{(c)} \tag{192}$$

$$\Omega_c = \tfrac{1}{2}[D - h'(H_{pp}^{(c)} + H_{\phi\phi}^{(c)})] \tag{193}$$

with

$$D^2 = (h')^2[(H_{pp}^{(c)} - H_{\phi\phi}^{(c)})^2 - 4a^{-2}H_{pp}^{(c)} H_{\phi\phi}^{(c)}] \tag{194}$$

$$a = \frac{h'}{g} = \left(\frac{\gamma}{(1 + a^2)M_s}\right)^{-1} \frac{a\gamma}{(1 + a^2)M_s} = \eta\gamma M_s \tag{195}$$

Equation (187) is now of the same form as the Kramers IHD formula, Eq. (169); consequently it is subject to the same limitations as that formula regarding the range of values of a for which it is applicable. Equation (187) is also of similar mathematical form to Eq. (5.60) of Geoghegan et al. (1997), originally derived by Brown in 1979. The details of the calculation, which is very lengthy, are given by Geoghegan et al. (1997). No discussion of the range of values of a in which Eq. (5.60) is valid was given there, however. The equation in question, considering for convenience only one escape path and assuming that the stationary points are coplanar is

$$\kappa = \kappa_1 = \tau^{-1} = \frac{b(-c_1 - c_2 + \sqrt{(c_2 - c_1)^2 - 4a^{-2}c_1c_2})}{4\pi\sqrt{-c_1c_2}}$$
$$\times \sqrt{c_1^{(1)}c_2^{(1)}} \, e^{-\beta(V_0 - V_1)} \tag{196}$$

where c_1 and c_2 are the coefficients in the Taylor series expansion of the potential function, in terms of the direction cosines at the saddle point c where $V = V_0$ and $c_1^{(1)}c_2^{(1)}$ are the coefficients of the truncated Taylor series expansion of the potential function at the bottom of the well A where $V = V_1$ in terms of the direction cosines at the bottom of the well and

$$b = h' \tag{197}$$

a, called α in Geoghegan et al. (1997) has the same meaning as before. In order to identify Eqs. (187) and (196), we must have

$$\omega_A^2 = \frac{\gamma^2}{M_s^2}\, c_1^{(1)} c_2^{(1)} \tag{198}$$

$$\omega_c^2 = \frac{\gamma^2}{M_s^2}\,(-c_1 c_2) \tag{199}$$

$$\Omega_c = \frac{b}{2}\left(-c_1 - c_2 + \sqrt{(c_2 - c_1)^2 - 4a^{-2}c_1 c_2}\right) \tag{200}$$

Equations (187) and (191)–(194) are essentially a special case of the IHD Kramers escape rate for a multidimensional system with *additive* white noise considered in detail by Brinkman (1956), Landauer and Swanson (1961), Langer (1969) and Dygas et al. (1986), reviewed in Hänggi et al. (1990), which is the underlying reason for the similarity with the Kramers IHD formula, Eq. (161), for a mechanical system with canonical variables (q, p). Additive white noise arises in the present context because the *linearized* form of the FPE in the vicinity of the stationary points, Eq. (5.24), of Geoghegan et al. (1997) and various equations in Brown (1979), corresponds to a Langevin equation driven by *additive* white noise rather than the *multiplicative* noise of the Gilbert or Landau–Liftshitz equations before the linearization procedure. The references quoted above appear to have been unknown to Smith and de Rozario (1976) and Brown (1979). They derived Eq. (187) from first principles in the context of magnetic relaxation.

F. Limits of Validity of the Magnetic Version of the IHD Formula

We shall repeat the procedure of Kramers just described for the mechanical problem by considering the various limiting cases of Eq. (187) and establish in like manner a range of validity for that formula. The behavior of Eq. (187) as a function of the dimensionless damping factor a is entirely controlled by Eq. (194) or equivalently Eq. (200). It will be convenient for the purpose of comparison with the work of Geoghegan et al. (1997) to use the notation of Eq. (200). In the very highly damped situation where $\eta \to \infty$ $(a = \eta\gamma M_s)$, we have

$$\Omega_c = -bc_1 = \frac{-a\gamma}{(1 + a^2)M_s}\, c_1 \approx \frac{-\gamma}{aM_s}\, c_1 \tag{201}$$

Equation (187) with this value of Ω_c is the magnetic analogue of the Kramers very high damping formula Eq. (174). We reiterate that Brown's axially symmetric formula, Eq. (184), is *not* the analogue of the Kramers strong damping formula, Eq. (174), as Eq. (184) is valid for all values of the damping a; in order to construct a true analogue, one must use Eq. (196). The second limiting case of Eq. (196) is the transition-state theory result where $a \to 0$; here

$$\kappa = \frac{\gamma}{2\pi M_s} \sqrt{c_1^{(1)} c_2^{(1)}} \, e^{-\beta(V_1 - V_0)} \qquad (202)$$

or

$$\kappa = \frac{\omega_A}{2\pi} e^{-\beta(V_1 - V_0)} \qquad (203)$$

which is independent of a. Equation (202) is the magnetic analogue of Eq. (175) in the mechanical problem. Consequently it has the same limitations as that equation, as it does not yield the frictional dependence of the prefactor. Once again the transition from small a to large a values is not uniform. Thus for small a we have to determine the magnetic analogue of the Kramers low-damping formula, Eq. (183a). Before doing this we state the range of validity of the IHD formula, Eq. (187) and its associated equations. First in Eq. (187)

$$\Omega_c = \frac{1}{2} \left(\frac{a}{1 + a^2} \frac{\gamma}{M_s} \right) \left(-c_1 - c_2 + \sqrt{(c_2 - c_1)^2 - 4a^{-2} c_1 c_2} \right) \qquad (204)$$

while in the Kramers IHD formula, Eq. (169)

$$\Omega_c = \frac{\eta}{2} \left[\left(4 \frac{\omega_c^2}{\eta^2} + 1 \right)^{1/2} - 1 \right] = \frac{\alpha \omega_c}{2} [(4\alpha^{-2} + 1)^{1/2} - 1] \qquad (205)$$

The condition for validity of this formula is, as we have seen,

$$\frac{\eta}{\omega_c} \gg \frac{kT}{V_1}, \qquad \frac{V_1}{kT} \gg 1$$

that is

$$\alpha \beta V_1 \gg 1 \qquad (206)$$

Now Eq. (204) may be rearranged as

$$\Omega_c = \frac{1}{2} \frac{a}{1+a^2} \, \omega_c \left(\frac{-c_1 - c_2}{\sqrt{-c_1 c_2}} + \sqrt{\frac{(c_2 - c_1)^2}{-c_1 c_2} + \frac{4}{a^2}} \right) \tag{207}$$

since

$$\omega_c^2 = \frac{\gamma^2}{M_s^2} (-c_1 c_2) \tag{208}$$

The damped angular frequency, Eq. (207), depends on the dimensionless friction factor a in two ways: one through the factor

$$\frac{a}{1+a^2} \tag{209}$$

which has its origin simply in the bare diffusion relaxation time τ_N and secondly the dependence via the square root in Eq. (207). The latter constitutes a dependence on a arising from the geometry of the problem which represents the coupling between the transverse and longitudinal relaxation modes of the system. We remark that the gyromagnetic term in the Gilbert–Langevin equation

$$\frac{d\mathbf{M}(t)}{dt} = g' M_s (\mathbf{M} \times \mathbf{H}) + h'(\mathbf{M} \times \mathbf{H}) \times \mathbf{M} \tag{210}$$

although it does not of course arise from mechanical inertia nevertheless plays the role of an 'effective' inertia. The inertial effects in the Kramers problem are at the root of the separate formulae Eqs. (169) and (183) which are used to describe the IHD and very low damping cases. If the inertial effects are neglected all that is required to fully describe the escape rate is the very high damping limit of Eq. (169), viz. Eq. (174). Likewise in the magnetic problem one has separate formulae, Eqs. (187) and (218), below depending on the range of a values chosen. Thus the gyromagnetic (g') term in Eq. (210) mimics the effect of inertia which is the fundamental reason for the similarity of the non axially symmetric magnetic problem and the Kramers problem. There exist of course certain detail differences between the escape rate formulae prefactors for the Kramers problem and the magnetic problem because the magnetic system is driven by vector multiplicative noise which by a linearization procedure, Brown (1979), Geoghegan et al. (1997) is transformed into additive white noise rather than the scalar additive noise of the Kramers problem. Furthermore, the magnetic problem is governed by two reaction coordinates (ϑ, ϕ) and so has two degrees of freedom. Nevertheless the underlying FPE, Eq. (6), in the configuration space (ϑ, ϕ) has still only *two* coordinates just as in the inertial Kramers problem in (p, q) space as, by definition, there are no inertial effects in the magnetic problem; we reiterate that *apparent* inertial effects are due to the gyromagnetic term.

As far as the analogy with the IHD Kramers result, Eq. (205), is concerned, only the dependence on a inside the radical in Eq. (207) is of interest. In view of the similarity of Eqs. (205) and (207) we may thus infer (proved in Section G below) that the criterion for the validity of Eq. (207) involves just as Eq. (205), the barrier height and the parameter a only and is as in Eq. (171)

$$a\beta(V_0 - V_1) \gg 1 \tag{211}$$

If the condition embodied in Eq. (211) is not satisfied, Eq. (207) will overestimate the escape rate and so underestimate the Néel relaxation time, indeed Eq. (207) used in this manner will predict Néel relaxation in the absence of dissipation which is impossible.

G. Formula for Very Low Damping in Magnetic Relaxation

We remark that just as in the calculation of Mel'nikov and Meshkov (1986) the transition between high and low damping formulae will be in the range of damping such that

$$a\beta(V_0 - V_1) \approx 1 \tag{212}$$

Thus we need a formula which will be valid for values of a such that

$$a\beta(V_0 - V_1) \ll 1$$

with

$$\beta(V_0 - V_1) \gg 1$$

Such a formula would be the magnetic analogue of the small damping Kramers formula, Eq. (183a) above. A low damping formula for the magnetic problem using the uniform expansion of the mean first passage time proposed by Matkowsky et al. (1984) (which was used by them to determine inter alia the underdamped Kramers escape rate, Eq. (183)) has been derived by Klik and Gunther in 1990 by extending the result of Matkowsky et al., their Eq. (2.15), which pertains of course to translational motion in the phase space (q, p)—the inertial Kramers problem) to rotational Brownian motion governed by Gilbert's equation in the space of the angles (ϑ, ϕ) or equivalently (p, ϕ) where p now means cos ϑ. The uniform expansion of the MFPT approach of Matkowsky et al. should be compared with that of Kramers (1940) who obtained his result by finding the asymptotic solution of the Fokker–Planck equation when that equation is written in angle-action variables and an average is taken over the fast angle variable.

Klik and Gunther (1990), state that the low-dissipation limit is distinguished by the fact that the contour of critical energy E_c lies within the boundary layer near the separatrix Γ between the domain of attraction of the stable equilibrium in the well and the rest of the phase space (p, ϕ). Thus, they introduce a domain Q such that $E < E_c$ in the interior of Q and $E = E_c$ on the boundary $\partial(Q)$. By assumption, $\partial(Q)$ lies so near to Γ that the passage time from $\partial(Q)$ to Γ is negligible (to leading order in kT) in comparison to the mean time $\tau(\phi, p)$ required to reach $\partial(Q)$ starting from a point $(\phi, p) \in Q$. The mean first passage time $\tau(\phi, p)$ is a solution of the adjoint FPE (Section III) given by the operator of Eq. (52) with $z' = p$, $\phi' = \phi$ subject to the boundary condition $\tau = 0$ on the boundary $\partial(Q)$.

Following Matkowsky et al. (1984), they set

$$\tau(\phi, p) = \tau(Q)u(\phi, p) \tag{213}$$

$$\tau(Q) = -\int_Q e^{-\beta H}\, dp\, d\phi \Big/ \int_Q e^{-\beta H} L^\dagger u(\phi, p) dp\, d\phi$$

where $\tau(Q)$ is an *exponentially large* quantity independent of the initial point (ϕ, p) and max $u(\phi, p) = 1$ in Q. Then to leading order in kT, Eq. (52) implies that $du^{(0)}/dt = 1$ along any noiseless trajectory (denoted by (0)) within Q so that by their normalization $u^{(0)} = 1$ within Q. In order to satisfy the boundary condition, introduce now a boundary layer by the stretching transformation, Matkowsky et al. (1984),

$$s = (E_c - E)/kT.$$

Equation (52) then yields to leading order in kT

$$u = 1 - \exp(-s) \tag{214}$$

which satisfies the boundary condition $u = 0$ on $\partial(Q)$ and the matching condition $u = 1$ within Q that is as $s \to \infty$. Using the fact that

$$\int_Q e^{-\beta H} L^\dagger u(\phi, p)\, dp\, d\phi = \beta^{-1} h' \int_Q (\nabla \times \mathbf{A}) \cdot \mathbf{e}_r\, dp\, d\phi$$

they now employ Green's theorem to obtain the constant $\tau(Q)$ in integral form (cf. Eq. (2.15) of Matkowsky et al.)

$$\tau(Q) = \frac{\displaystyle\int_Q dp \, d\phi \, \exp(-\beta H)}{\beta^{-1} h' \displaystyle\oint_{\partial(Q)} \exp(-\beta H)\left[(1 - p^2)\frac{\partial u}{\partial p} \, d\phi - (1 - p^2)^{-1}\frac{\partial u}{\partial \phi} \, dp\right]}$$

The integral in the numerator may be evaluated in the asymptotic limit and the denominator may be expressed in terms of the Hamiltonian H using Eq. (214). Thus, they finally arrive at $\tau(Q)$ in terms of the barrier height etc as

$$\tau(Q) \approx \frac{2\pi}{\beta \omega_A \Delta E} \exp \beta(E_c - E_1)$$

so that the Kramers escape rate is given by (in the very low damping limit)

$$\kappa \approx \frac{\omega_A \beta \, \Delta E}{2\pi} \exp[-\beta(V_0 - V_1)] \tag{215}$$

where ΔE is the energy loss per cycle of the almost periodic motion at the saddle point energy $E = V_0$. Note that Eq. (215) as quoted by them (1990a, b) is missing a factor 2. It has been checked with and agreed to by Dr. Klik (1997), see also Klik et al. (1993). According to Klik and Gunther (1990a,b) to first order in the damping factor a

$$\Delta E = a \oint_{E = V_0}\left[(1 - p^2)H_p \, d\phi - \frac{H_\phi}{1 - p^2} \, dp\right] \tag{216}$$

$$\sim a(V_0 - V_1) = a \times \text{barrier height} \tag{217}$$

for very small a so that

$$\kappa \sim \frac{a\omega_A \beta}{2\pi} (V_0 - V_1) \, e^{-\beta(V_0 - V_1)} \tag{218}$$

which is similar to the Kramers low damping formula given by Eq. (183a). The discussion leading to Eq. (216) should be compared with that of

Section 49 of Landau and Lifshitz (1976). We also remark that although Eq. (218) should not theoretically hold in the crossover region

$$a\beta(V_0 - V_1) \approx 1$$

as discussed by Matkowsky et al. (1984), in the context of the Kramers problem, nevertheless it often provides a reasonable practical approximation to the escape rate in this region (Coffey et al., 1998)

Equation (218) when combined with the transition state theory result, Eq. (203) namely

$$\kappa_{TS} = \frac{\omega_A}{2\pi} e^{-\beta(V_0 - V_1)} \tag{219}$$

yields (Klik and Gunther, 1990a) in the crossover region in accordance with the Kramers method described at the end of Section VC

$$\frac{\kappa}{\kappa_{TS}} = a\beta(V_0 - V_1) \approx 1 \tag{220}$$

so establishing that a is the appropriate damping parameter entirely analogous to α_κ of the Kramers problem. Thus the criteria for applicability of the IHD and LD formulas (Eqs. 196 and 218 respectively) are

$$a\beta(V_0 - V_1) > 1, \qquad a\beta(V_0 - V_1) < 1, \qquad \beta(V_0 - V_1) \gg 1 \tag{220a}$$

We remark that starting from Eq. (215) as given by Jannssen (1988), Eq. (216) may also be justified by writing Gilbert's equation as

$$\dot{\mathbf{M}} = g'M_s(\mathbf{M} \times \mathbf{H}) + h'(\mathbf{M} \times \mathbf{H}) \times \mathbf{M} \tag{221}$$

with

$$\mathbf{u} = \frac{\mathbf{M}}{M_s} \tag{222}$$

we then have

$$\dot{\mathbf{P}} = \frac{\dot{\mathbf{u}}}{g'} = \mathbf{u} \times (-) \frac{\partial V}{\partial \mathbf{u}} + a\left(\mathbf{u} \times (-) \frac{\partial V}{\partial \mathbf{u}}\right) \times \mathbf{u} \tag{223}$$

which is in the form for a particle of unit volume

Time rate of change of angular momentum = impressed torque (224)

since the gyromagnetic ratio

$$\gamma = \frac{\text{magnetic moment}}{\text{angular momentum}}.$$ (225)

Thus the dimensionless friction a plays the role of the friction η in the inertial Langevin equation

$$\dot{p} + \eta p + V'(\theta) = \lambda(t)$$ (226)

$$p = I\dot{\theta}$$ (227)

for a fixed axis rotator specified by the angular coordinate θ to which all the Kramers formulae of Section V.B, V.C will apply. We reiterate that the noise in Eq. (226) is an additive noise while in the Gilbert equation the noise term is a multiplicative noise which by a linearization procedure, Brown (1979), at the stationary points is reduced to an equation with additive noise. Furthermore the system described by Eq. (226) has only one degree of freedom while the Gilbert equation describes a system with two degrees of freedom. The overall effect of all this however, is merely to produce small differences of detail in the various Kramers formulae. In order to evaluate the energy loss per cycle for the lightly damped motion, namely ΔE of Eq. (215) we evaluate the energy per cycle of the undamped motion namely

$$E = \oint \dot{\mathbf{P}} \cdot d\mathbf{u}$$ (228)

where $\dot{\mathbf{P}}$ is evaluated from the undamped Gilbert Equation namely

$$\dot{\mathbf{P}} = \frac{\mathbf{u}M_s}{\gamma} = \mathbf{u} \times \left(\frac{-\partial V}{\partial \mathbf{u}}\right)$$ (229)

Now in spherical polar coordinates with $\mathbf{e}_9$ a unit vector in the direction of 9 increasing; $\mathbf{e}_\phi$ a unit vector in the direction of ϕ increasing; the rate of change of **u** is

$$\dot{\mathbf{u}} = \mathbf{e}_\vartheta \dot{\vartheta} + \mathbf{e}_\phi \dot{\phi} \sin \vartheta \tag{230}$$

where the line element is

$$d\mathbf{u} = \mathbf{e}_\vartheta \, d\vartheta + \mathbf{e}_\phi \sin \vartheta \, d\phi \tag{231}$$

Thus the undamped Gilbert equation (229) becomes in terms of its components

$$\frac{M_s}{\gamma} \dot{\vartheta} \equiv \frac{1}{\sin \vartheta} \frac{\partial V}{\partial \phi} = \dot{P}_\vartheta \tag{232}$$

$$\frac{M_s}{\gamma} \dot{\phi} \sin \vartheta = \frac{-\partial V}{\partial \vartheta} = \dot{P}_\phi \tag{233}$$

since

$$\mathbf{u} \times \left(\frac{-\partial V}{\partial \mathbf{u}} \right) = \mathbf{e}_\vartheta \frac{1}{\sin \vartheta} \frac{\partial V}{\partial \phi} - \mathbf{e}_\phi \frac{\partial V}{\partial \vartheta} \tag{234}$$

so that with Eq. (228), we have for the energy of the undamped motion

$$\begin{aligned} E &= \oint \dot{\mathbf{P}} \cdot d\mathbf{u} = \oint [\dot{P}_\vartheta \, d\vartheta + \dot{P}_\phi \sin \vartheta \, d\phi] \\ &= \oint \left[\frac{1}{\sin \vartheta} \frac{\partial V}{\partial \phi} \, d\vartheta - \frac{\partial V}{\partial \vartheta} \sin \vartheta \, d\phi \right]. \end{aligned} \tag{235}$$

The dissipative (that is the alignment) term in Gilbert's equation is directly proportional to the dimensionless damping factor a for small a so that in the lightly damped case in the first order of perturbation theory in a, we have, for the energy loss per cycle for a particle of unit volume.

$$\Delta E = a \oint \left[\frac{1}{\sin \vartheta} \frac{\partial V}{\partial \phi} \, d\vartheta - \frac{\partial V}{\partial \vartheta} \sin \vartheta \, d\phi \right] \tag{236}$$

or with $p = \cos \vartheta$ and substituting H for V

$$\Delta E = a \oint \left[(1 - p^2)H_p \, d\phi - \frac{H_\phi}{1 - p^2} \, dp \right] \qquad (237)$$

which is Eq. (216). Equation (217) may now be justified by supposing that the action I is approximately that of a harmonic oscillator so that

$$I \approx \frac{V_0 - V_1}{f}$$

whence with

$$E = If$$

$$\Delta E \approx a(V_0 - V_1)$$

so justifying Eq. (217).

Numerical calculations, Coffey et al. (1997b, 1998 in press), of the smallest non-vanishing eigenvalue λ_1 of the FPE for the particular non axially symmetric problem of a uniform field applied at an oblique angle to the easy axis suggest that Eqs. (215) (the LD limit) and (196) (the IHD limit) suitably rewritten to take into account the two escape paths from the bistable potential render a good approximation of λ_1 in their respective ranges of applicability given by Eqs. (220a). Moreover, both of them provide a reasonable approximation even in the cross-over region given by Eq. (212) as was conjectured by Kramers (1940) in the context of the mechanical problem of translational Brownian motion with inertial effects considered by him (see the discussion in p. 299 of his paper). Thus, the criteria for the applicability of the various formulae given in this section are justified by the results of the numerical solution for λ_1 in this case.

A more detailed analysis uniformly valid for all values of a can be given by adapting the methods of Matkowsky et al. (1984) and Mel'nikov and Meshkov (1986) for the translational problem to the magnetic problem. Such an analysis is not however essential in the present Section, the purpose of which is merely to establish a range of validity for the various Kramers formulae in the magnetic problem and to set that problem in its proper place in the statistical mechanical literature, (Langer 1969, Hängii et al. 1990) on multidimensional relaxing systems.

H.   Analogy with Dielectric Relaxation in Non-axially-Symmetric Potentials

Part of the analysis we have given may be applied to dielectric relaxation in non-axially-symmetric potentials. Due acknowledgment, however, is made of the fact that in the electric case the origin of the parameter a or α_κ is the actual moment of inertia of the molecules. Thus if inertial effects are included, a new analysis staring from first principles, that is, the appropriate Fokker–Planck equation in configuration-angular velocity space (Klein–Kramers equation) must be made. If, however, strong damping of the angular momentum is postulated, so that the evolution of the configuration space distribution is described by a rotational Smoluchowski equation, then that equation will be similar in form to Brown's Fokker–Planck equation, omitting of course the gyromagnetic term in a^{-1}. This means that the strong damping limit, Eq. (201), of the magnetic IHD formula Eq. (187) also applies to the dielectric relaxation, excluding the inertial effects in non-axially-symmetric potentials when the appropriate changes in notation have been made.

I.   Conclusion

This section brings to an end our long discussion of the problems that arise when one wishes to calculate reaction rates and (conversely) mean first passage times for orientational relaxation processes governed by more than one variable. It is apparent from the foregoing sections (see also Geoghegan et al., 1997) that all the treatments rely on reducing the problem to an effective one-dimensional problem (which allows the underlying partial differential equation to be integrated). A direct approach based on the adjoint Fokker–Planck operator in the manner outlined in Section III (ie an attempt to determine the exact solution as in the ID problem) manifestly fails in this instance due to the problem of integrating the adjoint Fokker–Planck equation in two or more variables. Nevertheless it appears that the asymptotic methods reviewed in Section V when combined with the Wiener–Hopf method of Mel'nikov and Meshkov (1986) or the uniform expansion of the first passage time of Matkowsky et al. (1984) are sufficiently powerful to treat almost all problems of practical interest in dielectric and magnetic overbarrier relaxation. We remark that for many problems the evaluation of the various constants c_1, c_2, etc. in the curvilinear Kramers formulae is a difficult and tedious process (Geoghegan et al., 1997; Coffey et al., 1997b; Kennedy, 1997; Coffey et al., 1998). The methods we have outlined here refer, of course, to the longest relaxation time of the system and so are of interest in problems that involve *overbarrier* relaxation. If *alternating current* measurements are made, on the other hand, the investigation of a particular relaxation problem will also involve the correlation time as outlined in Section III. Thus the particular experimental situation will determine which of the characteristic times the mean first passage time or the correlation time is relevant.

ACKNOWLEDGMENTS

I thank Professor Yu. P Kalmykov, Professor J. L. Dormann, Professor D. S. F. Crothers, Dr. I. Klik, Dr. E. Kennedy, Mr. D. McCarthy Dr. W. Wernsdorfer; and Dr. John Waldron for helpful conversations. I also acknowledge the financial support of the Forbairt Research Collaboration Fund and the Provost's Development Fund for this work. I thank Mrs. A. Raben for her careful preparation of the manuscript.

REFERENCES

Abramowitz, M., Stegun, I., *Handbood of Mathematical Functions*, eds., Dover, New York, 1964.

Aharoni, A., *Phys. Rev.* **135A**, 447 (1964).

Aharoni, A., *Phys. Rev.* **177**, 793 (1969).

Aharoni, A., *An Introduction to the Theory of Ferromagnetism*, Oxford University Press, London, 1996.

Bean, C. P., Livingston, J. D., *J. Appl. Phys.* **30**, 120S, (1959).

Benderskii, V. A., Makarov, D. E., Wight, C. A., *Adv. Chem. Phys.* **88**, (1994).

Brown, W. F., Jr. *Phys. Rev.* **130**, 1677 (1963).

Brown, W. F., Jr. *IEEE. Trans. Mag.* **15**, 1197 (1979).

Brinkman, H. C., *Physica* **22**, 29, 149 (1956).

Chandrasekhar, S., *Rev. Mod. Phys.* **15**, 1, (1943a).

Coffey, W. T., *Proc. 2nd International Conference on Fine Particle Magnetism*, Bangor, Wales, 1996, *J. Molec. Struct.* **416**, 221 (1997).

Coffey, W. T., Crothers, D. S. F., *Phys. Rev. E* **54**, 4768 (1996).

Coffey, W. T., Crothers, D. S. F., Dormann, J. L., Geoghegan, L. J., Kalmykov, Yu. P., Waldron, J. T., Wickstead, A. W., *Phys. Rev. B* **52**, 15951 (1995b).

Coffey, W. T., Crothers, D. S. F., Dormann, J. L., Geoghegan, L. J., Kennedy, E. C., *J. Mag. Mag. Mat.* **173**, L219 (1997b).

Coffey, W. T., Crothers, D. S. F., Kalmykov, Yu. P., *Phys. Rev. E* **55**, 4812 (1997a).

Coffey, W. T., Crothers, D. S. F., Kalmykov, Yu. P., Massawe, E. S., Waldron, J. T., *Phys. Rev. E* **49**, 1869 (1994b).

Coffey, W. T., Crothers, D. S. F., Kalmykov, Yu. P., Waldron, J. T., *Physica A* **213**, 551 (1994a).

Coffey, W. T., Crothers, D. S. F., Kalmykov, Yu. P., Waldron, J. T., *Phys. Rev. B* **51**, 15947 (1995a).

Coffey, W. T., Déjardin, J. L., Kalmykov, Yu. P., Titov, S. V., *Phys. Rev. E* **54**, 6462 (1996b).

Coffey, W. T., Kalmykov, Yu. P., Waldron, J. T., *The Langevin Equation* World Scientific, Singapore, 1996a.

Coffey, W. T., Crothers, D. S. F., Dormann, J. L., Geoghegan, L. J., Kennedy, E. C., *Phys. Rev. B.* (1998), to be published.

Crothers, D. S. F., unpublished work, (1996).

Dygas, M. M., Matkowsky, B. J., Schuss, Z., SIAM *J. Appl. Math* **46**, (1986).

Dormann, J. L., Fiorani, D., Tronc, E., *Adv. Chem. Phys.* **98**, 283 (1997).

Garanin, D. A., *Phys. Rev. E* **54**, 3250 (1996).

Garanin, D. A., Ischenko, V. V., Panina, L. V., *Teor. Mat. Fiz.* **82**, 242 (1990).[*Theor. Math. Phys.* **82**, 169 (1990)].

Gardiner, C. W., *Handbook of Stochastic Methods*, 2nd Edition, Springer Verlag, Berlin, 1985.

Geoghegan, L. J., Coffey, W. T., Mulligan, B., *Adv. Chem. Phys.* **100**, 475, (1997).

Goldstein, S., *Classical Mechanics*, 2nd Edition, Addison Wesley, New York, 1980.

Hänggi, P., Talkner, P., Borovec, M., *Rev. Mod. Phys.* **62**, 251, (1990).

Janssen L., *Physica A* **152**, 145 (1988).

Kalmykov, Yu. P., Coffey, W. T., Waldron, J., *J. Chem. Phys.* **105**, 2112 (1996).

Kalmykov, Yu. P., Déjardin, J. L., Coffey, W. T., *Phys. Rev. E* **55**, 2509 (1997).

Kennedy, E. C., Ph.D. Thesis, The Queen's University of Belfast, 1997.

Kim, G. H., Hwang, D. S., *Phys. Rev. B* **55**, 5918, (1997).

Klein, G., *Proc. R. Soc. Lond A* **211**, 431 (1952).

Klik, I., Gunther, L., (a) *J. Stat. Phys.* **60**, 473 (1990); (b) *J. Appl. Phys.* **67**, 4505, (1990).

Klik, I., personal communication (1997).

Klik, I., Chang, C. R., Huang, H. L., *Phys. Rev. B* **48**, 15823, (1993).

Kramers, H. A., *Physica* **7**, 284 (1940).

Landau, L. D., Lifshitz, E. M., A *Course of Theoretical Physics*, Vol. 1, *Mechanics*, 3rd Edition, Pergamon, London, 1976.

Landauer, R., Swanson, J. A., *Phys. Rev. B.* **121**, 1668 (1996).

Langer, J. S., *Ann. Phys.* **54**, 258 (1969).

Martin, A. J., Meier, G., Saupe, A., *Symp. Faraday Soc.* **5**, 119 (1971).

Matkowsky, B. J., Schuss, Z., Tier, C., *J. Stat Phys.* **35**, 443 (1984).

Mel'nikov, V. I., Meshkov, S. V., *J. Chem. Phys.* **85**, 1018, (1986).

Moro, G., Nordio, P. L., *Mol. Phys.* **56**, 255 (1985).

Perico, A., Pratolongo, R., Freed, K. F., Pastor, R. W., Szabo, A., *J. Chem. Phys.* **98**, 564 (1993).

Pfeiffer, H., *Phys. Status Solidi* **122**, 377 (1990).

Praestgaard, E., van Kampen, N. G., *Molec. Phys.* **43**, 33 (1981).

Risken, H., *The Fokker–Planck Equation*, Springer, Berlin, 2nd Edition, 1989.

Scaife, B. K. P., *Principles of Dielectrics*, Oxford University Press, London, 1989.

Smith, D. A., de Rozario, F. A., *J. Mag. Mag. Mat.* **3**, 219, (1976).

Szabo, A., *J. Chem. Phys.* **72**, 4620 (1980).

Wernsdorfer, W., Bonet Orozco, E., Hasselbach, K., Benoit, A., Barbara, B., Demoncy, N., Loiseau, A., Boivin, D., Pascard, H., Mailly, D., *Phys. Rev. Lett.* **78**, 1791, (1997a).

Wernsdorfer, W., Bonet Orozco, E., Hasselbach, K., Benoit, A., Mailly, D., Kubo, O., Barbara, B., *Phys. Rev. Lett.* **79**, 4014 (1997b).

Zwanzig, R. W., *Phys. Fluids* **2**, 12 (1959).

LATTICE CLUSTER THEORY OF MULTICOMPONENT POLYMER SYSTEMS: CHAIN SEMIFLEXIBILITY AND SPECIFIC INTERACTIONS

K. W. FOREMAN* AND KARL F. FREED

James Franck Institute and Department of Chemistry, The University of Chicago, Chicago, IL

CONTENTS

* Present address: Department of Pharmaceutical Chemistry, University of California at San Francisco, San Francisco, CA 94118

Advances in Chemical Physics, Volume 103, Edited by I. Prigogine and Stuart A. Rice.
ISBN 0-471-24752-9   © 1998 John Wiley & Sons, Inc.

I. INTRODUCTION

The goal of developing predictive theories for the thermodynamic properties of multicomponent polymer systems requires theoretical methods that
readily translate microscopic scale information into predictions for macroscopic properties. At one extreme, simple, easily applied, generic
(macroscopic) models, such as Flory–Huggins theory, provide general qualitative, semiempirical descriptions for a multitude of different phenomena.
The other extreme involves realistic, but computationally intensive, fully
atomistic simulations, which are consequently limited in scope to very particular systems. These simulations provide insights into microscopic phenomena that are indispensable for developing the intermediate class of
theories. This intermediate class of theories retains a large degree of realism
while enjoying greater analytic tractability, computational simplicity, or
both, as well as maintaining the ability to provide microscopic explanations
for qualitative trends in broad ranges of systems. The integral equation
methods (Schweizer and Curro, 1997) and the lattice cluster theory (LCT);
Freed and Dudowicz, 1995) fall into this intermediate class of theories, but
incur different tradeoffs between analytical tractability, computational simplicity, ease of usage, and range of applicability. Understandings gleaned
from one particular approach are available for testing or enhancing the
powers of other approaches.

The LCT builds on the long-recognized mathematical simplification to
the description of polymer systems conferred by lattice models that retain
sufficient realism for predicting new phenomena and explaining existing
data. The LCT uses an extended lattice model (Nemirovsky et al., 1987;
Dudowicz and Freed, 1991a) in which monomers are given explicit structures that occupy several lattice sites. This essential feature of the LCT
enables studying the influence of monomer size and shape disparities on the
thermodynamic properties of polymer melts and blends. Since simplified
treatments of the lattice model, such as those in Flory–Huggins and Guggenheim theories, cannot distinguish between different extended monomer
structures, the LCT has systematically developed more accurate solutions
to both the standard and extended lattice models. These LCT solutions
describe thermodynamic properties as strongly dependent on the different
monomer structures. Tests of the LCT against available Monte Carlo simulations for the identical lattice model display excellent agreement in certain
limits, but further improvements to the LCT solutions are required in
others (Madden et al., 1990; Dudowicz et al., 1990; Foreman et al., 1997).

The LCT already provides microscopic explanations for a number of
previously enigmatic observations, including the existence of an entropic
contribution to the effective Flory interaction parameter χ (Freed and Pesci,

1987, 1989; Freed and Dudowicz, 1992), the composition, temperature, and molecular weight dependence of χ (Dudowicz and Freed, 1991b), a monomer sequence dependence to χ for random copolymer systems (Dudowicz and Freed, 1996a), the extremely weak dependence of the critical temperature on molecular weight for blends of poly(isobutylene) with other polyolefins (Foreman and Freed, 1997b), and even the different miscibilities of poly(propylene) and head-to-head poly(propylene) in their blends with poly(ethylene propylene) (Freed et al., in press). The LCT explains the partial failure of a rule of thumb relating the critical temperature of polyolefin blends to the chemical composition of its components and provides a generalization based on simple structural considerations (Foreman and Freed, 1997b; Dudowicz and Freed, 1996b). In addition, the LCT has been used to predict interesting new phenomena, including the pressure dependence of χ (Dudowicz and Freed, 1995) and the possibility for observing a block copolymer order–disorder transition upon heating (Dudowicz and Freed, 1993). These features combine with the computational ease of implementing the LCT to motivate improving the theory further.

Several improvements are desirable because they increase the general realism and, hence, the applicability of the theory, the general accuracy of the theory, or both. Given the simplifications posed by any lattice model description, the improvements are all designed to enable representing general physical trends. For example, the computation of higher-order contributions, perhaps accompanied by series summation methods, would enhance the accuracy of the solutions to the extended lattice model, especially for treating highly structured monomers whose semiflexibility can drive liquid crystalline ordering. Another improvement is one focus of the present chapter, namely, endowing the lattice model with even greater realism by allowing the polymers to be semiflexible.

While polymer chains with structured monomers have a measure of stiffness arising from the application of excluded volume constraints to structured monomer lattice models, real polymer chains have steric interactions that impart additional chain stiffness whose description requires further extension of the lattice model. Thus we follow a preliminary treatment by Bawendi and Freed (1987) of describing chain stiffness by introducing energy differences between the lowest-energy *trans* and several higher-energy *gauche* conformations of bonds on the lattice. The earlier treatment only considers the lowest-order contribution to the athermal limit entropy, a contribution that is inadequate to distinguish between different monomer structures. Therefore, the current chapter develops the theory necessary for describing the influence of the *trans–gauche* energy differences along with the monomer structures, the local packing (i.e., nonrandom mixing), and the interaction induced correlations incorporated in previous developments of the LCT.

This semiflexible chain LCT has already been applied to consider several generic aspects of polymer blend thermodynamics and miscibilities (Foreman and Freed, 1997; 1997b; Foreman et al., 1997). These applications consider:

1. The influence of stiffness disparities on miscibilities;
2. Tests of simple models for polyolefin miscibilities that are based on stiffness disparities alone;
3. The reproduction of PRISM phase diagrams for athermal limit semi-flexible chain blends;
4. Tests of a thermodynamically equivalent semiflexible chain model;
5. The analysis of "anomalous" lower critical solution temperature phase diagrams observed in some polyolefin blends;
6. The explanation of the general pressure dependence for the phase behavior of polymer blends;
7. The description of the influence of semiflexibility on nearest-neighbor pair distribution functions; . . . .

Specific applications to polyolefin blends demonstrate a remarkable correlation between the ratio of the critical temperature to the molecular weight and a structural parameter that depends only on the fractions of tri- and tetrafunctional carbon atoms in the chains and on the *trans–gauche* bending energies, provided all CH_n united atom groups are taken as interacting with identical van der Waals energies. Computations with differing van der Waals energies explain the occurrence of lower critical phase diagrams for blends of poly(isobutylene) with other polyolefins, as well as the enigmatic lack of a molecular weight dependence to these critical temperatures. The present chapter provides the theoretical details for these improvements to the LCT that have already proven valuable in understanding the properties of polymer blends and various other theories of blend miscibilities.

In addition, the present chapter also extends the LCT to describe group-specific interactions. Previous applications of the LCT employ averaged van der Waals interactions in which each fundamental unit (called monads) from species s occupies a single lattice site and interacts equally with every monad from species s'; that is, each interaction yields the same energy $\varepsilon_{ss'}$. The phase behavior of polymer blends, however, is extremely sensitive to small changes in individual energies $\varepsilon_{ss'}$ (even in Flory–Huggins theory), as well as to changes in monomer structure (Foreman and Freed, 1997b; Dudowicz and Freed, 1996b). For example, typical off-lattice Lennard–Jones energy parameters for $(CH_3)–(CH_3)$ and $(CH_2)–(CH_2)$ united atom group interactions are 90 and 70 K, respectively (Mondello and Grest,

1995; Jorgensen, 1981; Toxvaerd, 1990). These energies imply rather large variations in monomer averaged interaction energies even among polyolefin blends, variations that invariably affect their phase behavior. Moreover, Yoon and co-workers (Smith et al., 1996) provide a unique analysis, based on quantum-chemical calculations and atomistic molecular-dynamics simulations, to demonstrate how specific interactions affect intramolecular conformations in poly(ethylene oxide) melts and dilute solutions. The present chapter provides a general representation of the LCT in which all monads in the system interact with distinct energies, a model of group-specific united atom interactions. The monomer averaged (unspecific) interactions and group-specific monad interaction approaches both emerge as special cases from the general representation derived here.

The semiflexible chain LCT is developed here by briefly reviewing some details from the previous LCT treatments in order to provide required notation. Since the early version of the LCT employs a field-theoretic formulation that is far less accessible, the full treatment here follows the significantly simpler algebraic derivation of the LCT (Dudowicz and Freed, 1991a). The brief review is then followed by describing the nontrivial modifications necessary to include semiflexibility and specific interactions. Section II begins by defining the basic variables in the LCT lattice model description of polymer systems and then continues by describing the partition function for a multicomponent mixture of interacting, semiflexible linear chains. (The full theory is likewise applicable to small molecule mixtures as well as to mixtures of polymers and solvent.) The derivation of the linear chain partition function is relatively straightforward and serves to illustrate the fundamental concepts necessary for including chain semiflexibility. Section II then concludes with the nontrivial extension to describe semiflexible polymer chains with structured monomers and specific interactions. Because the general partition function derived in Section II cannot be evaluated exactly, Section III begins by extracting a zeroth order mean-field component from the partition function in such a way that a cluster expansion emerges, as described in Sections III.B and III.C. Section III concludes by summarizing the diagrammatic rules for representing each term in the cluster expansion, including the new diagrams arising from chain semiflexibility. Earlier work (Dudowicz and Freed, 1991a) illustrates the treatment of all diagrams in the flexible chain limit; so the focus here is on the generalization required to treat semiflexibility and specific interactions. Since evaluating each diagram is a complicated process, Section IV details the method for decomposing a diagram into a series of easily calculated "contracted" diagrams along with the process of computing combinatorial prefactors, which becomes rather involved for chains of semiflexible structured monomers with specific interactions. Thus Section

IV, along with Appendices A and B, describe the new diagrams and new counting indices that are necessary for treating semiflexible chains. Detailed expressions for the Helmholtz free energy are derived in Section V and are provided in Appendix C for both the monomer averaged and group-specific interaction models.

II. THEORY

Although the theory may be developed using many different Bravais lattices, a d-dimensional cubic lattice model is employed because the cubic lattice introduces significant mathematical simplifications without loss of any essential physics (Nemirovsky et al., 1987). The lattice contains N_l sites, each with $z = 2d$ nearest neighbors. In general, calculations are performed with $d = 3$ dimensions, but the formal definition of z is required to develop the $1/z$ expansions (see Sections IV and V for more details).

Lattice chains represent the actual (atomistic) polymer chain architectures as composed of simplified fundamental units (called "monads"), which may comprise several atoms within a monomer or even whole monomers. Each monad resides on one lattice site whose volume v_c corresponds roughly to the volume occupied by the atoms represented by the monad. In principle, separate parts of the same chain may have monads with unequal volumes, but, if the lattice is regular, the model endows different monads with equal volumes. The monads in the traditional lattice model representations of polymers correspond to one or more monomers, but this choice generally yields monad volumes that vary considerably with species. Approximate combining rules are then employed when the monad volumes differ between species.

Previous applications of the LCT indicate that differences in monomer structures strongly influence the blend thermodynamics (Foreman and Freed, 1997b; Dudowicz and Freed, 1996b; Freed and Dudowicz, 1996; Foreman et al., 1997; Freed et al., to be published). This strong influence indicates that the lattice chain description should distinguish between different groups with monomers and between the monomer structures of various species. Therefore, the monad is chosen to represent the simplest combinations of atoms such that all monad volumes are kept as similar as possible to each other. For example, the CH_n ($n = 0$–3) groups of polyolefin chains are of similar sizes and are represented (Dudowicz and Freed, 1996b; Freed and Dubowicz, 1996) as united atom groups that become the monads on the lattice. Hence the united atom model of polyolefins requires each monomer to occupy several lattice sites.

We consider a multicomponent system in which species s has C_s chains with M_s monads per chain, producing a volume fraction $\phi_s = C_s M_s / N_l$.

Solvent molecules may be included as one or more components in the system by restricting $M_{solvent}$ to small positive integers for these species (Madden et al., 1990; Dudowicz et al., 1990). The treatment of chain connectivity constraints for models of polymers with structured monomers follows as a simple generalization of the linear chain case. Thus Subsections A and B use the example of linear chains to establish the notation and concepts necessary for treating both the theory and the general case of structured monomers in Subsection C.

A. Chain Connectivity Constraints

The monads in a linear chain may be numbered sequentially from end to end. The vector $\mathbf{r}_\alpha^{s,\,c}$ designates the position of the αth monad in linear chain c of species s on the lattice. Consecutive monads along the linear chain are constrained to lie on nearest-neighbor lattice sites, with the bond between monads α and $\alpha + 1$ equal to one of the lattice vectors $\mathbf{a}_\beta$, $\beta = 1, \ldots, z$. (All lattice vectors have the same length $|\mathbf{a}_\beta| = \sqrt[3]{v_c}$ for a cubic lattice.) Thus the connectivity between united atom groups α and $\alpha + 1$ within the chain translates into the condition,

$$\mathbf{r}_\alpha^{s,\,c} - \mathbf{r}_{\alpha+1}^{s,\,c} = \mathbf{a}_\beta \tag{1}$$

for some lattice direction $\beta = 1, \ldots, z$. The bonding constraint in Eq. (1) may be recast as a Kronecker delta,

$$\sum_{\beta=1}^{z} \delta(\mathbf{r}_\alpha^{s,\,c}, \mathbf{r}_{\alpha+1}^{s,\,c} + \mathbf{a}_\beta) = \begin{cases} 1, & \alpha,\, \alpha+1 \text{ nearest neighbors} \\ 0, & \text{otherwise} \end{cases} \tag{2}$$

where $\delta(x, y)$ is unity when $x = y$ and is zero otherwise. A one or a zero on the right-hand side of Eq. (2) indicates the presence or absence of a bond, respectively. Equation (2) is a mathematical representation of the constraint indicated by Eq. (1). The form of Eq. (2) proves more useful in developing an analytical representation for the partition function.

While many models of polymer thermodynamics introduce the mathematically convenient assumption of incompressibility, including a description of the system's compressibility is *necessary* to investigate the experimentally relevant pressure dependence of thermodynamic properties. Compressibility qualitatively alters the calculated thermodynamic properties of all polymer systems, and many workers now realize that only a compressible theory can provide a meaningful *microscopic* analysis (Schweitzer and Curro, 1997; Dudowicz and Freed, 1995, 1996b; Freed and Dudowicz, 1996). Compressible blends contain excess free volume, described here by including a volume fraction ϕ_v of unoccupied sites, called "voids." In computations for constant-pressure systems, the fraction ϕ_v is determined from the equation of state as a function of pressure, temperature, blend composition, and unit cell volume. The voids occupy single lattice sites, $M_{\text{void}} = 1$, but are *not* true thermodynamic components since they are strictly noninteracting.

B. The Partition Function for Linear Chains

The configurational partition function W is evaluated by counting all possible placements of all monads on the lattice, subject to the nontrivial constraints of chain connectivity and excluded volume. The connectivity constraints of Eqs. (1) and (2) are nearly sufficient for succinctly describing athermal limit multicomponent systems. Section II.B.1 presents the theory for the partition function of athermal limit multicomponent linear chain systems in order to develop some fundamental concepts necessary to formulate the theory of the full partition function for interacting structured monomer systems. Sections II.B.2 and II.B.3 detail the modifications of the athermal limit partition function that are required to describe, respectively, chain semiflexibility and attractive van der Waals interactions.

1. Partition Function for Multicomponent Blends with Only Hard-Core Interactions

If the placement of monads on the lattice were constrained only by the bonding restrictions of Eqs. (1) and (2), the model polymers would appear on the lattice as a set of independent random walk chains. Thus the constraint that a single linear chain c from species s has a particular random walk configuration $\{\mathbf{r}_\alpha^{s,\,c}\}$ can be represented in the notation of Eqs. (1) and (2) as

$$\prod_{\alpha=1}^{M_s-1} \left[\sum_{\beta_\alpha^{s,c}=1}^{z} \delta(\mathbf{r}_\alpha^{s,\,c}, \mathbf{r}_{\alpha+1}^{s,\,c} + \mathbf{a}_{\beta_\alpha^{s,c}}) \right] \tag{3}$$

where the product has α run over all but the final monad in the chain, or equivalently, over all the bonds in the chain and where the sequential bonding constraints are readily apparent. The bond direction label β is endowed with super- and subscripts in order to specify the directions for each of the different bonds in the chain.

Unlike model random walk chains, real polymers experience effective hard-core repulsions at short interatomic separations. Thus the partition function W incorporates excluded volume constraints that prevent more than one monad from simultaneously occupying a lattice site. The indistinguishability of all polymers from species s and the individual chain symmetries contribute factors of $(C_s!)^{-1}$ and of $\sigma_s^{-C_s}$, respectively, to the partition function, where σ_s is the chain symmetry number. Linear chains have symmetry about the midpoint, yielding the symmetry numbers $\sigma_s = 2$, but structured solvent molecules and branched network chains may require different σ_s. The partition function W for a p-component linear chain blend on a hypercubic lattice with only hard-core repulsive interactions is written with perfect generality as

$$W(\{C_s\}, \{M_s\}) = \sum_{\{\mathbf{r}_\alpha^{s,c}\}}' \left[\prod_{s=1}^{p} (C_s! \, \sigma_s^{C_s})^{-1} \prod_{c=1}^{C_s} \left(\prod_{\alpha=1}^{M_s-1} \sum_{\beta_\alpha^{s,c}=1}^{z} \delta(\mathbf{r}_\alpha^{s,c}, \mathbf{r}_{\alpha+1}^{s,c} + \beta_\alpha^{s,c}) \right) \right]$$

(4)

where the symbol $\sum_{\{\mathbf{r}_\alpha^{s,c}\}}'$ represents a restricted sum over all positions of every monad such that no two monads occupy the same lattice site and the notation $\beta_\alpha^{s,c}$ inside the Kronecker delta is shorthand for the lattice vector $\mathbf{a}_{\beta_\alpha^{s,c}}$ connecting monads α and $\alpha + 1$ on chain c from species s. The terms in the large parentheses on the right-hand side of Eq. (4) describe the connectivity constraints for chain c of species s as in Eq. (3). The remaining products in Eq. (4) run over the C_s chains from each species s.

2. Inclusion of Bending Energies

The hard-core repulsions in real polymers not only prevent the simultaneous occupation of space by more than one atom but also contribute to the steric hindrances for rotation about individual bonds. For example, the steric interactions in olefins typically produce a lowest-energy *trans* and several higher-energy, possibly degenerate, *gauche* rotational states. As the difference between *trans* and *gauche* conformations grows large, the chain stiffens considerably. Thus the difference between *trans* and *gauche* energies provides one measure of overall chain flexibility. The semiflexibility of linear chains is treated in the lattice model by introducing a bending energy penalty $B_\alpha^{s,c}$ (in units of K) for the pair of consecutive bonds α and $\alpha + 1$ in chain c from species s whenever the pair of bonds lies along orthogonal lattice directions (called the "*gauche*" conformation). The "*trans*" conformation corresponds to consecutive collinear bonds and is ascribed a vanishing bending energy.

The condition that only *gauche* conformations for each pair of consecutive bonds α and $\alpha + 1$ in chain c from species s incur the bending energy $B_\alpha^{s,c}$ translates into the bending energy factor

$$\exp\left(-\frac{B_\alpha^{s,c}}{T} \right) + \left[1 - \exp\left(-\frac{B_\alpha^{s,c}}{T} \right) \right] \delta(\beta_\alpha^{s,c}, \beta_{\alpha+1}^{s,c})$$

$$\equiv E_\alpha^{s,c} + [1 - E_\alpha^{s,c}] \, \delta(\beta_\alpha^{s,c}, \beta_{\alpha+1}^{s,c}) \quad (5)$$

When bonds α and $\alpha + 1$ are collinear (i.e., $\beta_\alpha^{s,\,c} = \beta_{\alpha+1}^{s,\,c}$), Eq. (5) yields a factor of unity, while any other configuration yields the Boltzmann factor $E_\alpha^{s,\,c}$ (for $\beta_\alpha^{s,\,c} \neq \beta_{\alpha+1}^{s,\,c}$). The bending energy factor $\beta_\alpha^{s,\,c}$ of Eq. (5) arises from purely intramolecular interactions. Hence the partition function W for a p-component blend with hard-core repulsions and semiflexibility only requires replacing the bonding constraint inside the large parentheses on the right-hand side of Eq. (4) with the extension to semiflexible, hard-core interaction chains,

$$\sum_{\{\beta_\alpha^{s,\,c}\}=1}^{z} \left(\prod_{\alpha=1}^{M_s-1} \delta(\mathbf{r}_\alpha^{s,\,c}, \mathbf{r}_{\alpha+1}^{s,\,c} + \beta_\alpha^{s,\,c}) \prod_{\gamma=1}^{M_s-2} [E_\gamma^{s,\,c} + (1 - E_\gamma^{s,\,c})\, \delta(\beta_\gamma^{s,\,c}, \beta_{\gamma+1}^{s,\,c})] \right) \quad (6)$$

where the sum is over the bond orientations $\beta_\alpha^{s,\,c}$ of every bond in chain c of species s. Although different indices α and γ are employed in the products inside the large parentheses, both terms range over all bonds in the system; so both products yield factors associated with the same bonds.

3.   Inclusion of van der Waals Interactions

Real polymers also contain longer-range attractive interactions that are necessary for them to attain liquidlike densities at atmospheric pressure. These longer-range interactions are modeled with nearest-neighbor attractive van der Waals energies $e(\{s, c, \alpha\}, \{s', c', \alpha'\})$ (in units of K), where α and α' designate the specific interacting pair of monads within chains c and c' of species s and s', respectively. This form of the interaction energies permits representing the chemically diverse monads within a monomer in accord with customary procedures for detailed phenomenological molecular modeling. The partition function W must include the Boltzmann factor for these van der Waals interactions

$$\exp\left(-\frac{1}{2!} \sum_{s,\,s'=1}^{p} \sum_{c=1}^{C_s} \sum_{c'=1}^{C_{s'}} \sum_{\alpha=1}^{M_s-1} \sum_{\alpha'=1}^{M_{s'}-1} [1 - \delta(s, s')\, \delta(c, c')\, \delta(\alpha, \alpha')] \right.$$
$$\left. \times \sum_{\zeta_\alpha^{s,\,c}=1}^{z} \delta(\mathbf{r}_\alpha^{s,\,c}, \mathbf{r}_{\alpha'}^{s',\,c'} + \zeta_\alpha^{s,\,c})\, \frac{e(\{s, c, \alpha\}, \{s', c', \alpha'\})}{T} \right) \quad (7)$$

where $\zeta_\alpha^{s,\,c}$ is shorthand for the lattice vector $\mathbf{a}_{\zeta_\alpha^{s,\,c}}$ separating the nearest-neighbor interacting monads α and α'. Contributions to Eq. (7) arise even if the nearest-neighbor monads α and α' are not bonded. The factor of 1/2! in Eq. (7) is necessary to prevent the double counting of the interactions between each pair of monads. The product $\delta(s, s')\, \delta(c, c')\, \delta(\alpha, \alpha')$ of these three Kronecker deltas (the "exclusion deltas") inside the large parentheses is necessary to ensure that no monad interacts with itself. The product of the exclusion deltas with the nearest-neighbor interaction condition $\delta(\mathbf{r}_\alpha^{s,\,c}, \mathbf{r}_{\alpha'}^{s',\,c'} + \zeta_\alpha^{s,\,c})$ always vanishes, allowing Eq. (7) to be rewritten as,

$$\prod_{s,\,s'=1}^{p} \prod_{c=1}^{C_s} \prod_{c'=1}^{C_{s'}} \prod_{\alpha=1}^{M_s-1} \prod_{\alpha'=1}^{M_{s'}-1} \exp\left(\sum_{\zeta_\alpha^{s,\,c}=1}^{z} \frac{\delta(\mathbf{r}_\alpha^{s,\,c}, \mathbf{r}_{\alpha'}^{s',\,c'} + \zeta_\alpha^{s,\,c})\varepsilon_{\alpha\alpha'}}{2} \right) \tag{8}$$

where the shorthand notation $\varepsilon_{\alpha\alpha'} = -e(\{s, c, \alpha\}, \{s', c', \alpha'\})/T$ is used.

Expanding a schematic representation of an individual exponential in Eq. (8) as a Taylor series produces

$$\exp\left(k \sum_\zeta \delta(\mathbf{r}, \mathbf{r}' + \zeta) \right) = 1 + k \sum_\zeta \delta(\mathbf{r}, \mathbf{r}' + \zeta)$$

$$+ \frac{k^2}{2!} \sum_\zeta \sum_{\zeta^*} \delta(\mathbf{r}, \mathbf{r}' + \zeta)\,\delta(\mathbf{r}, \mathbf{r}' + \zeta^*) + \cdots \tag{9}$$

Because the positions $\mathbf{r}$ and $\mathbf{r}'$ are fixed in Eq. (9), only one bond direction simultaneously satisfies every Kronecker delta in the quadratic and higher-order products of Eq. (9). Thus the expansion in Eq. (9) may be reduced to

$$\exp\left(k \sum_\zeta \delta(\mathbf{r}, \mathbf{r}' + \zeta) \right) = 1 + k \sum_\zeta \delta(\mathbf{r}, \mathbf{r}' + \zeta) + \frac{k^2}{2!} \sum_\zeta \delta(\mathbf{r}, \mathbf{r}' + \zeta) + \cdots \tag{10}$$

$$= 1 + \left(\sum_\zeta \delta(\mathbf{r}, \mathbf{r}' + \zeta) \right)\left(1 + k + \frac{k^2}{2!} + \cdots - 1 \right) \tag{11}$$

Resumming the term in the rightmost bracket in Eq. (11) and applying the result to Eqs. (7) and (8) yields the final form for the Boltzmann factor as

$$\prod_{s,\,s'=1}^{p} \prod_{c=1}^{C_s} \prod_{c'=1}^{C_{s'}} \prod_{\alpha=1}^{M_s-1} \prod_{\alpha'=1}^{M_{s'}-1} \left(1 + f_{\alpha\alpha'} \sum_{\zeta_\alpha^{s,\,c}=1}^{z} \delta(\mathbf{r}_\alpha^{s,\,c}, \mathbf{r}_{\alpha'}^{s',\,c'} + \zeta_\alpha^{s,\,c}) \right) \tag{12}$$

where the Mayer f function $f_{\alpha\alpha'}$ is

$$f_{\alpha\alpha'} = \exp\left(\frac{\varepsilon_{\alpha\alpha'}}{2} \right) - 1 \tag{13}$$

Since the products over s, c, and α in Eq. (12) are identical to the products in Eq. (4), only the remainder of Eq. (12) must be inserted into Eq. (4) in order to provide the general expression for the partition function W of chains with nearest-neighbor van der Waals interactions. Thus the partition function for a multicomponent linear chain blend with hard-core repulsions, semiflexibility, and nearest-neighbor van der Waals attractions may be written as

$$W(\{C_s\}, \{M_s\}) = \sum_{\{\mathbf{r}_\alpha^{s,c}\}}{}' \left\{ \prod_{s=1}^{p} (C_s! \, \sigma_s^{C_s})^{-1} \prod_{c=1}^{C_s} \sum_{\{\beta_\alpha^{s,c}\}=1}^{z} \left[\prod_{\alpha=1}^{M_s-1} \delta(\mathbf{r}_\alpha^{s,c}, \mathbf{r}_{\alpha+1}^{s,c} + \beta_\alpha^{s,c}) \right. \right.$$

$$\times \left[\prod_{\gamma=1}^{M_s-2} [E_\gamma^{s,c} + (1 - E_\gamma^{s,c}) \, \delta(\beta_\gamma^{s,c}, \beta_{\gamma+1}^{s,c})] \right.$$

$$\times \left. \left. \prod_{s'=1}^{p} \prod_{c'=1}^{C_{s'}} \prod_{\alpha'=1}^{M_{s'}-1} \left(1 + f_{\alpha\alpha'} \sum_{\zeta_\alpha^{s,c}=1}^{z} \delta(\mathbf{r}_\alpha^{s,c}, \mathbf{r}_{\alpha'}^{s',c'} + \zeta_\alpha^{s,c}) \right) \right] \right\} \qquad (14)$$

C. The Partition Function for Structured Monomer Chains

The extended lattice model differentiates between monomer structures by representing various monomers with different connectivities between the monads, producing multiply branched structures. Although many numbering schemes equally well identify the monads in a structured monomer or branched chain, we employ a scheme that reduces the structure into a series of connected linear portions called *branches*. The procedure for assigning specific bonds to the different linear branches of a polymer chain is illustrated, beginning with the simple example of homopolymer united atom model polyolefin chains and then passing on to the description for chains whose monomers contain any structures without closed loops.

Since homopolymer polyolefin chains are composed of a regular, repeating array of monomers, the chain is decomposed into branches by defining the first branch as containing all backbone monads, and then the remaining branches may be derived solely from the monomer structures. Distinct branches meet only at tri- or higher functional monads, called *branch points*. The linear monad model for a polyethylene (PE) chain contains two monofunctional monads at the chain ends, while the interior chain monomers contain a pair of difunctional monads. Clearly, no branch points exist, and hence the chain yields only one single branch. In contrast, the monad model for the isobutyl monomer in Fig. 5.1 contains one tetrafunctional branch point. An n-functional branch point has either $n/2$ (n even) or $(n+1)/2$ (n odd) independent, linear branches, called *subchains*, that all contain the n-functional monad. Thus the isobutyl monomer of Fig. 5.1 contains two subchains, one subchain composed of the two backbone bonds eminating from the branch point and the other subchain composed of the remaining two (side-chain) bonds. The monad model for the propyl monomer contains one fewer side chain monad than the isobutyl monomer,

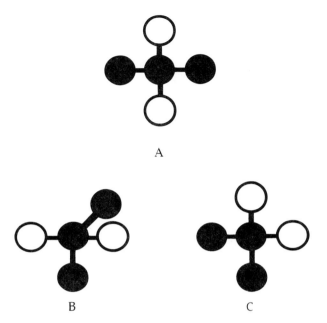

Figure 5.1. The *trans–trans* (A), *trans–gauche* (B), and *gauche–gauche* (C) conformations of the isobutylene monomer. Filled and open circles designate backbone and side-chain monads, respectively. The central monad is the tetrafunctional branch point. In each configuration, bends between any pair of connected bonds are assigned an energy penalty E_b, except where the pair consists of one backbone bond and one side-chain bond (those pairs that lie along the different subchains). This four-bond structure has two (linear) subchains, each containing a pair of consecutive bonds. *Gauche* conformations of these pairs are assigned a bending energy penalty of E_b. This structure contains four pairs of consecutive bonds, which lie on separate subchains and thus incur no bending energy penalty for the *gauche* conformation. Hence, configurations A, B, and C have energy penalties of 0, E_b, and $2E_b$, respectively.

but still comprises two subchains: The backbone bonds form one subchain, and the side-chain bond is the other subchain. This definition for the number of subchains per n-functional branch point is sufficient to describe the monomer structures of nearly all united atom model polyolefins. The first subchain is always defined as composed of all the backbone bonds.

The description of more complex monomer structures (containing no closed loops) employs a decomposition of the full monomer architecture into a minimum set S^c of linear subchains that share no bonds but that may share branch point monads. For example, the set S^c for the united atom model of an isobutyl monomer is composed of the two (the backbone and the side-chain) subchains. The members of S^c are chosen as follows: All backbone bonds compose the first member of S^c. In order to define the remaining members of S^c, all the backbone bonds are removed from the original monomer structure along with the monads (called *associated monads*) that are no longer attached to any of the remaining bonds. The longest linear, connected group of bonds (called a *connecting path*) in the remaining structure is extracted as the next member of S^c. If more than one connecting path is present, any one of these is selected, with no loss of generality, as the next member of S^c. The bonds from this new member of S^c, along with their associated monads, are then removed from the remaining monomer structure. The next connecting paths (and their associated groups) are then selected successively until no bonds remain from the original monomer structure.

Including a description of branching in the partition function is now straightforward, requiring only a suitable notation. The set S^c has L_1 members. The monads along each of the $L_1^{s,\,c}$ linear subchains in chain c from species s can be labeled sequentially from one end to the other, with monads shared by different subchains acquiring multiple (redundant) labels. A single branched chain contributes to the partition function a factor similar to that in Eq. (3)

$$\prod_{\lambda=1}^{L_1^{s,\,c}} \prod_{\alpha=1}^{n_\lambda^{s,\,c}-1} \left(\sum_{\beta_{\lambda,\alpha}^{s,\,c}=1}^{z} \delta(\mathbf{r}_{\lambda,\,\alpha}^{s,\,c}, \mathbf{r}_{\lambda,\,\alpha+1}^{s,\,c} + \mathbf{a}_{\beta_{\lambda,\alpha}^{s,\,c}}) \right) \prod D_0(s) \qquad (15)$$

where $n_\lambda^{s,\,c}$ is the total number of monads in the λth subchain of chain c from species s and where $\Pi\, D_0(s)$ represents a factor enforcing the equivalence of the multiple labels assigned to monads that are shared by different subchains. No specific form is required here for the product over the $D_0(s)$ since the expression is dependent on the monomer structure. The treatment of $D_0(s)$, however, becomes clearer when developing the cluster expansion in Section III. The first product in Eq. (15) runs over the subchains in chain c of species s, while the second product runs over the bonds along a specific subchain. The quantity $n_\lambda^{s,\,c}$, by definition, has a minimum value of two, corresponding to a single bond.

Structured monomers, and more generally, branched polymers also introduce an additional complication because only some pairs of bonds should be assessed a bending energy penalty. For example, if a rigid rod has flexible side branches, the side-chain bonds directly attached to the rod *must*, in the lattice model, lie along a direction orthogonal to the rod. Thus the bending energy is omitted when the pair of consecutive bonds lies along different subchains in the same polymer. Hence, for example, a united atom monad model for the propyl monomer contains one trifunctional branch point and is assigned one bending energy only for *gauche* conformations of the two consecutive backbone bonds. The united atom monad model for the isobutyl monomer contains a tetrafunctional branch point and is, therefore, assigned two different bending energies, one for *gauche* conformations of the pair of backbone bonds and one for the *gauche* conformations of the consecutive pair of side-chain bonds emanating from the tetrafunctional branch point.

The condition that only *gauche* conformations for each pair of consecutive bonds α and $\alpha + 1$ on the same subchain λ in chain c from species s incur the bending energy $B_{\lambda, \alpha}^{s, c}$ translates into the generalization of Eq. (6)

$$
\sum_{\{\beta_{\lambda, \alpha}^{s, c}\} = 1} \prod_{\lambda = 1}^{L_1^{s, c}} \left(\prod_{\alpha = 1}^{n_\lambda^{s, c} - 1} \delta(\mathbf{r}_{\lambda, \alpha}^{s, c}, \mathbf{r}_{\lambda, \alpha+1}^{s, c} + \beta_{\lambda, \alpha}^{s, c}) \right.
$$

$$
\left. \times \prod_{\gamma = 1}^{n_\lambda^{s, c} - 2} [E_{\lambda, \gamma}^{s, c} + (1 - E_{\lambda, \gamma}^{s, c}) \, \delta(\beta_{\lambda, \gamma}^{s, c} \beta_{\lambda, \gamma+1}^{s, c})] \right) \prod D_0(s) \quad (16)
$$

When $n_\lambda^{s, c} = 2$, only one bond is present in the λth subchain, and the bending energy factor in Eq. (16) is meaningless. Therefore, the product over γ in Eq. (16) is defined as unity when $n_\lambda^{s, c} = 2$. A comparison of Eqs. (6) and (16) reveals that the description of a branched chain merely introduces an additional counting index and some additional constraints, which define the connectivity of the subchains, but otherwise does not affect the mathematical representation of the partition function. Therefore, the partition function in Eq. (14) is readily generalized to branched structures as

$$
W(\{C_s\}, \{M_s\}) = \sum_{\{\mathbf{r}_{\lambda, \alpha}^{s, c}\}}' \left\{ \prod_{s = 1}^{p} (C_s! \, \sigma_s^{C_s})^{-1} \right.
$$

$$
\times \prod_{c = 1}^{C_s} \sum_{\{\beta_{\lambda, \alpha}^{s, c}\} = 1}^{z} \prod_{\lambda = 1}^{L_1^{s, c}} \left[\prod_{\alpha = 1}^{n_\lambda^{s, c} - 1} \delta(\mathbf{r}_{\lambda, \alpha}^{s, c}, \mathbf{r}_{\lambda, \alpha+1}^{s, c} + \beta_{\lambda, \alpha}^{s, c}) \right.
$$

$$
\times \prod_{\gamma = 1}^{n_\lambda^{s, c} - 2} [E_{\lambda, \gamma}^{s, c} + (1 - E_{\lambda, \gamma}^{s, c}) \, \delta(\beta_{\lambda, \gamma}^{s, c} \beta_{\lambda, \gamma+1}^{s, c})]
$$

$$
\left. \times \prod_{s' = 1}^{p} \prod_{c' = 1}^{C_{s'}} \prod_{\lambda' = 1}^{L_1^{s', c'}} \prod_{\alpha' = 1}^{n_\lambda^{s, c}} \left(1 + f_{\alpha\alpha'} \sum_{\zeta_{\lambda, \alpha}^{s, c} = 1}^{z} \delta(\mathbf{r}_{\lambda, \alpha}^{s, c}, \mathbf{r}_{\lambda', \alpha'}^{s', c'} + \zeta_{\lambda, \alpha}^{s, c}) \right) \right]
$$

$$
\left. \times \prod D_0(s) \prod D_0(s') \right\} \quad (17)
$$

III. DEVELOPMENT OF CLUSTER EXPANSION FOR THE PARTITION FUNCTION

A.   Extraction of a Zeroth-Order Mean-Field Contribution

The partition function W in Eq. (17) cannot be evaluated exactly in all but the most trivial cases. Thus we extract a convenient zeroth-order approximation and then develop methods for calculating systematic corrections to this contribution. An approximate Flory-type mean-field contribution W^{MF} may be extracted as a zeroth-order approximation by removing an average portion from each Kronecker delta in Eq. (17) as follows: The Kronecker deltas in Eq. (17) either arise from connectivity constraints between monads $[\delta(\mathbf{r}_{\lambda,\alpha}^{s,c}, \mathbf{r}_{\lambda,\alpha+1}^{s,c} + \beta_{\lambda,\alpha}^{s,c})]$ or conformational (*trans–gauche*) restrictions on bond orientations $[\delta(\beta_{\lambda,\gamma}^{s,c}, \beta_{\lambda,\gamma+1}^{s,c})]$ for pairs of consecutive bonds. Consider the $\delta(\beta_{\lambda,\gamma}^{s,c}, \beta_{\lambda,\gamma+1}^{s,c})$ bending terms first. If $\beta_{\lambda,\gamma}^{s,c}$ and $\beta_{\lambda,\gamma+1}^{s,c}$ are the only two bonds in the system and if the direction of $\beta_{\lambda,\gamma}^{s,c}$ is specified, only one out of the z possible orientations for $\beta_{\lambda,\gamma+1}^{s,c}$ satisfies the Kronecker delta, thus yielding a "mean-field" average of $1/z$ for the $\delta(\beta_{\lambda,\gamma}^{s,c}, \beta_{\lambda,\gamma+1}^{s,c})$ factor. Likewise, if only two monads are placed on the lattice at the locations $\mathbf{r}_{\lambda,\alpha}^{s,c}$ and $\mathbf{r}_{\lambda,\alpha+1}^{s,c}$, the first monad can be found in only one out of N_l possible lattice sites, and the bonding constraint Kronecker delta $\delta(\mathbf{r}_{\lambda,\alpha}^{s,c}, \mathbf{r}_{\lambda,\alpha+1}^{s,c} + \beta_{\lambda,\alpha}^{s,c})$ is satisfied when the second monad occupies the nearest-neighbor lattice site to the first monad along the bond direction $\beta_{\lambda,\alpha}^{s,c}$. Thus $1/N_l$ represents the average "mean-field" contribution from the bonding constraint $\delta(\mathbf{r}_{\lambda,\alpha}^{s,c}, \mathbf{r}_{\lambda,\alpha+1}^{s,c} + \beta_{\lambda,\alpha}^{s,c})$. Consequently, each Kronecker delta δ in the partition function of Eq. (17) is rewritten below as the sum of the average portion A and a remainder $R \equiv \delta - A$; that is, $\delta = A + R$.

The approximate zeroth-order Flory mean-field expression follows by employing physically identical assumptions to those used for deriving the average portions A above. Thus the zeroth-order mean-field term is extracted by neglecting all remainders R. The zeroth-order contribution from the average terms in Eq. (17) takes the form

$$W(\{C_s\}, \{M_s\}) = \sideset{}{'}\sum_{\{\mathbf{r}_{\lambda,\alpha}^{s,c}\}} \left[\prod_{s=1}^{p} (C_s! \sigma_s^{C_s})^{-1} \prod_{c=1}^{C_s} \prod_{\lambda=1}^{L_1^{s,c}} \left\{ \prod_{\alpha=1}^{n_\lambda^{s,c}-1} \sum_{\beta_{\lambda,\alpha}^{s,c}=1}^{z} \frac{1}{N_l} \right. \right.$$

$$\times \prod_{\gamma=1}^{n_\lambda^{s,c}-2} \left(E_{\lambda,\gamma}^{s,c} + \frac{(1 - E_{\lambda,\gamma}^{s,c})}{z} \right) \prod_{s'=1}^{p} \prod_{c'=1}^{C_{s'}} \sum_{\lambda'=1}^{L_1^{s',c'}} \sum_{\alpha'=1}^{n_\lambda^{s,c}-1} \left(1 + f_{\alpha\alpha'} \sum_{\zeta_{\lambda,\alpha}^{s,c}=1}^{z} \frac{1}{N_l} \right) \right\}$$

$$\times \prod D_0(s) \prod D_0(s') \Bigg] \quad (18)$$

The monad positions and bond directions, which appear in the summation indices of Eq. (18), are absent in the summand. Consequently, the sums in Eq. (18) are readily evaluated. The sums over β and ζ each yield a factor of z, while the sum over positions produces the combinatorial factor $N_l!/(N_l - \sum_{s=1}^{p} C_s M_s)!$ (see below for an explanation). The products over $D_0(s)$ and $D_0(s')$ do not influence the evaluation of the remaining factors in Eq. (18) and are thus dropped. Since chains c and c' from species s must have identical numbers of subchains ($L_1^{s,\,c} = L_1^{s,\,c'} \equiv L_1^s$), the total number of subchains in all chains from species s must equal $C_s L_1^s$. Combining the factors of z leads to the final expression for the zeroth-order mean-field expression,

$$
W^{\mathrm{MF}} = \frac{N_l!}{(N_l - \sum_{s=1}^{p} C_s M_s)!} \left(\prod_{s=1}^{p} (C_s! \, \sigma_s^{C_s})^{-1} \right) \left(\frac{z^{N_{\mathrm{sc}}}}{N_l^{N_b}} \right)
$$

$$
\times \left(\prod_{s=1}^{p} \prod_{c=1}^{C_s} \prod_{\lambda=1}^{L_1^s} \prod_{\gamma=1}^{n_\lambda^{s,c}-2} [(z-1)E_{\lambda,\gamma}^{s,\,c} + 1] \right)
$$

$$
\times \prod_{s=1}^{p} \prod_{c=1}^{C_s} \prod_{\lambda=1}^{L_1^s} \prod_{\alpha=1}^{n_\lambda^{s,c}-1} \prod_{s'=1}^{p} \prod_{c'=1}^{C_{s'}} \prod_{\lambda'=1}^{L_1^s} \sum_{\alpha'=1}^{n_{\lambda'}^{s',c'}-1} \left(1 + \frac{zf_{\alpha\alpha'}}{N_l} \right) \tag{19}
$$

where the total numbers N_{sc} and N_b of subchains and of bonds in the system are, respectively, $\sum_{s=1}^{p} C_s L_1^s$ and $\sum_{s=1}^{p} C_s(M_s - 1)$. Equation (19) reduces to the previous LCT zeroth-order mean-field partition function for flexible chains when $E_{\lambda,\gamma}^{s,\,c} \to 1$, while the limit $L_1^s \to 1$ reduces Eq. (19) to the linear semiflexible chain case. The Flory theory semiflexible chain partition function differs from the LCT expression in Eq. (19) in two ways: The Flory theory uses $z - 1$ instead of z in all but the rightmost term in Eq. (19) and retains only the leading term ($\varepsilon_{\alpha\alpha'}$) from the Mayer f function $f_{\alpha\alpha'}$ in calculations of thermodynamics quantities.

The voids each occupy single lattice sites and do not interact with each other or with the polymer chains. Although this implies that the voids are not components in the thermodynamic sense, some mathematical convenience emerges by treating the C_v voids as an additional component in Eq. (19). The *only* contribution to Eq. (19) from the voids is an overall factor of $1/C_v!$, but then the lattice is taken as completely occupied by the system of polymers and voids. In this case, the sums over all positions for both polymer and voids in Eq. (18) yields $N_l!$, and the ratio $N_l!/C_v!$ of these two factors is identical to the factor of $N_l!/(N_l - \sum_{s=1}^{p} C_s M_s)!$ in Eq. (19). Introducing the voids as an additional component ascribes absolutely no physical character to the voids but is convenient in automatically producing the correct combinatorial prefactor.

B.   General Form of the Cluster Expansion

Reintroducing all remainders into Eq. (17) yields the partition function as a product of a zeroth-order contribution and a correction term; that is, $W = W^{\mathrm{MF}}$ (1 + *corrections*). Equations (18) and (19) provide the expression for W^{MF}, thus enabling the determination of the form of the corrections from Eq. (17). The expression for W in Eq. (17) contains Kronecker deltas in three separate portions of the partition function. The first portion of the partition function treats the connectivity (bonding) within chains; the second involves the bending constraints between pairs of consecutive bonds along a given subchain; and the third considers the van der Waals interactions between nearest-neighbor monads. Because the myriad of different indices, products, and summations in the equations obscure some of the simple physical concepts, the theory is stripped of these complications whenever exploring basic concepts, leaving only a representation employing a schematic notation. Thus the partition function of Eq. (17) can be rewritten in this schematic notation as

$$W = \sum_{\{\mathbf{r}\}}{}' \left[\prod_{\text{bonds}} \left(\sum_{\beta_{\text{bond}}=1}^{z} \delta_{\text{bond}} \right) \prod_{\text{bending constraints}} [E_{\text{bend}} + (1 - E_{\text{bend}}) \delta_{\text{bend}}] \right.$$

$$\left. \times \prod_{\text{pairs of monads}} \left(1 + f_{\text{pair}} \sum_{\beta_{\text{pair}}=1}^{z} \delta_{\text{pair}} \right) \right] \tag{20}$$

where the product over bonds includes all bonds between monads, the bending constraint products run over all pairs of consecutive bonds in each and every subchain, and the pair product runs over all nearest-neighbor pairs of monads.

The cluster expansion is now generated by re-expressing each of the three Kronecker deltas in Eq. (20) in the schematic form

$$\delta = A + (\delta - A) = A\left(1 + \frac{\delta - A}{A} \right) \tag{21}$$

where A represents the (average) contribution to δ. We employ the averages calculated in Section III.A in order to extract the mean-field component of Eq. (19) (i.e., $A_{\text{bond}} = A_{\text{pair}} = 1/N_l$ and $A_{\text{bend}} = 1/z$). The partition function in Eq. (20) is the product of three portions, each of which is decomposed into the product of its contribution to the zeroth-order mean-field approximation and a term again of the form $(1 + corrections)$. More explicitly, the first factor of Eq. (20) from bonding constraints includes the sum $\sum_{\beta_{bond}=1}^{z} \delta_{bond}$ over the z possible orientations of the bond. Applying the decomposition Eq. (21) to this term yields

$$\sum_{\beta_{bond}=1}^{z} \delta_{bond} = A_{bond}\left(z + \sum_{\beta_{bond}=1}^{z} \frac{\delta_{bond} - A_{bond}}{A_{bond}}\right)$$

which, after algebraic manipulation, is,

$$= zA_{bond}\left(1 + \frac{1}{zA_{bond}} \sum_{\beta_{bond}=1}^{z} (\delta_{bond} - A_{bond})\right)$$

$$= zA_{bond}(1 + X_{bond}) \tag{22}$$

thereby defining X_{bond}. Equation (22) is a product of the zeroth-order mean-field portion zA_{bond} and a term in the desired form $(1 + corrections)$, with the single-bond correlation correction X_{bond} corresponding to the previously derived correlation correction (Dudowicz and Freed, 1991a) for the bond between monads α and $\alpha + 1$ along subchain λ of chain c from species s. More specifically, this bond correction term is

$$X_{bond} \equiv X_{\lambda,\alpha}^{s,c} = \frac{N_l}{z} \sum_{\beta_{\lambda,\alpha}^{s,c}=1}^{z} \left(\delta(\mathbf{r}_{\lambda,\alpha}^{s,c}, \mathbf{r}_{\lambda,\alpha+1}^{s,c} + \beta_{\lambda,\alpha}^{s,c}) - \frac{1}{N_l}\right) \tag{23}$$

The remaining portions of the partition function in Eq. (20) are decomposed following similar analyses. For example, the third portion of Eq. (20) from nearest-neighbor van der Waals interactions is nearly identical to the sum in the first portion of Eq. (20), except for the extra factor of unity. The results from Eq. (22) can be substituted into the third portion of Eq. (20), producing

$$1 + f_{pair} \sum_{\beta_{pair}=1}^{z} \delta_{pair} = 1 + zA_{pair} f_{pair}(1 + X_{pair}) \tag{24}$$

where X_{pair} is the interaction correction, which is identical to X_{bond} except that X_{pair} constrains the pair of monads to be nearest-neighbors even if the pair is not connected by a bond. Rearranging Eq. (24) yields the desired form as

$$1 + f_{pair} \sum_{\beta_{pair}=1}^{z} \delta_{pair} = (1 + zA_{pair} f_{pair})\left(1 + \frac{zA_{pair} f_{pair} X_{pair}}{1 + zA_{pair} f_{pair}}\right)$$

$$= (1 + zA_{pair} f_{pair})(1 + G_{pair} X_{pair}) \tag{25}$$

where the factor G_{pair} is written in formal notation as,

$$G_{pair} = G_{\alpha\alpha'} = \frac{zf_{\alpha\alpha'}/N_l}{1 + zf_{\alpha\alpha'}/N_l} \tag{26}$$

Applying Eq. (21) to the bending constraint portion of the partition function in Eq. (20) generates an expression analogous to Eq. (24):

$$E_{bend} + (1 - E_{bend}) \delta_{bend} = E_{bend} + (1 - E_{bend})A_{bend}(1 + Y_{bend}) \tag{27}$$

where the bending energy correction is

$$Y_{bend} = \frac{\delta_{bend} - A_{bend}}{A_{bend}} = z \, \delta(\beta_{\lambda, \gamma}^{s, c}, \beta_{\lambda, \gamma+1}^{s, c}) - 1 \tag{28}$$

Algebraic rearrangement in Eq. (27) yields:

$$E_{bend} + (1 - E_{bend}) \delta_{bend} = [E_{bend} + (1 - E_{bend})A_{bend}]$$

$$\times \left(1 + \frac{(1 - E_{bend})A_{bend} Y_{bend}}{E_{bend} + (1 - E_{bend})A_{bend}}\right)$$

$$= \frac{z_{b1}(E_{bend})}{z} (1 + K_{bend} Y_{bend}) \tag{29}$$

where the bending factor K_{bend} is written explicitly as

$$K_{bend} = K_{\lambda, \alpha}^{s, c} = \frac{(1 - E_{\lambda, \alpha}^{s, c})}{z_{b1}(E_{\lambda, \alpha}^{s, c})} \tag{30}$$

$$z_{b1}(E_{\lambda, \alpha}^{s, c}) = (z - 1)E_{\lambda, \alpha}^{s, c} + 1 \tag{31}$$

The bending factor $K_{\lambda, \alpha}^{s, c}$ is related to the semiflexibility scaling factor $(g \equiv 1 - K)$ used by Bawendi and Freed (1987) and is treated as formally of order unity.

C. Cluster Expansion

Comparison of Eqs. (18) and (19) with Eqs. (22), (24), (25), (27), and (28) enables rewriting the partition function of Eq. (20) schematically as:

$$
W = \sum_{\{\mathbf{r}\}}' \frac{W^{\mathrm{MF}}(N_l - \sum_{s=1}^{p} C_s M_s)!}{N_l!}
$$

$$
\times \left(\prod_{\mathrm{bonds}} (1 + X_{\mathrm{bond}}) \prod_{\mathrm{bending\ constraints}} (1 + K_{\mathrm{bend}}\, Y_{\mathrm{bend}}) \right.
$$

$$
\left. \times \prod_{\mathrm{pairs\ of\ monads}} (1 + G_{\mathrm{pair}}\, X_{\mathrm{pair}}) \right) \tag{32}
$$

Multiplying the terms from all products within the large parentheses in Eq. (32) produces a leading factor of unity plus various classes of corrections. Since the representation of these corrections is somewhat complicated, we briefly review enough features of the cluster expansion produced by the products in Eq. (32) in order to describe how the bending terms are treated. Each of the products is individually of the form $\prod_i (1 + t_i)$, where t_i is either X_{bond}, $K_{\mathrm{bend}}\, Y_{\mathrm{bend}}$, or $G_{\mathrm{pair}}\, X_{\mathrm{pair}}$. Multiplying these products out produces a leading factor of unity along with clusters of increasing powers in the correction factors t_i,

$$
\prod_i (1 + t_i) = 1 + \sum_i t_i + \sum_{i,\,i'} t_i t_{i'} + \cdots \tag{33}
$$

The cluster expansion for the bending constraints differs from the other two cluster expansions because a factor of Y_{bend} only contributes to the partition function when multiplied by the correlation corrections X_{bond} for the two bonds affected by Y_{bend}. Thus, if $Y_{1,2}$ is the bending restriction involving the consecutive bonds 1 and 2 and X_i is the correlation correction for bond i, then linear terms in $Y_{1,2}$ must also contain a factor of $X_1 X_2$ in order to contribute in the expansion. The quadratic contributions in Y_{bend} therefore appear in two forms, $X_1 X_2 X_3\, Y_{1,2}\, Y_{2,3}$ or $X_1 X_2 X_3 X_4\, Y_{1,2}\, Y_{3,4}$. Higher-order terms in Y_{bend} follow similarly.

Extracting a factor of W^{MF} from the right-hand side of Eq. (32) and applying Eq. (33) lead to nearly the desired representation of the partition function, which in schematic form is:

$$
W = W^{\mathrm{MF}} \left\{ \sum_{\{\mathbf{r}\}}' \frac{(N_l - \sum_{s=1}^{p} C_s M_s)!}{N_l!} \left(1 + \sum_i t_i + \sum_{i,\,i'} t_i t_{i'} + \cdots \right) \right\} \tag{34}
$$

where the sum involving i implies that t_i ranges over the X_{bond}, K_{bend} Y_{bend}, and $G_{\text{pair}} X_{\text{pair}}$ type correction terms. Terms involving a bending restriction Y_{bend} yield vanishing contributions to the partition function unless accompanied by the requisite pair of X_{bond} factors as described above. Thus no corrections from Y_{bend} appear until third or higher order in t_i. Applying the sum over monad positions $\{\mathbf{r}\}$ separately to the factor of unity and to the remaining factors in Eq. (34) yields the final schematic form for the partition function,

$$W = W^{\text{MF}}\left\{1 + \sum_{\{\mathbf{r}\}}' \frac{(N_l - \sum_{s=1}^p C_s M_s)!}{N_l!}\left(\sum_i t_i + \sum_{i,i'} t_i t_{i'} + \cdots\right)\right\} \quad (35)$$

As mentioned previously, including voids as an extra "component" in Eq. (35) is equivalent to removing the factor of $(N_l - \sum_{s=1}^p C_s M_s)!$.

D. Diagrammatic Representation of the Cluster Expansion Contributions

The systematic calculation of contributions from the cluster corrections in Eqs. (34) and (35) is simplified by representing each t_i factor in Eqs. (34) and (35) diagrammatically. We begin by describing the diagrams produced when the t_i involve only X_{bond} factors, since the treatment of the other diagrams follows as an extension of these procedures. Each factor of a correlation correction X_{bond} is depicted by a solid straight line connecting a pair of consecutive, bonded monads, represented by circles. Figure 5.2 illustrates all contributing diagrams involving only bond correlation corrections, called *bond diagrams*, with up to four bonds. The diagram with two bonded monads is not depicted in Fig. 5.2 because it does not contribute in the thermodynamic limit (see Dudowicz and Freed, 1991a, for details). Connected bonds within the same chain may belong to the same or to different subchains, but the bond diagrams in Fig. 5.2 are intended to represent all possible unique combinations of the connected bonds that produce the depicted arrangement of bonds and monads. For example, two equivalent diagrams (o1 and o2) are presented for bond diagram o in Fig. 5.2 as a reminder that the connectivity depicted in bond diagram o arises either from three bonds along one subchain and one bond along another subchain (o1) or from two bonds along two separate subchains (o2). Both diagrams yield identical numerical results, but, as becomes apparent below, the introduction of additional bending restrictions into these diagrams produces a new set of diagrams whose evaluation produces different numerical results. Wavy lines (in, for example, bond diagram j of Fig. 5.2) imply that one or more intervening bonds separate the explicitly depicted bonds in the diagram, with the intervening bonds attached to any one of the monads from each of the two sets of explicitly depicted bonds in the diagram.

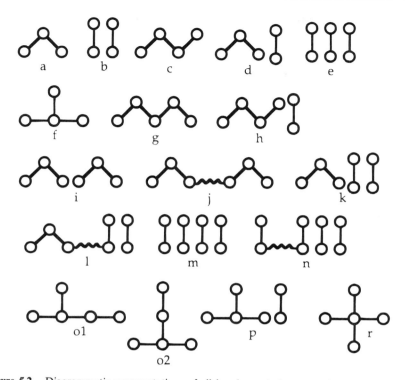

Figure 5.2. Diagrammatic representations of all bond correlation corrections to the zeroth-order mean-field approximation from four or fewer correlating bonds. These athermal limit bond diagrams contribute through $O(z^{-2})$ and are depicted with open circles representing monads, solid straight lines representing bond correlation corrections, and wavy lines indicating the presence of one or more intervening bonds between the bonds depicted. All other diagrams employed in the current model can be derived by operations applied to these athermal limit diagrams.

The bond diagrams in Fig. 5.2 provide the essential building blocks for constructing all remaining diagrams, which involve van der Waals interactions, bending restrictions, bond correlation corrections, and combinations thereof (see Fig. 5.3). The nearest-neighbor van der Waals interaction correction factors $G_{pair} X_{pair}$ are pictured as solid curved interaction lines. Diagrams involving only bond correlation corrections and van der Waals interactions are called *bond/pair diagrams*. Dudowicz and Freed (1991a) provide all relevant bond/pair diagrams containing up to a total of four bond and/or interaction lines; so we only describe those aspects of generating bond/pair diagrams that are required for the extension to include the bending energies.

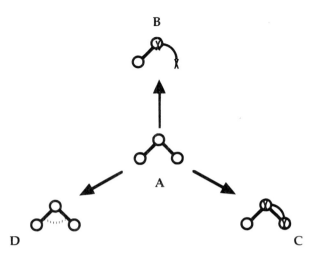

Figure 5.3. All cluster correction diagrams are derivable from the bond diagrams of Fig. 5.2. The circles, straight lines, solid curved lines, and dashed curved lines represent, respectively, monads, bonds, van der Waals interactions, and bending restrictions. Manipulation of diagram a of Fig. 5.2A produces diagrams with van der Waals interactions (B and C) or with bending restrictions (D).

Bond/pair diagrams are derived from the bond diagrams in Fig. 5.2 in a stepwise procedure, with two distinct parts to each step. The first step converts all bond diagrams in Fig. 5.2 into a set of bond/pair diagrams, each with only one curved interaction line. Successive steps convert the previous generated set of diagrams into a new set whose individual diagrams have one more interaction line than the diagrams from the previous set. The first part of the nth step involves the replacement, in all possible ways, of a straight line with a curved interaction line in each diagram from the $(n - 1)$st set. (The first step operates on the diagrams from Fig. 5.2.) Figure 5.4A uses diagram b from Fig. 5.2 to illustrate the repeated application of this process. Once a bond/pair diagram is constructed, the two monads connected by the curved interaction line need not reside on the same chain.

The second part of the nth step involves creating loops that are closed upon insertion of an interaction line (see diagrams B and C in Fig. 5.4). Loops can be closed only when the monads to be connected lie on nearest-neighbor lattice sites. Hypercubic lattices require a minimum of one bond and one interaction line (as in diagram B of Fig. 5.4) to form a closed loop, with larger loops produced by adding an interaction line to a connected, linear series of three, five, etc. bonds. Therefore, only diagrams with an odd number of consecutive bonds can form a closed loop upon the addition of a single interaction line. Hence, diagrams B and C of Fig. 5.4 are the only

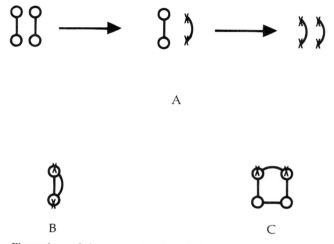

A

B   C

Figure 5.4. Illustrations of the process for introducing an interaction line into individual diagrams. Successive applications of this process generate all bond/pair diagrams. Each application requires two distinct steps. In the first step, an interaction line replaces a bond in every possible way in each diagram. (A) depicts the results of repeated use of this first step on diagram b of Fig. 5.2. The second step involves the construction of diagrams with closed loops containing, for instance, one (B), three (C), etc., bonds and one interaction line, as generated from the original set of diagrams.

relevant classes of closed loops for diagrams that arise in the first step for diagrams with a total of four or fewer bonds. The second part of the nth step considers all possible ways of making a closed loop from each diagram produced in the previous set (e.g., those from Fig. 5.2 for the first step). The thermodynamically relevant diagram B in Fig. 5.4 arises from the bond diagram involving a single bond and is not depicted in Fig. 5.2. Only diagrams containing $N_e \leq 2$ interaction lines are included in the present LCT. In addition, only diagrams with four or fewer total bond and/or interaction lines, that is, the numbers N_b and N_e of, respectively, bond corrections and interaction lines, are restricted to $N_b + N_e \leq 4$ and $N_e \leq 2$. Finally, any pair of monads can have at most one interaction line connecting them.

Once all the bond and bond/pair diagrams are known, the diagrams including bending restrictions are generated by adding curved dashed lines connecting pairs of consecutive bonds in a diagram as depicted in Fig. 5.5. The curved dashed line connecting a pair of consecutive bonds α and $\alpha + 1$ serves as a reminder that the bending restriction factor $Y_{\alpha, \alpha+1}$ must be multiplied by $X_\alpha X_{\alpha+1}$ for the two bonds. Figure 5.5A illustrates the application of this process to a four bond portion from a single subchain. Recall that a bending restriction only applies when the pair of consecutive bonds

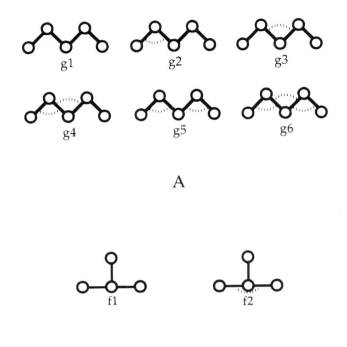

A

B

Figure 5.5. The process of adding bending restriction factors to all bond and bond/pair diagrams. Bending restriction factors are applied to all pairs of consecutive bonds that lie on the same subchain. All possible combinations of the bending factors are retained, as illustrated in A for a four-bond segment from a single subchain. When consecutive bonds lie along different subchains, no bending restriction factor is applied, and, hence, no curved dashed line is introduced between these bonds. The propyl monomer (whose united atom model corresponds to the three right most monads of diagram f1 in B) has two bonds along the backbone subchain and one bond along the side subchain, thus permitting only one bending restriction factor to be added to the pair along the backbone (diagram f2).

lies on the same subchain. Thus only two possible bond or bond/bend diagrams are derivable from diagram f in Fig. 5.2: one diagram with no bending restriction (f1 in Fig. 5.5B) and one with a bending restriction for the pair of bonds along same subchain (f2 in Fig. 5.5B). No additional bending restrictions can be added to the diagram f2 in Fig. 5.5B since any pair of bonds, other than the pair connected by the dashed line in diagram f2, must lie on separate subchains. The diagrammatic representation for the partition function of Eqs. (34) and (35), therefore, contains all possible diagrams with combinations of (straight) bond correlation lines, curved interaction lines, and curved (dashed) bending constraint lines.

IV. EVALUATION OF DIAGRAMS FROM THE CLUSTER EXPANSION

Dudowicz and Freed (1991a) and Nemirovsky et al. (1987) describe the general rules for evaluating the bond and bond/pair diagrams. The extension of the LCT to include a description of semiflexible chains requires nontrivial alterations of the diagrammatic rules for evaluating bond and bond/pair diagrams, but retains many basic features. Therefore, we outline the process for evaluating diagram a in Fig. 5.2 to provide the necessary details and notation for describing the generalization to treat the corrections in Eqs. (34) and (35) that contain bending energies.

The pair of consecutive bonds in diagram a of Fig. 5.2 can lie either on the same or different subchains within the same chain c from species s. Each bond in diagram a contributes a factor of the associated correlation correction X_{bond}, producing for this diagram a quadratic term in the t_i of Eqs. (34) and (35), namely, a factor of $X_{bond, 1} X_{bond, 2}$. The contribution to Eqs. (34) and (35) from all such terms due to correlation corrections between consecutive pairs of bonds is written schematically as

$$\frac{(N_l - \sum_{s=1}^{p} C_s M_s)!}{N_l!} \sum_{\{\mathbf{r}_{\lambda,\alpha}^{s,c}\}}' \sum_{s=1}^{p} \sum_{c=1}^{C_s} \sum_{\substack{\text{all consecutive} \\ \text{pairs of bonds}}} X_{bond, 1} X_{bond, 2} \qquad (36)$$

The prime on the summation in Eq. (36) implies that the sum is subject to the excluded volume constraints as in, for example, Eq. (4). The restricted sum over monad positions in Eq. (36) may be separated into two sums, one for the three monads appearing in the $X_{bond, 1} X_{bond, 2}$ factor and the other for all the remaining monads. Let α, $\alpha + 1$, and $\alpha + 2$ label the three monad positions in the correlation factor $X_{bond, 1} X_{bond, 2}$. For each specified position $(\mathbf{r}_\alpha, \mathbf{r}_{\alpha+1}, \mathbf{r}_{\alpha+2})$ of the three monads α, $\alpha + 1$, and $\alpha + 2$, the remaining restricted sums yield the simple combinatorial factor $(N_l - 3)!/(N_l - \sum_{s=1}^{p} C_s M_s)!$. A restricted sum over the three correlating monad positions remains, and Eq. (36) is thereby reduced to the form,

$$\frac{1}{N_l(N_l - 1)(N_l - 2)} \sum_{s=1}^{p} \sum_{c=1}^{C_s} \sum_{\substack{\text{all consecutive} \\ \text{pairs of bonds}}} \sum_{\mathbf{r}_{\alpha}^{s,c} \neq \mathbf{r}_{\alpha+1}^{s,c} \neq \mathbf{r}_{\alpha+2}^{s,c}}' X_{bond, 1} X_{bond, 2} \qquad (37)$$

where the prime implies excluded volume constraints for the three monad positions and the factor of $1/N_l(N_l - 1)(N_l - 2)$ is the remainder of the combinatorial portions after cancellation. More generally, this combinatorial portion contributes an overall factor from any diagram of $[\prod_{j=0}^{N_v - 1} (N_l - j)]^{-1}$, where N_v is the number of monads (or vertices) in the diagram.

Substituting the definition of X_{bond} from Eq. (23) facilitates further evaluation of Eqs. (36) and (37). The present work employs a momentum space representation (Dudowicz and Freed 1991a) for calculating the correction terms, but an alternative coordinate-space (Baker et al., 1993) method may be used. The Fourier transform of the Kronecker delta $\delta(\mathbf{r}_k, \mathbf{r}_j + \mathbf{a}_\beta)$ replaces $\delta(\mathbf{r}_k, \mathbf{r}_j + \mathbf{a}_\beta)$ with a sum over wave vectors $\mathbf{q}$ using the lattice identity

$$\delta(\mathbf{r}_k, \mathbf{r}_j + \beta) = \frac{1}{N_l} \sum_{\mathbf{q}} \exp[i\mathbf{q} \cdot (\mathbf{r}_k - \mathbf{r}_j - \mathbf{a}_\beta)] \tag{38}$$

where the wave vector summation index $\mathbf{q}$ runs over the reciprocal space unit cell. (The reciprocal-space unit cell for the three-dimensional cubic lattice is defined by the wave vectors $\mathbf{q}$ whose individual components obey the relation $q_x = 2\pi n_x/a_x$, with $n_x = 0, 1, \ldots, N_l^{1/3} - 1$ and $a_x = \sqrt[3]{N_l}$ the magnitude of the lattice vector.) The wave vector summation index for the bond between monads α and $\alpha + 1$ on subchain λ of chain c from species s is denoted as $\mathbf{q}_{\lambda, \alpha}^{s, c}$. The average contribution $(1/N_l)$ to the connectivity constraint $\delta(\mathbf{r}_k, \mathbf{r}_j + \mathbf{a}_\beta)$ between monads $\mathbf{k}$ and $\mathbf{j}$ arises from the $\mathbf{q} = 0$ term on the right-hand side of Eq. (38). Since the average contribution is already subtracted from the Kronecker delta representing the connectivity constraint in the definition for X_{bond} in Eq. (23), only the nonvanishing wave vectors contribute to X_{bond}. Thus X_{bond} may be rewritten as:

$$X_{\text{bond}} = \frac{N_l}{z} \sum_{\beta=1}^{z} \frac{1}{N_l} \sum_{\mathbf{q} \neq 0} \exp[i\mathbf{q} \cdot (\mathbf{r}_k - \mathbf{r}_j - \mathbf{a}_\beta)]$$

$$= \frac{1}{z} \sum_{\mathbf{q} \neq 0} f(\mathbf{q}) \exp[i\mathbf{q} \cdot (\mathbf{r}_k - \mathbf{r}_j)] \tag{39}$$

where the nearest-neighbor structure factor $f(\mathbf{q})$ is

$$f(\mathbf{q}) = \sum_{\beta=1}^{z} \exp(-i\mathbf{q} \cdot \mathbf{a}_\beta) \tag{40}$$

Substituting the definitions for X_{bond} from Eq. (39) into Eqs. (36) and (37) produces the equivalent expression for diagram a of Fig. 5.2 as

$$\frac{1}{N_l(N_l - 1)(N_l - 2)} \sum_{s=1}^{p} \sum_{c=1}^{C_s} \sum_{\substack{\text{all consecutive} \\ \text{pairs of bonds}}} \sum_{\mathbf{r}_\alpha^{s,c} \neq \mathbf{r}_{\alpha+1}^{s,c} \neq \mathbf{r}_{\alpha+2}^{s,c}} \frac{1}{z^2}$$

$$\times \sum_{\mathbf{q}_{\alpha}^{s,\,c} \neq 0}^{z} \sum_{\mathbf{q}_{\alpha+1}^{s,\,c} \neq 0}^{z} f(\mathbf{q}_{\alpha}^{s,\,c}) f(\mathbf{q}_{\alpha+1}^{s,\,c}) \exp\{i[\mathbf{q}_{\alpha}^{s,\,c} \cdot (\mathbf{r}_{\alpha}^{s,\,c} - \mathbf{r}_{\alpha+1}^{s,\,c})$$

$$+ \mathbf{q}_{\alpha+1}^{s,\,c} \cdot (\mathbf{r}_{\alpha+1}^{s,\,c} - \mathbf{r}_{\alpha+2}^{s,\,c})]\} \tag{41}$$

The evaluation of contributions from each diagram in the cluster expansion follows the above treatment of diagram a in Fig. 5.2, with the sums first performed over the positions of all monads not explicitly represented in the diagram. This process leaves the summations over the bonded (or otherwise interacting) cluster of monads in the diagram, subject to the usual excluded volume constraints within the cluster. Thus each diagram contribution corresponds to an exact partition function for the specific cluster of monads, depicted by the diagram, in a Flory-like mean field of the surroundings.

A. Contracted Diagrams

The excluded volume constraints on the summation over the monad positions in the cluster are now converted to a series of unconstrained summations. For example, excluded volume constraint between the two monads i and j prevents these two monads from occupying the same lattice site, a requirement made explicit by the condition $1 - \delta(i, j)$. The set of excluded volume constraints for a single diagram is schematically represented by

$$\sum_{\{k\}}' h(\{k\}) = \sum_{\{k\}} \prod_{i > j} [1 - \delta(i, j)] h(\{k\}) \tag{42}$$

where $\{k\} \equiv \{\mathbf{r}_{\lambda,\,\alpha}^{s,\,c}\}$ is the set of positions for the monads represented in the diagram and where i (j) represent the ith $(j$th$)$ element in the set $\{k\}$. Multiplying the product on the right-hand side of Eq. (42) enables conversion of a restricted sum into a series of unrestricted summations with reduced numbers of distinct indices. For example, the leading term of unity in the product produces a set of monads that are not subjected to excluded volume constraints, and hence the resultant unconstrained sum overcounts the number of allowable configurations (as restricted by the original excluded volume constraints). Successive terms in the expansion of the product in Eq. (42) include increasing numbers of Kronecker deltas and thereby correct for the configurations overcounted by terms with fewer deltas.

Equation (42) converts the restricted summation in Eq. (41) to a series of unrestricted sums, written symbolically as

$$\sum_{k_1,\,k_2,\,k_3}' h(\{k\}) = \sum_{k_1,\,k_2,\,k_3} [1 - \delta(k_1, k_2)][1 - \delta(k_1, k_3)][1 - \delta(k_2, k_3)] h(\{k\})$$

$$\tag{43}$$

The operation of each factor $\delta(i, j)$ upon the LCT diagrams is represented diagrammatically by coalescing the pair of monads i and j from the original diagram, producing what we term *contracted diagrams*. Thus each original diagram with n restricted summations is converted through the process of Eq. (42) into a series of 2^n contracted diagrams, each with unrestricted summations over all remaining monad positions.

The removal of the excluded volume restrictions on the summations over monad positions greatly simplifies the evaluation of the resulting contracted diagram. The overwhelming majority of the 2^n contracted diagrams, produced from an original diagram with n monads, vanish identically, (Nemirovsky et al., 1987; Brazhnik and Freed, 1996) leaving a small subset of easily recognizable diagrams. For example, the identity from lattice Fourier transforms,

$$\sum_{\mathbf{r}_j} \exp(i\mathbf{q} \cdot \mathbf{r}_j) = N_l \, \delta(\mathbf{q}, 0) \tag{44}$$

implies that contracted diagrams vanish identically if they contain dangling monads, which are defined to be monads that are connected to the remainder of the contracted diagram by only one bond. Thus the contributing contracted diagrams must contain only closed loops (see Fig. 5.6).

As described in previous developments of the LCT, the contribution from any diagram can be decomposed into the product of two components, a monomer-structure-dependent (lattice-independent) combinatorial factor γ_D and a lattice-dependent (monomer-structure-independent) connectivity factor D_B. We again illustrate this process for diagram a of Fig. 5.2 to provide the necessary ingredients for including the bending energies. Equation (41) is thus represented symbolically as

$$\frac{1}{\prod_{j=0}^{2} (N_l - j)} \left(\sum_{s=1}^{p} C_s N_2(s) \right) \sum_{d} f_{B, d} R_{B, d} \tag{45}$$

where $N_2(s)$ is the total number of pairs of consecutive bonds in a chain from species s, d is a sequential counting label designating the unique contributing contracted diagrams, and B is the number of bonds in the original diagram. The combinatorial factor γ_D is $\gamma_D = (\sum_{s=1}^{p} C_s N_2(s))/\prod_{j=0}^{2} (N_l - j)$, while the connectivity factor $D_B = \sum_d f_{B, d} R_{B, d}$ contains the remainder of Eq. (45). The numerator of γ_D is specific to the architecture of the original diagram and is simply the total number of ways of selecting a set of monads with the original diagram architecture from all the polymers in the system. The values of diagrams involving wavy lines, such as diagram

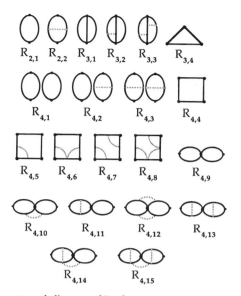

$$R_{2,1} \quad R_{2,2} \quad R_{3,1} \quad R_{3,2} \quad R_{3,3} \quad R_{3,4}$$

$$R_{4,1} \quad R_{4,2} \quad R_{4,3} \quad R_{4,4}$$

$$R_{4,5} \quad R_{4,6} \quad R_{4,7} \quad R_{4,8} \quad R_{4,9}$$

$$R_{4,10} \quad R_{4,11} \quad R_{4,12} \quad R_{4,13}$$

$$R_{4,14} \quad R_{4,15}$$

Figure 5.6. All the contracted diagrams $\{R_{B,d}\}$ necessary for evaluating the diagrams in or derived from Fig. 5.2. Nonvanishing contracted diagrams always contain closed loops, such that at least two solid lines connect each distinct vertex (black dots). Upon contracting the three- and four-bond diagrams of Fig. 5.2, many more contracted diagrams are produced than depicted, but only the contracted diagrams that contribute to leading order in N_i are presented. The structure of a cubic lattice implies that all contracted diagrams involving odd cycles (e.g., $R_{3,4}$) do not contribute to the free energy, thus simplifying many calculations.

j in Fig. 5.2, contain counting factors, such as the number $N_{x,y}$ of ways to decompose a chain into two disconnected pieces with x and y consecutive bonds in each respective piece and one or more intervening bonds between them. Nemirovsky and co-workers (1992) provide relations between the $N_{x,y}$ and the set $\{N_i\}$ of numbers of i consecutive bonds in a chain. Appendix A describes some generalized relations between the $N_{x,y}$ and the $\{N_i\}$ for describing semiflexible chains.

The coefficient $f_{B,d}$ in Eq. (45) is a combination of the numerical coefficients $N_{B,d}$ and $C_{B,d}$ (defined below) that arise in the steps of producing contracted diagrams from original diagrams. $N_{B,d}$ is the number of unique ways that the contracted diagram can be generated from the original diagram through the contraction process. The combinatorial contribution is $C_{B,d} = \prod_{j=1}^{N'_v} [(-1)^{k_j - 1}(k_j - 1)!]$, where N'_v vertices are present in the contracted diagram and k_j of the original vertices are coalesced onto the jth vertex of the contracted diagram (i.e., k_j monads from the original diagram are forced to be at the same position on the lattice in the contracted diagram). This expression for $C_{B,d}$ emerges from expanding the $\prod_{i>j}[1-\delta(i,j)]$ to convert the restricted sums into unrestricted sums in Eq. (42). Both $N_{B,d}$ and $C_{B,d}$ depend on the specific initial diagram and on the final contracted diagram.

The values $R_{B,d}$ of the contracted diagrams are lattice structure dependent but are independent of the specific monomer structures. As described in more detail by previous references (Nemirovsky et al., 1987), the $R_{B,d}$ are evaluated by summing over the nonzero wave vectors (or momenta) to yield the symbolic form:

$$R_{B,d} = \sum_{\{\mathbf{q}\}\neq 0} N_l^{N_v} \prod_{j=1}^{N_v'} \delta_j(\sum \mathbf{q}, 0) \prod_{t=1}^{B} \frac{f(\mathbf{q}_t)}{z} \tag{46}$$

where the set $\{\mathbf{q}\}$ contains one wave vector $\mathbf{q}$ for each of the B bonds in the contracted diagram and δ_j is the Kronecker delta representing the momentum conservation condition ($\sum \mathbf{q} = \mathbf{0}$) for the momenta converging on the jth vertex. Recall that application of this momentum conservation condition to a dangling monad yields $\delta(\mathbf{q}, \mathbf{0})$ where $\mathbf{q}$ is the momentum vector for the bond connected to this monad. Thus the $\mathbf{q} \neq 0$ condition on the summand in Eq. (46) implies that the corresponding $R_{B,D}$ vanishes. For example, diagram a in Fig. 5.2 has $B = 2$ bonds but only one contributing contracted diagram [$R_{2,1}$ of Fig. 5.6]. Therefore, diagram a of Fig. 5.2 yields $N_{2,1} = 1$, $C_{2,1} = -1$, and

$$R_{2,1} = \sum_{\mathbf{q}_1, \mathbf{q}_2 \neq 0} N_l^2 \, \delta(\mathbf{q}_1 - \mathbf{q}_2, 0) \frac{f(\mathbf{q}_1)f(\mathbf{q}_2)}{z^2} \tag{47}$$

Table I presents the values of all contributing contracted diagrams $R_{B,d}$ from Fig. 5.6, including the value of Eq. (47). Substituting the values for $N_{2,1}$, $C_{2,1}$, and $R_{2,1}$ into Eq. (45) yields the contribution from diagram a of Fig. 5.2 as

$$\frac{1}{\prod_{j=0}^{2}(N_l - j)} \left(\sum_{s=1}^{P} C_s N_2(s)\right)(-1)\left(\frac{N_l^3}{z}\right) \tag{48}$$

B. Bending Energies

The inclusion of bending energy corrections requires modification of the above treatment for flexible chains. Consider now the bond/bend diagram D is Fig. 5.3. The combinatorial factor γ_D for the bond/bend diagram is similar to the γ_D for diagram a in Fig. 5.2, with only two differences. First,

TABLE I
All Contributing Contracted Diagrams from the
Diagrams in Fig. 5.2

Diagram Label in Fig. 5.6	Diagram Value
$R_{2,1}$	$\dfrac{N_i^3}{z}$
$R_{2,2}$	$-R_{2,1}$
$R_{3,1}$	$\dfrac{N_i^4}{z^2}$
$R_{3,2}$	$-R_{3,1}$
$R_{3,3}$	$-R_{3,1}$
$R_{3,4}$	0
$R_{4,1}$	$R_{2,1}^2$
$R_{4,2}$	$R_{2,1}R_{2,2}$
$R_{4,3}$	$R_{2,2}^2$
$R_{4,4}$	$\dfrac{3N_i^5}{z^2}$
$R_{4,5}$	$\dfrac{-2N_i^5}{z^2}$
$R_{4,6}$	$-R_{4,4}$
$R_{4,7}$	$\dfrac{N_i^5}{z}\left(1+\dfrac{1}{z}\right)$
$R_{4,8}$	$-R_{4,4}$
$R_{4,9}$	$\dfrac{N_i^5}{z^2}$
$R_{4,10}$	0
$R_{4,11}$	$-R_{4,9}$
$R_{4,12}$	$\dfrac{N_i^5}{z}\left(1-\dfrac{1}{z}\right)$
$R_{4,13}$	$R_{4,9}$
$R_{4,14}$	$-R_{4,9}$
$R_{4,15}$	$-R_{4,9}$

diagram a in Fig. 5.2 represents all possible pairs of consecutive bonds, regardless of which subchains contain the bonds, while the bond/bend diagram arises only when the two consecutive bonds lie on the same subchain. In general, diagrams without bending restrictions include contributions from all possible combinations of bonds that produce the cluster architecture depicted in the diagram, while the bending restrictions Y_{bend} limit the number of combinations by constraining the affected pair of consecutive bonds to lie on the same subchain. Second, a factor of K is present for each bending restriction Y_{bend} in the diagram. Thus γ_D for the bond/

bend diagram D of Fig. 5.3 is given by:

$$\gamma_D = \frac{1}{N_l(N_l - 1)(N_l - 2)} \sum_{s=1}^{p} \sum_{c=1}^{C_s} \sum_{\lambda=1}^{L_1^s} \sum_{\alpha=1}^{n_\lambda^{s,c} - 2} K_{\lambda, \alpha}^{s, c} \tag{49}$$

where $K_{\lambda, \alpha}^{s, c}$ is defined in Eq. (30). Note that γ_D in Eq. (49) cannot readily be expressed as multiples of C_s and M_s because the factor of $K_{\lambda, \alpha}^{s, c}$, in practice, may differ for topologically distinct pairs of consecutive bonds along the same subchain. Previous applications of the LCT (Foreman and Freed, 1997a,b; Foreman et al., 1997) consider chains in which all bending energies for species s are identical (i.e., $K_{\lambda, \alpha}^{s, c} = K_s$), allowing γ_D in Eq. (49) to be written as the product of K_s with a γ_D that is similar to the one obtained for diagram a in Fig. 5.2:

$$\gamma_D(\text{bond/bend}) = \frac{\sum_{s=1}^{p} C_s N_2^{(0)}(s) K_s}{\prod_{j=0}^{2} (N_l - j)} \tag{50}$$

where $N_2^{(0)}(s)$ is the total number of pairs of consecutive bonds that each lie along the same subchain in a chain from species s.

Diagram a in Fig. 5.2 and its derivative bond/bend diagram in Fig. 5.3D contribute contracted diagrams with identical connectivities to their respective D_B, but the presence of the bending restrictions in the bond/bend diagrams affects the evaluation of the contracted diagrams. Upon addition of the bending factors, the contracted diagram $R_{2,1}$, produced from diagram a of Fig. 5.2, is converted to the corresponding $R_{2,2}$ (see Fig. 5.6) associated with the bond/bend diagram D in Fig. 5.3: The evaluation of the contracted diagram $R_{2,2}$ proceeds by replacing the product over nearest-neighbor structure factors in Eq. (47) as follows:

$$\frac{f(\mathbf{q}_1) f(\mathbf{q}_2)}{z^2} \to \left(\frac{f(\mathbf{q}_1 + \mathbf{q}_2)}{z} - \frac{f(\mathbf{q}_1) f(\mathbf{q}_2)}{z^2} \right). \tag{51}$$

Appendix B provides a general discussion of how to convert the contracted diagrams involving only correlation corrections to the corresponding contracted diagrams with bending restrictions.

Semiflexible chains lie intermediate between the purely flexible ($E_{\lambda, \alpha}^{s, c} \to 1$) and rigid rod ($E_{\lambda, \alpha}^{s, c} \to 0$) limits studied in the previous applications of the LCT (Dudowicz and Freed, 1996b; Freed and Dudowicz, 1996; Li and Freed and 1995; Houston et al., 1993). Different classes of diagrams contribute to $O(1/z^n)$ in these two limits, requiring the semiflexible chain LCT to incorporate diagrams appropriate to both limits. The rod-limit LCT partition function involves contributions from large numbers of correlating bonds (Huston et al., 1993) because of the simplifications accorded in requiring only one bond direction to specify the direction of all bonds within a single linear chain. The semiflexible chain LCT description does not enjoy such a simplification, and some compromise in retaining rod-limit diagrams is necessary. Particularly, the semiflexible chain LCT retains all diagrams necessary to obtain the exact flexible chain limit through a certain order in $1/z$ [$O(1/z^2)$ here]. This selection of diagrams yields expressions for the Helmholtz free energy in the rod limit that are nearly identical to those obtained previously, except for a single (ill-behaved and therefore undesirable) term from the rod LCT.

C. Interaction Energies

The evaluation of diagrams with interaction lines introduces a slightly different complication. The similarity between the X_{pair} and X_{bond} factors allows the van der Waals interaction corrections to be treated just as the bonding corrections, but with a different overall prefactor. The mathematical representation of diagram b in Fig. 5.2 and the intermediate diagram in Fig. 5.4A are identical, except that the intermediate diagram in Fig. 5.4A has a factor of G_{pair} [see Eqs. (24) and (25)], which multiplies the sums over wave vectors, and the connectivity constraints for the two diagrams differ. Both of these features only affect the value of γ_D, as explained below. The connectivity factor D_B, on the other hand, is evaluated by treating the interaction lines in the contracted diagrams as bonds; that is, B now is the total number of lines, either from bonds or from interactions, in the contracted diagram. For example, diagram a of Fig. 5.2 produces only one bond/pair diagram with one interaction line and one bond (diagram B in Fig. 5.3). This diagram yields a D_B identical to the D_B of Eq. (48) ($= -N_l^3/z^2$).

The γ_D for a particular bond/pair diagram is a function of G_{pair}. The thermodynamic limit ($N_l \to \infty$) produces the simplification that the denominator of G_{pair} in Eq. (26) may be replaced with a factor of unity: that is, $G_{pair} \to z f_{pair}/N_l$ as $N_l \to \infty$. For example, the combinatorial factor γ_D for the bond/pair diagram B of Fig. 5.3 is calculated by enumerating the total number of G_{pair} contributions from all interacting pairs of monads such that one of the paired monads is also bonded to another monad. This total number may be calculated by separately summing over all bonds and all pairs of interacting monads in the system and by restricting the positions of one of the bonded monads and of one of the interacting monads to be identical. The two monad positions k_0 and k_1 (in shorthand notation) from the bond and the positions of the two interacting monads k_2 and k_3 can be paired in four distinct ways, leading to the symbolic representation for γ_D as

$$\gamma_D = \frac{z}{N_l} \sum_{\{k_0, k_1\}} \sum_{k_2} \sum_{k_3}' [\delta(k_0, k_2) + \delta(k_0, k_3) + \delta(k_1, k_2) + \delta(k_1, k_3)] f_{k_2, k_3} \quad (52)$$

where the first sum is over all bonds in the system, the remaining sums run over the positions of the interacting monads, and f_{k_2, k_3} refers to a Mayer f function as in Eq. (13). The prime over the sums indicates that the position of the noninteracting monad in the bond is distinct from that of the non-bonded interacting monad.

The treatment of γ_D follows similarly for other diagrams with bending constraints. In general, the presence of n solid curved lines in a diagram leads to $2n$ sums over the locations of single monads in the expression corresponding to the diagram. The Kronecker deltas, analogous to those in Eq. (52), guarantee that the restrictions on the monad positions are appropriate to describe the original diagram. These deltas are nearly sufficient to produce the symmetry factors described in earlier work (Dudowicz and Freed, 1991a), provided that each pair of interacting monads is distinct; otherwise a trivial symmetry factor appears. The present work permits each monad to interact differently with every other monad in the system. This general treatment of the specific interactions between the monads representing different chemical groups prevents the reduction of the sums in Eq. (52) to expressions containing factors of C_s and M_s. Previous applications of the LCT take all monads from species s and s' as interacting with a single averaged van der Waals energy $\varepsilon_{ss'}$. Introduction of this simplification into Eq. (52) allows the sums to be evaluated, generating functions of C_s and M_s.

V. THERMODYNAMIC QUANTITIES

The LCT Helmholtz free energy F follows by applying the standard thermodynamic definition to the partition function from Eqs (34) and (35)

$$\frac{F}{k_B T} = -\ln W(\{n_\mu\}, \{M_\mu\}) = -\ln W^{MF} - \ln(1 + \mathscr{C}) \qquad (53)$$

where $\mathscr{C}$ represents the sum of all corrections to the zeroth-order mean-field approximation in Eqs. (34) and (35). The $\ln(1 + \mathscr{C})$ term in Eq. (51) may be expanded in a Taylor series:

$$\ln(1 + \mathscr{C}) = \mathscr{C} - \frac{\mathscr{C}^2}{2} + \frac{\mathscr{C}^3}{3} - \cdots \qquad (54)$$

Because the Helmholtz free energy F is extensive, the left-hand side of Eq. (54) must scale as the first power of N_l. Individual terms contributing to $\mathscr{C}$ are hyperextensive, proportional to an integer power of N_l greater than unity. An example is diagram i of Fig. 5.2, which scales as N_l^2. Rearranging the series on the right-hand side of Eq. (54) into cumulants (Kubo, 1962), each of which is a sum of several original diagrams, identically eliminates all hyperextensive contributions. The contribution from diagram i in Fig. 5.2 that scales as N_l^2 is exactly cancelled by the square of the contribution from diagram a in Fig. 5.2 that emerges from the $\mathscr{C}^2/2$ term on the right-hand side of Eq. (54). The sum of diagrams for each cumulant is designated by the term contributing from $\mathscr{C}$ in Eq. (54). This sum may also be ordered as an expansion in $1/z$. We retain all diagrams through second order in $1/z$. In addition, we follow previous applications of the LCT in using a high-temperature expansion of the Meyer f functions $f_{ss'} = \varepsilon_{ss'} + \varepsilon_{ss'}^2/2 + \varepsilon_{ss'}^3/3!$ $+ \cdots$, in which only powers of $\varepsilon_{ss'}$ less than or equal to two are retained and $\varepsilon_{ss'}$ is treated formally as order $1/z$.

The general schematic form of the Helmholtz free energy F may thus be written from Eqs. (53) and (54) as

$$\frac{F}{k_B T} = -\ln W(\{n_\mu\}, \{M_\mu\}) = -\ln W^{MF} - \sum_d [\gamma_D D_B]_d^c \tag{55}$$

where the superscript c indicates that the sum runs over the cumulants for each diagrammatic structure d in $\mathscr{C}$. The zeroth-order mean-field contribution to the free energy in Eq. (55) is given by,

$$-\ln W^{MF} = N_l \left\{ \phi_v \ln \phi_v + \sum_{s=1}^p \frac{\phi_s}{M_s} \left[\ln\left(\frac{\sigma_s \phi_s}{z^{L_1^s} M_s}\right) + N_1(s) \right] \right.$$
$$\left. + \frac{1}{N_l} \left[\sum_{k_1, k_2} \ln\left(1 + \frac{z f_{k_1 k_2}}{N_l}\right) + \sum_{(s, c, \lambda, \gamma)} \ln(\{z-1\}E_{\lambda, \gamma}^{s, c} + 1) \right] \right\} \tag{56}$$

where ϕ_v is the void volume fraction and the last sum in Eq. (56) is over all pairs of consecutive bonds in every subchain λ from each chain c of every species s. The logarithm containing the Mayer f function $f_{k_1 k_2}$ is expanded using Eq. (54). Only the lowest-order $z f_{k_1 k_2}/N_l$ term contributes in the thermodynamic limit, transforming Eq. (56) into

$$-\ln W^{MF} = N_l \left\{ \phi_v \ln \phi_v + \sum_{s=1}^p \frac{\phi_s}{M_s} \left[\ln\left(\frac{\sigma_s \phi_s}{z^{L_1^s} M_s}\right) + N_1(s) \right] \right.$$
$$\left. + \frac{1}{N_l} \left(\sum_{k_1, k_2} \frac{z f_{k_1 k_2}}{N_l} + \sum_{(s, c, \lambda, \gamma)} \ln(\{z-1\}E_{\lambda, \gamma}^{s, c} + 1) \right) \right\} \tag{57}$$

It is straightforward to prove that if all monads from species s and s' interact identically (that is, $f_{k_1k_2} = f_{ss'}$ for $k_1 \in s$, $k_2 \in s'$) and if only the first term in the expansion of $f_{k_1k_2}$ is retained (i.e., the $\varepsilon_{ss'}$ term), the Flory semiflexible chain Helmholtz free energy is recovered. After reordering the cumulant contributions to Eq. (55) as an expansion in $1/z$ and inserting Eq. (57) into Eq. (55), the LCT Helmholtz free energy takes the final form

$$\frac{F}{N_l k_B T} = \phi_v \ln \phi_v + \sum_{s=1}^{p} \frac{\phi_s}{M_s}\left[\ln\left(\frac{\sigma_s \phi_s}{z^{L_1^s} M_s}\right) + N_l(s)\right]$$
$$+ \frac{1}{N_l}\left[\sum_{(s,\,c,\,\lambda,\,\gamma)} \ln(\{z-1\}E_{\lambda,\,\gamma}^{s,\,c} + 1)\right] + \sum_{t=1}^{6} C(t) \tag{58}$$

where the correction terms $C(t)$ are given in Appendix C and where the mean-field term from the $\sum_{k_1k_2} f_{k_1k_2}$ in Eq. (57) is included in the correction $C(2)$ of Eq. (58).

VI. DISCUSSION

The lattice cluster theory (LCT) is extended to describe polymer systems containing semiflexible chains and specific interactions. This extension is designed to facilitate the development of a more robust qualitative (and often quantitative) theory that explains how various thermodynamic properties of multicomponent polymer systems depend on general structural and energetic features of the individual monomers. The semiflexible chain LCT employs a lattice model and thereby sacrifices some microscopic-scale details in order to achieve both analytic tractability and an enormous increase in computational simplicity. The LCT, however, retains enough realism for readily translating these microscopic scale details into predictions of macroscopic properties. Previous applications of the LCT have provided illumination for a variety of microscopic aspects of polymer system thermodynamics, including the establishment of a microscopic basis for changes in the Flory χ parameter with alterations of the system properties (Freed and Pesci, 1987, 1989; Freed and Dudowicz, 1992; Dudowicz and Freed, 1991b), explanations for the successes and failures of several phenomenological models (Foreman and Freed, 1997b; Foreman et al., 1997), and the description of the pressure dependence of polymer blends (Dudowicz and Freed, 1995). These successful applications of the LCT motivate the development of improved, more realistic descriptions of the polymers by the LCT, without significant loss of computational speed. The semiflexible LCT represents just such an improvement over previous versions of the LCT and has already provided insights (some of which are summarized in the introduction) into the properties of polyolefin blends as well as into various models proposed to predict their phase behavior (Foreman and Freed, 1997b; Foreman et al., 1977; Freed et al., to be published). The recent applications of the semiflexible LCT only provide the final form of the free energy for specific limiting cases. The present work provides a detailed account of the derivation, including the generalizations necessary for a treatment of group-specific interactions.

The previous LCT description of polymers does include some measure of semiflexibility, due to the local excluded volume constraints, especially in models with extended monomer structures. The chain flexibility in these models, however, is not adjustable as is necessary to represent the influences of structure and steric hindrances in realistic polymers. The semiflexible LCT permits the treatment of adjustable chain stiffness by assigning to bent configurations of pairs of consecutive bonds the positive energy difference (called the bending energy) between *trans* and *gauche* conformations. Collinear configurations are assigned a vanishing bending energy.

The inclusion of semiflexibility in structured monomer chains introduces several additional difficulties in calculating the thermodynamic properties beyond those already overcome by the LCT. For example, different bending energies may apply to chemically distinct portions of the monomers, backbone, etc. Therefore, rather involved additional bookkeeping is necessary to specify the position and arrangement of bond clusters within a chain, such that different clusters with formally identical topologies can be distinguished according to their occurrence in different places within the chain structure. Appendices A and B present the technical details associated with this bookkeeping. In addition, the inclusion of *trans–gauche* energy differences introduces a new category of LCT diagrams whose evaluation requires sophisticated modifications of previous procedures for treating flexible chain systems. Furthermore, the semiflexible chains, in principle, span the range of behavior from purely flexible chains to purely rigid rods, but different diagrams contribute to the LCT expansion in these two limits. We retain all diagrams that exactly reproduce the flexible chain limit when all bending energies vanish identically. This selection of diagrams nearly reproduces the rod-limit treatment (Huston et al., 1993), except for a single poorly behaved (and hence undesirable) term from the rod LCT.

The present chapter also devises a method to include specific interactions. The specific interactions are important in realistic models because large energetic differences generally exist for interactions between different pairs of united atom groups, even ones considered as being chemically similar. For example, typical off-lattice Lennard–Jones energy parameters for (CH_3)–(CH_3) and (CH_2)–(CH_2) united atom group interactions are 90 and 70 K, respectively (Mondello and Grest, 1995; Jorgensen, 1981; Toxvaerd, 1990). All theories of polymer systems (including Flory–Huggins theory) recognize that the phase behavior of polymer blends is, however, extremely sensitive to small changes in individual interaction energies. Previous LCT computations demonstrate a similar sensitivity to monomer structure (Dudowicz and Freed, 1996b; Freed and Dudowicz, 1996). Since the extended lattice model and the LCT both differentiate between monomer structures, the various chemical groups in the monomers should be permitted to interact with different energies, because these specific interactions surely affect the phase behavior of many polymer blends. Atomistic simulations by Yoon and co-workers (Smith et al., 1996) demonstrate how specific interactions affect poly(ethylene oxide) intramolecular conformations. Previous applications of the LCT employ the simplified model of monomer-averaged van der Waals interactions in which each united atom group from species s interacts with every united atom group from species s' through the same energy $\varepsilon_{ss'}$. The present chapter provides a general representation in which all united atom groups in the system interact with distinct energies. The monomer-averaged interaction model emerges as a special limit of the more general treatment of specific interactions.

The semiflexible chain LCT may be applied to a variety of other interesting topics beyond those recently considered (concerning the influences of semiflexibility and monomer structures on blend phase diagrams). For example, the Gibbs–DiMarzio theory (DiMarzio and Gibbs, 1958; Gibbs and DiMarzio, 1958) of polymer glasses considers chain semiflexibility as an important component of the configurational entropy that determines the location of the glass transition temperature T_g. The Gibbs–DiMarzio theory employs a Guggenheim-type model (1952), which is unable to distinguish between different chain architectures (Foreman and Freed, 1995). The present semiflexible chain LCT describes monomer structure and variable chain stiffness, features which should permit the semiflexible chain LCT to describe the relationship between the T_g for blends and those for the constituents. Another application involves determining the influence of chain stiffness upon the surface properties of polymers. Stiffer chains tend to order more easily than flexible chains near surfaces. The semiflexible chain LCT may be used to provide simple molecular-based models of the polymer–surface interactions.

ACKNOWLEDGMENTS

K. W. Foreman thanks the Department of Education for a GAANN fellowship. This research is supported, in part, by NSF Grant DMR-9530403. We appreciate helpful comments from Stuart Rice, Jacek Dudowicz, and Mary Jay.

APPENDIX A.   TOPOLOGY-INDEPENDENT RELATIONS BETWEEN COUNTING FACTORS CONTRIBUTING TO γ_D

The prefactor coefficient γ_D contains a sum associated with the number of ways to select the depicted set of bonds and monads from all those in a chain. The introduction of bending restrictions leads to special complications in treating diagrams with wavy lines. These diagrams yield a total of $N_{x_1, \ldots, x_n}$ identical contributions to γ_D. This number $N_{x_1, \ldots, x_n}$ of disconnected portions with $x_1, \ldots, x_n$ bonds in each of the n different portions of a chain may be calculated by taking the product of the total number of ways of finding each portion separately in the chain and then subtracting the overcounting. The overcounting arises from the number of ways that one or more of the portions at least partially overlap with each other (Nemirovsky et al., 1992). For example, the number $N_{1,1}$ of pairs of disconnected single bonds in a chain provides the simplest relation for flexible chains:

$$2! N_{1,1} = N_1^2 - N_1 - 2N_2 \tag{59}$$

where N_i is the number of i consecutive bonds in the chain. The N_1^2 term on the right-hand side of Eq. (59) is the product of the number of ways of finding each bond separately in the chain (independent of their mutual overlap), while the N_1 and N_2 terms in Eq. (59) count the total number of times when the two bonds coincide and when they are sequentially bonded (i.e., share a single monad), respectively. The factor of two multiplying N_2 in Eq. (59) appears because the second bond may be bonded to either end of the first bond. The expression in Eq. (59) and similar ones for $N_{2,1}$ and $N_{2,2}$ are all that are required here for treating the diagrams without bending constraints. Diagrams with bending restrictions, however, require the numbers of various subsets of the bonds that contribute to $N_{2,1}$ and $N_{2,2}$ and that are affected by bending energy constraints. Therefore, this appendix presents the analysis necessary to extract these new counting factors.

We begin with a derivation of Eq. (59) that develops the techniques necessary to treat the counting factors considered when, for example, bending constraints are operative. The N_1^2 term in Eq. (59) is identically cancelled by contributions from other diagrams, leading to a final free energy expression that is well behaved in the long-chain limit. This cancellation is more readily enforced by using a representation similar to the right-hand side of Eq. (59). The symmetry factor $\sigma = 2!$ on the left-hand side of Eq. (59) reflects the indistinguishability of the two separate bonds. Similarly, a symmetry factor is used to reflect the degeneracy of indices in $N_{x_1, \ldots, x_n}$. Thus, if $N_{x_1, \ldots, x_n}$ has k distinct indices, with the ith distinct index repeated g_i times, the symmetry factor is $\sigma = \prod_{i=1}^{k} g_i!$. (Note that $\sum_{i=1}^{k} g_i = n$.) For example, $N_{1,1}$ has a single ($k = 1$) distinct index that is twofold degenerate, leading to $\sigma = 2!$ in Eq. (59).

Equation (59) can be rewritten in terms of sums over possible monad locations within the chain. Begin with the symbolic representation for $N_{1,1}$:

$$N_{1,1} = \sum_{\{k_{1,2}\} < \{k_{1,2}\}'}^{'} 1 \tag{60}$$

where $\{k_{1,2}\}$ represents the locations for a pair of consecutive bonded monads on the chain, for example, the first and second monads on the backbone subchain. The notation $\{k_{1,2}\} < \{k_{1,2}\}'$ indicates that the summation in Eq. (60) runs over all distinct pairs of disconnected bonds in the system, and the prime on the sum refers to the restriction that the pair of bonds cannot be adjacent, that is, cannot have one (or more) monad(s) in common. Since the two bonds $\{k_{1,2}\}$ and $\{k_{1,2}\}'$ are indistinguishable, the sum in Eq. (60) may equally well be written as:

$$\sum_{\{k_{1,2}\} < \{k_{1,2}\}'}^{'} 1 = \frac{1}{2!} \sum_{\{k_{1,2}\} \neq \{k_{1,2}\}'}^{'} 1 \tag{61}$$

Removing the restriction that the two bonds cannot be identical (i.e., the $\{k_{1,2}\} \neq \{k_{1,2}\}'$ constraint) in the sum on the right-hand side of Eq. (61) and subtracting away the subsequent overcounting transform Eq. (61) into

$$\sum_{\{k_{1,2}\} < \{k_{1,2}\}'}^{'} 1 = \frac{1}{2!} \left[\sum_{\{k_{1,2}\}, \{k_{1,2}\}'}^{'} 1 - \sum_{\{k_{1,2}\}} 1 \right] \tag{62}$$

The N_1 term on the right-hand side of Eq. (59) emerges from the subtracted term on the right-hand side of Eq. (62) since this term counts the total number of pairs of bonded monads in the chain. No prime appears on the second sum on the right-hand side of Eq. (62) because the pair of bonds in this summation have been completely superimposed and, hence, are never adjacent. The first sum on the right-hand side of Eq. (62) may be transformed into a pair of sums by again subtracting the violations to the restriction from an unrestricted sum. The number of violations in this case is simply the N_2 possible combinations of pairs of consecutive (or equivalently, adjacent) bonds in the chain. In removing the restriction on the first summation on the right-hand side of Eq. (62), it is useful (in treating diagrams with bending restrictions) to separate the violations into categories where the adjacent bonds lie on the same or different subchains, yielding an equivalent form for Eq. (59):

$$
\sigma \sum_{\{k_{1,2}\}<\{k_{1,2}\}'}' 1 = \left\{ \sum_{\{k_{1,2}\},\{k_{1,2}\}'} - \sum_{\{k_{1,2}\}} - 2 \sum_{\{k_{1,2,3}\}:\text{same}} \right.
$$
$$
- \sum_{\{k_{1,2}\},\{k_{1,2}\}':\text{different}} [\delta(k_1, k_1') + \delta(k_1, k_2')
$$
$$
\left. + \delta(k_2, k_1') + \delta(k_2, k_2')] \right\} 1 \tag{63}
$$

where $\sigma = 2!$ and where the final summation on the right-hand side of Eq. (63) twice runs over all pairs of bonds that lie on separate subchains but share a common branch point (indicated by the Kronecker deltas). The factor of two appears in Eq. (63) for the same reason as on the right-hand side of Eq. (59). The sum over $\{k_{1,2,3}\}$ in Eq. (63) runs over all pairs of consecutive bonds that lie on the same subchain.

The proliferation of indices and steps necessary to describe even the simplest case $[N_{1,1}$ in Eq. (63)] motivates the use of a convenient shorthand notation. The establishment of the notation begins from the recognition that each sum in Eq. (63) corresponds to an overall counting factor in the γ_D appropriate to a diagram (without wavy lines) in Fig. 5.2. For example, the last two sums on the right-hand side of Eq. (63) both correspond to the counting factor in γ_D for diagram a of Fig. 5.2. The two sums differ only because the next to last sum is over pairs of consecutive bonds along the same subchain, while the last sum is over pairs of consecutive bonds along different subchains. Let $S(\Gamma, \{b_1, b_2, \ldots, b_n\})$ specify the set of structures in a chain containing b_1 consecutive bonds in the first subchain, b_2 consecutive bonds in the second subchain, etc., but with topology identical to the bond topology in diagram Γ of Fig. 5.2. In this notation, the set of structures specified by the indices (and Kronecker deltas) in the last two sums of Eq. (63) correspond to the sets $S(a, \{2\})$ and $S(a, \{1, 1\})$, respectively. The cumbersome descriptive indices in each sum are replaced by the more explicit labels $S(\Gamma, \{b_1, b_2, \ldots, b_n\})$ for the corresponding sets. This substitution permits Eq. (63) to be rewritten as

$$
\sigma \sum_{\{k_{1,2}\}<\{k_{1,2}\}'}' 1 = N_1^2 - N_1 - 2 \left\{ \sum_{S(a, \{2\})} + \sum_{S(a, \{1, 1\})} \right\} 1 \tag{64}
$$

where the factor of two, implicitly contained in the final sum of Eq. (63), is explicitly written in Eq. (64) and where N_1 is the number of bonds in the chain.

An identical approach produces the preferred representation [similar to Eqs. (63) and (64)] of other counting factors. The present LCT calculations require only $N_{2,1}$ and $N_{2,2}$ to be written in a form similar to Eqs. (63) and (64). The representation of these two counting factors is complicated by the possibility that the portions involving two bonds can originate from a pair on the same or different subchains, and different counting factors are required in these two cases when bending constraints appear in the diagram. The counting factors $N_{2,1}$ and $N_{2,2}$ are, therefore, decomposed into distinct portions according to whether the two bonds lie on the same (s) or different (d) subchains,

$$N_{2,1} = N_{2,1}(s) + N_{2,1}(d) \tag{65}$$

$$N_{2,2} = N_{2,2}(s, s) + N_{2,2}(s, d) + N_{2,2}(d, d) \tag{66}$$

where expressions for $N_{2,1}$ and $N_{2,2}$ are previously presented as Eq. (A2) and (A4), respectively, of Nemirovsky et al. (1992). The γ_D for diagrams (derived from Fig. 5.2) with wavy lines and bending energy restrictions each contain one of the five counting indices present on the right-hand side of Eqs. (65) and (66).

When evaluating diagrams containing bending energy factors for pairs of consecutive bonds from the same subchain, each sum over monad positions has a counting factor that depends on the bending restrictions in the diagram. The evaluation of these counting factors involves sums with indices that are associated with those on the bending factors K_{bend}. In order to keep track of the labels on K_{bend}, the shorthand notation must also specify which pair of bonds has labels corresponding to K_{bend}. In general, such an identification is not difficult, since each diagram considered here contains only one subchain composed of more than one bond. If this subchain contains only two bonds and a bending restriction, the two bonds must be associated with K_{bend}. Diagrams with three or more bonds in one subchain require an additional index i to describe the location of the first of the two bonds associated with a factor of K_{bend} within the depicted subchain. For example, diagram c in Fig. 5.2, a linear three-bond diagram, can have a factor of K_{bend} arising from bending restriction factors on either the first two or the last two consecutive bonds in the diagram. If K_{bend} is associated with the first two consecutive bonds along the diagram, the first bond from that pair is the first bond in the diagram, and this case is specified by the additional index $i = 1$. Likewise, if K_{bend} is associated with the next two consecutive bonds in the diagram, the first bond from that pair is the second bond in the diagram, and we use the index $i = 2$. The index i is indicated as a subscript to the number of bonds along the subchain. Thus

$S(c, \{3_1\})$ represents the set of all structures similar to diagram c in Fig. 5.2 in which all bonds lie along the same subchain and in which the first pair of consecutive bonds is labeled $(i = 1)$ because of bending constraints. Occasionally, it becomes necessary to specify the location where bonds of one subchain attach to those of another subchain. The index i then denotes the bond in the first subchain after which the second subchain begins. For example, consider a set of bonds with topology identical to diagram g in Fig. 5.2 for which the first three bonds lie on one subchain and the last bond lies on a separate subchain. The arrangement in this case is denoted as $\{3, 1_3\}$.

Following a procedure similar to the one that transforms Eq. (60) into Eqs. (63) and (64) yields the individual terms on the right-hand side of Eqs. (65) and (66) as

$$N_{2,1}(s) = \left(N_1 \sum_{S(a, \{2\})} - 2 \sum_{S(a, \{2\})} - \sum_{S(c, \{3_1\})} - \sum_{S(c, \{3_2\})} - \sum_{S(c, \{2, 1\})} - \sum_{S(f, \{2, 1\})} \right) 1 \tag{67}$$

$$N_{2,1}(d) = \left(N_1 \sum_{S(a, \{1, 1\})} - 2 \sum_{S(a, \{1, 1\})} - \sum_{S(c, \{2, 1\})} - 2 \sum_{S(c, \{1, 1, 1\})} \right.$$
$$\left. - 3 \sum_{S(f, \{1, 1, 1\})} - 2 \sum_{S(f, \{2, 1\})} \right) 1 \tag{68}$$

$$N_{2,2}(s, s) = \left(\sum_{S(a, \{2\})} \sum_{S(a, \{2\})} - \sum_{S(a, \{2\})} - 2 \sum_{S(c, \{3\})} \right.$$
$$\left. - 2 \sum_{S(g, \{4\})} - 2 \sum_{S(g, \{2, 2\})} - 2 \sum_{S(o2, \{2, 2\})} - 2 \sum_{S(r, \{2, 2\})} \right) 1 \tag{69}$$

$$N_{2,2}(s, d) = \left(\sum_{S(a, \{2\})} \sum_{S(a, \{1, 1\})} - \sum_{S(c, \{2, 1\})} - \sum_{S(g, \{3_1, 1_3\})} - \sum_{S(g, \{2, 1_2, 1_1\})} \right.$$
$$\left. - \sum_{S(o2, \{2, 1, 1\})} - \sum_{S(o1, \{3_1, 1_2\})} - 2 \sum_{S(f, \{2, 1\})} - \sum_{S(r, \{2, 1, 1\})} \right) 1 \tag{70}$$

$$N_{2,2}(d, d) = \left(\sum_{S(a, \{1, 1\})} \sum_{S(a, \{1, 1\})} - \sum_{S(a, \{1, 1\})} - 2 \sum_{S(c, \{1, 1, 1\})} - 2 \sum_{S(g, \{1, 2_1, 1_2\})} \right.$$
$$- 2 \sum_{S(g, \{1, 1, 1, 1\})} - 6 \sum_{S(f, \{1, 1, 1\})} - 2 \sum_{S(o2, \{1, 1, 1, 1\})}$$
$$- 2 \sum_{S(o1, \{1, 2_1, 1_1\})} - 6 \sum_{S(r, \{1, 1, 1, 1\})} - 2 \sum_{S(f, \{2, 1\})} - 4 \sum_{S(r, \{2, 2\})}$$
$$\left. - 4 \sum_{S(r, \{2, 1, 1\})} \right) 1 \tag{71}$$

The evaluation of diagrams with wavy lines and bending energy constraints requires a γ_D that contains all the same summations as those appearing in one of the equations in Eqs. (67)–(71) but with summands that differ from unity. Nevertheless, when these summands are independent of the summation indices, the sums may be replaced with the number $N(\Gamma, \{b_1, b_2, \ldots, b_n\})$ of times the diagram Γ of Fig. 5.2 occurs in a chain with b_1 consecutive bonds in the first subchain, etc. Thus, for example, the structures specified by the indices in the last two sums of Eqs. (63) and (64) occur a total of, respectively, $N(a, \{2\})$ and $N(a, \{1, 1\})$ times. Since the last two sums in Eqs. (63) and (64) represent sums over all pairs of consecutive bonds, we have the sum rule $N(a, \{2\}) + N(a, \{1, 1\}) = N_2$, as implied by detailed comparison between Eqs. (63), (64) and (59). Similar sum rules may be applied to the $N(\Gamma, \{b_1, b_2, \ldots, b_n\})$ from Eqs. (67)–(71).

APPENDIX B. METHOD FOR INCLUDING BENDING ENERGY CORRECTIONS IN D_B

Cluster diagrams that include bending energy corrections Y_{bend} must also contain the bond correlation correction factors $X_{\text{bond, 1}}$ and $X_{\text{bond, 2}}$ for the two bonds that Y_{bend} affects. The evaluation of these diagrams then requires applying the constraints in Y_{bend} to the product $X_{\text{bond, 1}} X_{\text{bond, 2}} Y_{\text{bend}}$ and then simplifying the expressions as much as possible. Below, we illustrate the partial evaluation of a bond/bend diagram for the simplest case (diagram D in Fig. 5.3) and then provide the results for more complex cases.

The summand for diagram D in Fig. 5.3 contains the product of factors $X_{\text{bond, 1}} X_{\text{bond, 2}} Y_{\text{bend}}$. Substituting the definitions for the X_{bond} [Eq. (39)] and Y_{bend} [Eq. (28)] into this product of factors yields the equivalent representation,

$$
X_{\text{bond, 1}} X_{\text{bond, 2}} Y_{\text{bend}} = \left(\frac{1}{z} \sum_{\mathbf{q}_1 \neq 0} f(\mathbf{q}_1) \exp([i\mathbf{q}_1 \cdot (\mathbf{r}_1 - \mathbf{r}_2)] \right)
$$
$$
\times \left(\frac{1}{z} \sum_{\mathbf{q}_2 \neq 0} f(\mathbf{q}_2) \exp[i\mathbf{q}_2 \cdot (\mathbf{r}_2 - \mathbf{r}_3)] \right) [z \, \delta(\beta_1, \beta_2) - 1]
$$

$$(72)$$

where $\mathbf{r}_i$ is the position of the ith monad in the diagram and β_j represents the bond direction of the jth bond in the diagram. When more factors of Y_{bend} are present in a diagram, a product P is taken for each Y_{bend}. Each factor yields terms analogous to those in brackets on the right-hand side of eq. (72). Multiplying out this product P yields a sum of terms with increasing numbers of Kronecker delta bending restrictions. Each resulting term in

P is multiplied by the product of all bond correlation correction factors X_{bond} for the bonds in the diagram. Since $P = z\,\delta(\beta_1, \beta_2) - 1$ in Eq. (72), no expansion of P is required in this case.

All factors in P are independent of the wave vectors $\{\mathbf{q}_i\}$ from the X_{bond} factors, permitting Eq. (72) to be rewritten as:

$$
X_{\text{bond},\,1} X_{\text{bond},\,2} Y_{\text{bend}} = \left(\frac{1}{z^2} \sum_{\mathbf{q}_{1,2} \neq 0} \exp[i\mathbf{q}_1 \cdot (\mathbf{r}_1 - \mathbf{r}_2) + i\mathbf{q}_2 \cdot (\mathbf{r}_2 - \mathbf{r}_3)] \right.
$$
$$
\left. \times \{f(\mathbf{q}_1)f(\mathbf{q}_2)[z\,\delta(\beta_1, \beta_2) - 1]\}\right)
\tag{73}
$$

Each term in P multiplies the product of nearest-neighbor structure factors $\{f(\mathbf{q}_i)\}$ on the right-hand side of Eq. (73):

$$
f(\mathbf{q}_1)f(\mathbf{q}_2)[z\,\delta(\beta_1, \beta_2) - 1] = f(\mathbf{q}_1)f(\mathbf{q}_2)z\,\delta(\beta_1, \beta_2) - f(\mathbf{q}_1)f(\mathbf{q}_2)
\tag{74}
$$

Substituting the definition for f [Eq. (40)] into the first term on the right-hand side of Eq. (74) and applying the bending restriction $\delta(\beta_1, \beta_2)$ yields the form:

$$
f(\mathbf{q}_1)f(\mathbf{q}_2)[z\,\delta(\beta_1, \beta_2) - 1] = z \sum_{\beta_1 = 1}^{z} \exp[-i a_\beta \cdot (\mathbf{q}_1 + \mathbf{q}_2)] - f(\mathbf{q}_1)f(\mathbf{q}_2)
\tag{75}
$$

The first term in Eq. (75) resembles the definition for f in Eq. (40), except that the argument is $(\mathbf{q}_1 + \mathbf{q}_2)$ instead of $\mathbf{q}$, allowing Eq. (75) to be rewritten as:

$$
f(\mathbf{q}_1)f(\mathbf{q}_2)[z\,\delta(\beta_1, \beta_2) - 1] = zf(\mathbf{q}_1 + \mathbf{q}_2) - f(\mathbf{q}_1)f(\mathbf{q}_2)
\tag{76}
$$

When Eq. (76) is multiplied by the factor of $1/z^2$ in Eq. (73), the final expression is identical to the factor presented in Eq. (51).

Similar expressions, presented below, are derived when more than one bending restriction factor Y_{bend} appears. Before presenting the final results, it is convenient to introduce a function $\Delta\{(\alpha_1, \alpha_1 + 1), (\alpha_2, \alpha_2 + 1), \ldots, (\alpha_n, \alpha_n + 1)\}$ that generalizes the form of Eq. (76):

$$
\Delta\{(\alpha_1, \alpha_1 + 1), (\alpha_2, \alpha_2 + 1), \ldots, (\alpha_n, \alpha_n + 1)\} \equiv z^{n-1} f\left(\sum_{i=1}^{n} \mathbf{q}_i\right) - \prod_{i=1}^{n} f(\mathbf{q}_i)
$$
$$
\tag{77}
$$

where the label $(\alpha_i, \alpha_i + 1)$ denotes the ith pair of consecutive bonds affected by some bending restriction Y_{bend} and where n is the total number of bending restrictions. For example, Eq. (76) now takes the shorthand form:

$$zf(\mathbf{q}_1 + \mathbf{q}_2) - f(\mathbf{q}_1)f(\mathbf{q}_2) = \Delta\{(1, 2)\} \qquad (78)$$

Upon multiplying the factors in P, the factor for a set of three consecutive bonds with two bending restrictions may be written as:

$$X_{bond, 1}X_{bond, 2}X_{bond, 3}Y_{bend(1, 2)}Y_{bend(2, 3)}$$

$$\rightarrow \Delta\{(1, 2), (2, 3)\} - \frac{f(\mathbf{q}_1)}{z}\Delta\{(2, 3)\} - \frac{f(\mathbf{q}_3)}{z}\Delta\{(1, 2)\} \quad (79)$$

while the factor for a set of four consecutive bonds with three bending restrictions is:

$$X_{bond, 1}X_{bond, 2}X_{bond, 3}X_{bond, 4}Y_{bend(1, 2)}Y_{bend(2, 3)}Y_{bend(3, 4)}$$

$$\rightarrow \Delta\{(1, 2), (2, 3), (3, 4)\} - \frac{f(\mathbf{q}_4)}{z}\Delta\{(1, 2), (2, 3)\} - \Delta\{(1, 2)\}\,\Delta\{(3, 4)\}$$

$$- \frac{f(\mathbf{q}_1)}{z}\Delta\{(2, 3), (3, 4)\} + \frac{f(\mathbf{q}_1)f(\mathbf{q}_4)}{z^2}\Delta\{(2, 3)\} \qquad (80)$$

where only the portions contributing in the thermodynamic limit are retained in Eqs. (79) and (80). When the factors on the right-hand side of Eq. (79) or (80) are substituted for the corresponding factors on the left-hand side into the expressions for the diagrams D_B, the resultant is a series of terms, each of which corresponds to a distinct contracted diagram depicted in Fig. 5.6. More explicitly, the terms in Eq. (79) from left to right produce contracted diagrams $R_{3, 3}$, $R_{3, 2}$, and $R_{3, 2}$, respectively.

APPENDIX C. CORRECTION TERMS TO THE ZEROTH-ORDER HELMHOLTZ FREE ENERGY

The correction terms $C(t)$ in Eq. (5.8) are functions of the chain topologies from each of the p different species, as well as functions of the bending and interaction energies; that is $C(t) = C(p, \{M\}, \{E_b\}, \{\varepsilon\}; t)$. The expressions given below are quite general, but are lengthy because each monad in a monomer may interact with different energies. Therefore, the general expressions are also specialized to the more commonly used case in which all bending energies within chains of the same species are all identical and all van der Waals energies between monads from species s and s' are identical (with the lone van der Waals energy parameter denoted as $\varepsilon_{ss'}$). The Mayer f functions in the latter case are evaluated in the high-temperature limit, retaining contributions only through order ε^2.

Before presenting the general expressions for the $C(t)$, we introduce some necessary shorthand notation. As in Appendix A, let $S(\Gamma, \{b_1, b, \ldots, b_n\})$ specify the set of structures in a chain containing b_1 consecutive bonds in the first subchain, etc., but with topology identical to the bond topology in diagram Γ of Fig. 5.2. Further, let $N(\Gamma, \{b_1, b_2, \ldots, b_n\})$ be the total number of elements in the set $S(\Gamma, \{b_1, b_2, \ldots, b_n\})$. When evaluating diagrams containing bending energy factors, the sums in the definition of the counting factor γ_D run over the monad positions in the chain but also contain factors of the bending term K_{bend}, where some of the monad summation indices coincide with the monad indices on the bending factors K_{bend}. Thus the shorthand notation must specify which pair of bonds has labels corresponding to those in K_{bend}.

In general, the required specification is straightforward, since each diagram typically possesses only one subchain that contains more than one bond. If the subchain is composed of only two bonds, the bonds must be associated with the K_{bend} factor. Diagrams with three or more bonds in one subchain require an additional index i to describe the location in the subchain of the first of the two bonds that is associated with the K_{bend} factor. For example, diagram c in Fig. 5.2 is a linear three-bond diagram, which can have a factor of K_{bend} arising from bending restriction on either the first two or the last two consecutive bonds in the diagram. If the K_{bend} factor applies to the first pair of consecutive bonds, the first bond from the first pair is designated as the first bond in the diagram, which is given the label $i = 1$. Likewise, if the K_{bend} factor applies to the next pair of consecutive bonds in the diagram, the first bond from that pair is the second bond in the diagram, and the label is $i = 2$. Thus the notation $S(c, \{3_1\})$ represents the set of all structures similar to diagram c in Fig. 5.2, in which all bonds lie along the same subchain and in which the first pair of consecutive bonds are labeled with $i = 1$. The bending factor K_{bend} (or $g_{\text{bend}} \equiv 1 - K_{\text{bend}}$) is similarly labeled with the subscripted values for i. For example, all contributions to $S(c, \{3_1\}$ from a chain of species s with a single bending restriction contain a factor of $K_1(s)$. Occasionally it becomes necessary to specify the location of junctions between subchains. If two subchains α and β meet at a branch point located at the end of the ith bond in subchain α, the arrangement is denoted as $\{\alpha, \beta_i\}$. For example, consider a set of bonds with topology identical to diagram g in Fig. 5.2 for which the first three bonds lie on one subchain and the last bond lies on a separate subchain. The arrangement in this case is denoted as $\{3, 1_3\}$.

Four collections of combinatorial indices frequently appear in the coefficients $C(t)$. Two collections involve sums over all possible ways of placing the bonds from either diagram a or c in Fig. 5.2 along various subchains. The remaining two collections are sums over the bonds and monads in a chain. Some collections are functions of the bending factor g_{bend}. When the collection is evaluated in the limit of vanishing bending energies, each collection yields the total number of times that the set of monads particular to the diagram occurs in the chain. Let $\tau_n[x]$ denote the collection of terms for an n-bond "linear" diagram (such as diagram a or c in Fig. 5.2), which operate on the summand factor represented as x. Therefore, in the flexible chain limit, $\tau_n[1] \to N_n$, where N_n is the number of times that an n-bond "linear" diagram appears within a single chain. The $\tau_n^{(s)}[x]$ corresponding to the two- and three-bond "linear" diagrams within a chain from species s are given by:

$$\tau_2^{(s)}[x] = \left(\sum_{S(a,\,\{1,\,1\})} + \sum_{S(a,\,\{2\})} g_1(s) \right) x \tag{81}$$

$$\tau_3^{(s)}[x] = \left(\sum_{S(c,\,\{1,\,1,\,1\})} + \sum_{S(c,\,\{2,\,1\})} g_1(s) + \sum_{S(c,\,\{3\})} g_1(s)g_2(s) \right) x \tag{82}$$

where the bending factor $g_i(s)$ is defined as $1 - K_i(s)$ and where the index i denotes the location in the subchain of the first of the two bonds that are associated with the $K_i(s)$ factor. The labels $\tau_1^{(s)}$ and $\tau_0^{(s)}$, respectively, denote the sums over the $M_s - 1$ bonds and M_s monads from a chain of species s. In the simplified limit where all van der Waals energies between monads from species s and s' are identical and where all bending energies for chains from species s are equal, the arguments x are always independent of the summation indices in Eqs. (81) and (82) defining the operators $\tau_n^{(s)}[x]$. Thus, for this simple limit, the operators $\tau_n^{(s)}[x]$ reduce to special counting indices, with the following correspondences:

$$\tau_0^{(s)} \to M_s \tag{83}$$

$$\tau_1^{(s)} \to N_1(s) \tag{84}$$

$$\tau_2^{(s)} \to [N(a, \{1, 1\}) + N(a, \{2\})g(s)] \equiv N_{t2}(s) \tag{85}$$

$$\tau_3^{(s)} \to [N(c, \{1, 1, 1\}) + N(c, \{2, 1\})g(s) + N(c, \{3\})g(s)^2] \equiv N_{t3}(s) \tag{86}$$

where $g(s)$ is the bending energy factor for any pair of consecutive bonds along the same subchain in a chain from species s.

A specification is required for the interacting monads in the Mayer f functions. Although branched diagrams (such as f or o in Fig. 5.2) appear in the LCT partition function, their contributions to excess thermodynamic properties are identically cancelled by contributions from diagrams (such as j in Fig. 5.2) with disconnected pieces. These cancellations occur for nearly all expressions from bond and bond/bend diagrams as well as for all derived bond/pair and bond/bend/pair diagrams. All interaction diagrams that remain in the final LCT expressions for $C(t)$ ($t > 1$) involve single monads, single bonds, and "linear" diagrams similar to diagrams a and c from Fig. 5.2. Hence, the monads in each linear diagram can be labeled sequentially from end to end. Every interacting monad from a diagram in Fig. 5.2 is then specified by the chemical species of the chain, followed by the diagram letter in Fig. 5.2 along with its sequential label in the diagram. For example, $(s, a2)$ specifies the middle monad of a two-bond segment in a chain from species s. The two monads from the single-bond diagram (not depicted in Fig. 5.2) are labeled by the chemical species of the chain and the word "bond" followed by either 1 or 2 (corresponding to the two monads in the bond), while the lone monad is labeled by its chemical species and the number "1."

The general expressions for the corrections $C(t)$ are:

$$
\begin{aligned}
C(1) = \sum_{s=1}^{p} \frac{\phi_s}{M_s} & \left(\frac{1}{z} \left\{ -\tau_2^{(s)}[1] + \sum_{S(r,\,\{2,\,2\})} K_1(s)K_2(s) \right\} \right. \\
& + \frac{1}{z^2} \left[-N_\perp(s) - 0.5 \sum_{S(a,\,\{2\})} g_1(s)^2 - 0.5 N(a, \{1,1\})(s) \right. \\
& - \sum_{S(g,\,\{4\})} g_1(s)g_2(s)g_3(s) - \sum_{S(g,\,\{3,\,1\})} g_1(s)g_2(s) \\
& - \sum_{S(g,\,\{2,\,2\})} [1 + K_1(s)K_2(s)] - \sum_{S(g,\,\{1,\,2_1,\,1_2\})} g_1(s) - \sum_{S(g,\,\{2,\,1_2,\,1_1\})} g_1(s) \\
& - N(g, \{1,1,1,1\})(s) + \left. \sum_{S(r,\,\{2,\,2\})} K_1(s)[1 - 2K_2(s)] \right] \\
& + 2\tau_1^{(s)}[f_{(s,\,\text{bond}1),\,(s,\,\text{bond}2)}] + \frac{2\tau_3^{(s)}[f_{(s,\,c1),\,(s,\,c4)}]}{z} \\
& \left. - 2\tau_1^{(s)}[f_{(s,\,\text{bond}1),\,(s,\,\text{bond}2)} f_{(s,\,\text{bond}1),\,(s,\,\text{bond}2)}] \right) \tag{87}
\end{aligned}
$$

$$
\begin{aligned}
C(2) = \sum_{s_1,\,s_2=1}^{p} \frac{\phi_{s_1}\phi_{s_2}}{M_{s_1}M_{s_2}} & \left(\frac{N_1(s_1)N_1(s_2)}{z} + \frac{2N_1(s_1)\tau_3^{(s2)}[1] + \tau_2^{(s1)}[1]\tau_2^{(s2)}[1]}{z^2} \right. \\
& + 2\tau_0^{(s1)}\tau_0^{(s2)}[f_{(s_1,\,1),\,(s_2,\,1)}] - 2\tau_1^{(s1)}\tau_0^{(s2)} \\
& \times [f_{(s_1,\,\text{bond}1),\,(s_2,\,1)} + f_{(s_1,\,\text{bond}2),\,(s_2,\,1)}] + \frac{1}{z} \{\tau_1^{(s1)}\tau_1^{(s2)}
\end{aligned}
$$

$$\times [f_{(s_1, \text{bond}1), (s_2, \text{bond}1)} + f_{(s_1, \text{bond}1), (s_2, \text{bond}2)} + f_{(s_1, \text{bond}2), (s_2, \text{bond}1)}$$

$$+ f_{(s_1, \text{bond}2), (s_2, \text{bond}2)}] - 2\tau_3^{(s1)}[\tau_0^{(s2)}[f_{(s_1, c1), (s_2, 1)} + f_{(s_1, c4), (s_2, 1)}]]]$$

$$- 2\tau_2^{(s1)}[\tau_1^{(s2)}[f_{(s_1, a1), (s_2, \text{bond}1)} + f_{(s_1, a1), (s_2, \text{bond}2)}$$

$$+ f_{(s_1, a3), (s_2, \text{bond}1)} + f_{(s_1, a3), (s_2, \text{bond}2)}]\}]$$

$$+ 4\tau_2^{(s1)}[\tau_0^{(s2)}[f_{(s_1, a1), (s_2, 1)} f_{(s_1, a3), (s_2, 1)}]] + 2\tau_1^{(s1)}\tau_1^{(s2)}$$

$$\times [f_{(s_1, \text{bond}1), (s_2, \text{bond}1)} f_{(s_1, \text{bond}2), (s_2, \text{bond}2)} + f_{(s_1, \text{bond}1), (s_2, \text{bond}2)}$$

$$\times f_{(s_1, \text{bond}2), (s_2, \text{bond}1)}] \Bigg) \tag{88}$$

$$C(3) = \sum_{s_1, s_2, s_3 = 1}^{p} \frac{\phi_{s_1}\phi_{s_2}\phi_{s_3}}{M_{s_1}M_{s_2}M_{s_3}} \left(\frac{\{\frac{2}{3}N_1(s_1) - 4\tau_2^{(s1)}[1]\}N_1(s_2)N_1(s_3)}{z^2} \right.$$

$$+ 2N_1(s_1)\tau_0^{(s2)}\tau_0^{(s3)}[f_{(s_2, 1), (s_3, 1)}]$$

$$+ \frac{1}{z} \{ -4N_1(s_1)\tau_1^{(s2)}\tau_0^{(s3)}[f_{(s_2, \text{bond}1), (s_3, 1)} + f_{(s_2, \text{bond}2), (s_3, 1)}]$$

$$+ 2\tau_3^{(s1)}[\tau_0^{(s2)}\tau_0^{(s3)}[f_{(s_2, 1), (s_3, 1)}]] + 4\tau_2^{(s1)}[\tau_1^{(s2)}\tau_0^{(s3)}[f_{(s_2, \text{bond}1), (s_3, 1)}$$

$$+ f_{(s_2, \text{bond}2), (s_3, 1)} + f_{(s_1, a1), (s_3, 1)} + f_{(s_1, a3), (s_3, 1)}]]$$

$$+ 2N_1(s_1)\tau_1^{(s2)}\tau_1^{(s3)}[f_{(s_2, \text{bond}1), (s_3, \text{bond}1)} + f_{(s_2, \text{bond}1), (s_3, \text{bond}2)}$$

$$+ f_{(s_2, \text{bond}2), (s_3, \text{bond}1)} + f_{(s_2, \text{bond}2), (s_3, \text{bond}2)}]\}$$

$$- 2z\tau_0^{(s1)}\tau_0^{(s2)}\tau_0^{(s3)}[f_{(s_1, 1), (s_2, 1)} f_{(s_2, 1), (s_3, 1)}]$$

$$+ 2\tau_1^{(s1)}\tau_0^{(s2)}\tau_0^{(s3)}[2f_{(s_2, 1), (s_3, 1)}\{f_{(s_1, \text{bond}1), (s_2, 1)} + f_{(s_1, \text{bond}2), (s_2, 1)}\}$$

$$+ f_{(s_1, \text{bond}1), (s_2, 1)} f_{(s_1, \text{bond}1), (s_3, 1)} + f_{(s_1, \text{bond}2), (s_2, 1)} f_{(s_1, \text{bond}2), (s_3, 1)}]$$

$$- 4\tau_2^{(s1)}[\tau_0^{(s2)}\tau_0^{(s3)}[f_{(s_1, a1), (s_2, 1)} f_{(s_1, a3), (s_3, 1)}$$

$$+ f_{(s_2, 1), (s_3, 1)}\{f_{(s_1, a1), (s_2, 1)} + f_{(s_1, a3), (s_2, 1)}\}]]$$

$$- 4\tau_1^{(s1)}\tau_1^{(s2)}\tau_0^{(s3)}[0.5\{f_{(s_1, \text{bond}1), (s_3, 1)} + f_{(s_1, \text{bond}2), (s_3, 1)}\}$$

$$\times \{f_{(s_2, \text{bond}1), (s_3, 1)} + f_{(s_2, \text{bond}2), (s_3, 1)}\} + f_{(s_1, \text{bond}1), (s_3, 1)}$$

$$\times \{f_{(s_1, \text{bond}2), (s_2, \text{bond}1)} + f_{(s_1, \text{bond}2), (s_2, \text{bond}2)}\}$$

$$+ f_{(s_1, \text{bond}2), (s_3, 1)}\{f_{(s_1, \text{bond}1), (s_2, \text{bond}1)} + f_{(s_1, \text{bond}1), (s_2, \text{bond}2)}\}] \Bigg) \tag{89}$$

$$C(4) = \sum_{s_1, s_2, s_3, s_4 = 1}^{p} \frac{\phi_{s_1}\phi_{s_2}\phi_{s_3}\phi_{s_4}}{M_{s_1}M_{s_2}M_{s_3}M_{s_4}} \left(\frac{2N_1(s_1)N_1(s_2)N_1(s_3)N_1(s_4)}{z^2} \right.$$

$$+ \frac{1}{z} \{4N_1(s_1)N_1(s_2)\tau_0^{(s3)}\tau_0^{(s4)}[f_{(s_3, 1), (s_4, 1)}]$$

$$- 8\tau_2^{(s1)}[\tau_1^{(s2)}\tau_0^{(s3)}\tau_0^{(s4)}[f_{(s3, 1), (s4, 1)}]] - 8N_1(s_1)N_1(s_2)$$

$$\times \ \tau_1^{(s3)}\tau_0^{(s4)}[f_{(s3, \text{bond}1), (s4, 1)} + f_{(s3, \text{bond}2), (s4, 1)}]\}$$

$$+ z\tau_0^{(s1)}\tau_0^{(s2)}\tau_0^{(s3)}\tau_0^{(s4)}[f_{(s1, 1), (s2, 1)}f_{(s3, 1), (s4, 1)}]$$

$$- 4\tau_1^{(s1)}\tau_0^{(s2)}\tau_0^{(s3)}\tau_0^{(s4)}[2f_{(s2, 1), (s3, 1)}f_{(s3, 1), (s4, 1)}$$

$$+ f_{(s2, 1), (s3, 1)}\{f_{(s1, \text{bond}1), (s4, 1)} + f_{(s1, \text{bond}2), (s4, 1)}\}]$$

$$+ 4\tau_2^{(s1)}[\tau_0^{(s2)}\tau_0^{(s3)}\tau_0^{(s4)}[f_{(s2, 1), (s3, 1)}\{f_{(s3, 1), (s4, 1)} + f_{(s1, a1), (s4, 1)}$$

$$+ f_{(s1, a3), (s4, 1)}\}]] + 4\tau_1^{(s1)}\tau_1^{(s2)}\tau_0^{(s3)}\tau_0^{(s4)}[f_{(s3, 1), (s4, 1)}$$

$$\times \ \{0.5[f_{(s1, \text{bond}1), (s2, \text{bond}1)} + f_{(s1, \text{bond}1), (s2, \text{bond}2)}$$

$$+ f_{(s1, \text{bond}2), (s2, \text{bond}1)} + f_{(s1, \text{bond}2), (s2, \text{bond}2)}]$$

$$+ 2[f_{(s1, \text{bond}1), (s3, 1)} + f_{(s1, \text{bond}2), (s3, 1)}]\}$$

$$+ f_{(s1, \text{bond}1), (s3, 1)}\{f_{(s2, \text{bond}1), (s4, 1)} + 2f_{(s2, \text{bond}2), (s4, 1)}$$

$$\left. + 2f_{(s1, \text{bond}2), (s4, 1)}\} + f_{(s1, \text{bond}2), (s3, 1)}f_{(s2, \text{bond}2), (s4, 1)}]\right) \tag{90}$$

$$C(5) = \sum_{s1, s2, s3, s4, s5 = 1}^{p} \frac{\phi_{s1}\phi_{s2}\phi_{s3}\phi_{s4}\phi_{s5}}{M_{s1}M_{s2}M_{s3}M_{s4}M_{s5}}$$

$$\times \left\{ \frac{8N_1(s_1)N_1(s_2)N_1(s_3)}{z} \tau_0^{(s4)}\tau_0^{(s5)}[f_{(s4, 1), (s5, 1)}] \right.$$

$$+ 6N_1(s_1)\tau_0^{(s2)}\tau_0^{(s3)}\tau_0^{(s4)}\tau_0^{(s5)}[f_{(s2, 1), (s3, 1)}f_{(s4, 1), (s5, 1)}]$$

$$- 4\tau_2^{(s1)}[\tau_0^{(s2)}\tau_0^{(s3)}\tau_0^{(s4)}\tau_0^{(s5)}[f_{(s2, 1), (s3, 1)}f_{(s4, 1), (s5, 1)}]]$$

$$- 8N_1(s_1)\tau_1^{(s2)}\tau_0^{(s3)}\tau_0^{(s4)}\tau_0^{(s5)}[2f_{(s3, 1), (s4, 1)}\{f_{(s2, \text{bond}1), (s5, 1)}$$

$$\left. + f_{(s2, \text{bond}2), (s5, 1)}\} + f_{(s3, 1), (s4, 1)}f_{(s4, 1), (s5, 1)}]\right\} \tag{91}$$

$$C(6) = \sum_{s1, s2, s3, s4, s5, s6 = 1}^{p} \frac{\phi_{s1}\phi_{s2}\phi_{s3}\phi_{s4}\phi_{s5}\phi_{s6}}{M_{s1}M_{s2}M_{s3}M_{s4}M_{s5}M_{s6}} 12N_1(s_1)N_1(s_2)$$

$$\times \ \tau_0^{(s3)}\tau_0^{(s4)}\tau_0^{(s5)}\tau_0^{(s6)}[f_{(s3, 1), (s4, 1)}f_{(s5, 1), (s6, 1)}] \tag{92}$$

where ϕ_s is the volume fraction of species s, z is the lattice coordination number, $N_1(s)$ is the number of bonds in a chain from species s, $N_\perp(s)$ is the number of times that diagram f in Fig. 5.2 appears in a chain from species s, and $f_{(), ()}$ is the Mayer f function. The expression for the $C(t)$ as given in Eqs. (87)–(92) involve a large number of parameters and require some group additive scheme for their specification.

Semiempirical applications of the LCT are based on the simplest case that retains all the desired qualitative features; namely, all bending energies for chains from species s are set equal and all van der Walls energies between monads from species s and s' are identical (with the lone van der Waals energy parameter denoted as $\varepsilon_{ss'}$). With the understanding that the correspondences in Eqs. (83)–(86) apply in Eqs. (87)–(92), the $C(t)$ for the simplified case are, after algebraic rearrangement,

$$
\begin{aligned}
C(1) = \sum_{s=1}^{p} \phi_s \Bigg\{ &\frac{1}{z} \left[-u_{t2}(s) + u(r, \{2, 2\})(s)K(s)^2 \right] \\
&+ \frac{1}{z^2} \left[-u_\perp(s) - 0.5u(a, \{2\})(s)g(s)^2 \right. \\
&- 0.5u(a, \{1, 1\})(s) - u(g, \{4\})(s)g(s)^3 - u(g, \{3, 1\})(s)g(s)^2 \\
&- u(g, \{2, 2\})(s)[1 + K(s)^2] - u(g, \{1, 2_1, 1_2\})(s)g(s) \\
&- u(g, \{2, 1_2, 1_1\})(s)g(s) - u(g, \{1, 1, 1, 1\})(s) \\
&+ u(r, \{2, 2\})(s)K(s)[1 - 2K(s)]] + \left(u_1(s) + \frac{u_{t3}(s)}{z} \right)\varepsilon_{ss} \Bigg\}
\end{aligned}
\tag{93}
$$

$$
\begin{aligned}
C(2) = \sum_{s_1, s_2=1}^{p} \phi_{s_1}\phi_{s_2} \Bigg(&\frac{u_1(s_1)u_1(s_2)}{z} + \frac{2u_1(s_1)u_{t3}(s_2) + u_{t2}(s_1)u_{t2}(s_2)}{z^2} \\
&+ \varepsilon_{s_1s_2} \Bigg\{ \frac{z}{2} - 2u_1(s_1) + \frac{2u_1(s_1)u_1(s_2) - 2u_{t3}(s_1) - 4u_{t2}(s_1)u_1(s_2)}{z} \\
&+ \varepsilon_{s_1s_2} \left[\frac{z}{4} - u_1(s_1) + u_{t2}(s_1) + u_1(s_1)u_1(s_2) \right] \Bigg\} \Bigg)
\end{aligned}
\tag{94}
$$

$$
\begin{aligned}
C(3) = \sum_{s_1, s_2, s_3=1}^{p} \phi_{s_1}\phi_{s_2}\phi_{s_3} \Bigg(&\frac{[\frac{2}{3}u_1(s_1) - 4u_{t2}(s_1)]u_1(s_2)u_1(s_3)}{z^2} \\
&+ \varepsilon_{s_2s_3} \Bigg\{ u_1(s_1) + \frac{1}{z} \left[-4u_1(s_1)u_1(s_2) + u_{t3}(s_1) \right. \\
&+ 4u_{t2}(s_1)u_1(s_2) + 4u_{t2}(s_2)u_1(s_1) + 4u_1(s_1)u_1(s_2)u_1(s_3)] \\
&+ \frac{\varepsilon_{s_2s_3}u_1(s_1)}{2} + \varepsilon_{s_1s_2} \left[-\frac{z}{2} + 2u_1(s_1) + u_1(s_2) \right. \\
&- 2u_{t2}(s_1) - u_{t2}(s_2) - 2u_1(s_1)u_1(s_3) - 4u_1(s_1)u_1(s_2)] \Bigg\} \Bigg)
\end{aligned}
\tag{95}
$$

$$
C(4) = \sum_{s_1, s_2, s_3, s_4=1}^{p} \phi_{s_1}\phi_{s_2}\phi_{s_3}\phi_{s_4} \left(\frac{2u_1(s_1)u_1(s_2)u_1(s_3)u_1(s_4)}{z^2} \right.
$$

$$+ \varepsilon_{s_3s_4} \left\{ \frac{1}{z} \left[2u_1(s_1)u_1(s_2) - 4u_{t2}(s_1)u_1(s_2) - 8u_1(s_1)u_1(s_2)u_1(s_3) \right] \right.$$

$$+ \varepsilon_{s_1s_2} \left[\frac{z}{4} - 2u_1(s_1) + 2u_{t2}(s_1) + 2u_1(s_1)u_1(s_2) + 4u_1(s_1)u_1(s_3) \right]$$

$$\left. + \varepsilon_{s_2s_3} \left[-2u_1(s_1) + u_{t2}(s_1) + 4u_1(s_1)u_1(s_2) + 2u_1(s_1)u_1(s_3) \right] \right\} \right)$$

(96)

$$C(5) = \sum_{s_1, s_2, s_3, s_4, s_5 = 1}^{p} \phi_{s_1} \phi_{s_2} \phi_{s_3} \phi_{s_4} \phi_{s_5} \left(\varepsilon_{s_4s_5} \left\{ \frac{4u_1(s_1)u_1(s_2)u_1(s_3)}{z} \right. \right.$$

$$+ \varepsilon_{s_2s_3} \left[1.5u_1(s_1) - u_{t2}(s_1) - 8u_1(s_1)u_1(s_4) \right]$$

$$\left. \left. - 2u_1(s_1)u_1(s_2)\varepsilon_{s_3s_4} \right\} \right)$$

(97)

$$C(6) = \sum_{s_1, s_2, s_3, s_4, s_5, s_6 = 1}^{p} \phi_{s_1} \phi_{s_2} \phi_{s_3} \phi_{s_4} \phi_{s_5} \phi_{s_6} \, 3u_1(s_1)u_1(s_2)\varepsilon_{s_3s_4} \varepsilon_{s_5s_6} \quad (98)$$

where the factor $u_x(s) \equiv N_x(s)/M_s$.

REFERENCES

Baker, D., Chan, H. S., Dill, K. A., *J. Chem. Phys.* **98**, 9951 (1993).

Bawendi, M. G., Freed, K. F., *J. Chem. Phys.* **86**, 3720 (1987).

Brazhnik, O. D., Freed, K. F., *J. Chem. Phys.* **105**, 837 (1996).

DiMarzio, E. A., Gibbs, J. H., *J. Chem. Phys.* **28**, 807 (1958).

Dudowicz, J., Freed, K. F., *Macromolecules* **24**, 5076 (1991a).

Dudowicz, J., Freed, K. F., *Macromolecules* **24**, 5112 (1991b).

Dudowicz, J., Freed, K. F., *Macromolecules* **26**, 213 (1993).

Dudowicz, J., Freed, K. F., *Macromolecules* **28**, 6625 (1995).

Dudowicz, J., Freed, K. F., *Macromolecules* **29**, 7826 (1996a).

Dudowicz, J., Freed, K. F., *Macromolecules* **29**, 8960 (1996b).

Dudowicz, J., Freed, K. F., Madden, W. G., *Macromolecules* **23**, 4803 (1990).

Foreman, K. W., Freed, K. F., *J. Chem. Phys.* **102**, 4663 (1995).

Foreman, K. W., Freed, K. F., *J. Chem. Phys.* **106**, 7422 (1997a).

Foreman, K. W., Freed, K. F., Ngola, I. M., *J. Chem. Phys.* **107**, 4688 (1997).

Foreman, K. W., Freed, K. F., *Macromolecules,* **30**, 7279 (1997b).

Freed, K. F., Dudowicz, J., *J. Chem. Phys.* **97**, 2105 (1992).

Freed, K. F., Dudowicz, J., *Trends Polym. Sci.* **3**, 248 (1995).

Freed, K. F., Dudowicz, J., Foreman, K. W., to be published.

Freed, K. F., Dudowicz, J., *Macromolecules* **29**, 625 (1996).

Freed, K. F., Pesci, A. I., *J. Chem. Phys.* **87**, 7342 (1987).

Freed, K. F., Pesci, A. I., *Macromolecules* **22**, 4048 (1989).

Gibbs, J. H., DiMarzio, E. A., *J. Chem. Phys.* **28**, 373 (1958).

Guggenheim, E. A., *Mixtures*, Oxford University, Oxford, 1952.

Huston, S. E., Nemirovsky, A. M., Freed, K. F., *J. Chem. Phys.* **99**, 2149 (1993).

Jorgensen, W. L., *J. Am. Chem. Soc.* **103**, 335 (1981).

Kubo, R., *J. Phys. Soc. Jpn.* **17**, 1100 (1962).

Li, W., Freed, K. F., *J. Chem. Phys.* **103**, 5693 (1995).

Madden, W. G., Pesci, A. I., Freed, K. F., *Macromolecules* **23**, 1181 (1990).

Mondello, M., Grest, G. S., *J. Chem. Phys.* **103**, 7156 (1995).

Nemirovsky, A. M., Bawendi, M. G., Freed, K. F., *J. Chem. Phys.* **87**, 7272 (1987).

Nemirovsky, A. M., Dudowicz, J., Freed, K. F., *Phys. Rev. A* **45**, 7111 (1992).

Schweizer, K. S., Curro, J. G., *Adv. Chem. Phys.* **98**, 1 (1997).

Smith, G. D., Yoon, D. Y., Jaffe, R. L., Colby, R. H., Krishnamoorti, R., Fetters, L. J., *Macromolecules* **29**, 3462 (1996).

Toxvaerd, S., *J. Chem. Phys.* **93**, 4290 (1990).

AUTHOR INDEX

Numbers in parentheses are reference numbers and indicate that the author's work is referred to although his name is not mentioned in the text. Numbers in *italic* show the pages on which the complete references are listed.

Abramcyzk, H., 52(1–2), 76–81(1) , 127(3), *134*
Abramowitz, M., 236(1), *256*, 281–282(1), 284(1), *332*
Ackroyd, R., 71(99), *137*
Aharoni, A., 288(2–3), 293–294(4), *332*
Asgharian, A., 190(11), 194–195(11), 201–202(11), 210(11), *215*
Asselin, M., 132(4–5), *134*

Badger, R., 12(6), *134*
Baker, D., 362(1), *389*
Bambini, A., 251(2), *256*
Bárány, A., 218(3–4), *256*
Barbara, B., 305(44–45), *333*
Barton, S., 18(7), *134*
Baryshanskaya, F., 7(42), *135*
Bates, D. R., 229(5–6), *256*
Bauer, S., 12(6), *134*
Bawendi, M. G., 336(30), 337(2), 340(30), 354(2), 361(30), 364(30), 366(30), *389–390*
Bayfield, J. E., 239(7), *256*
Bean, C. P., 305(5), *332*
Belch, A., 10(67), *136*
Belsely, M., 251(27), 254(27), *257*
Benderskii, V. A., 305(6), *332*
Benoit, A., 305(44–45), *333*
Bergren, M., 10(67), *136*
Berman, P. R., 232(30), 250(30), 251(2,40), 254(30), *256–257*
Berry, M. V., 220(8), 224(8), *256*
Beswick, J., 51(8), *134*
Blaise, P., 10(37), 28(12,14), 47(14), 53(14), 54(12), 123(96), 127(14), 166(12,14), 173(9), 183(10), *134–135, 137*
Boivin, D., 305(44), *333*
Bonet Orozco, E., 305(44–45), *333*

Borgis, D., 127(86), *136*
Born, M., 189(8), *215*
Borovec, M., 276(27), 309(27), 312–313(27), 315–316(27), 321(27), 330(27), *333*
Borstnik, B., 12(11), *134*
Boulil, B., 28(12–14), 53(13), 54(12), 125(13), 127(14), 166(12–14), *134*
Bournay, J., 71(15), *135*
Bratos, S., 8(17), 10(36), 12(16), 52(16,19), 63–61(16), 125(16,18), 126(16), *135*
Brazhnik, O. D., 364(3), *389*
Brinkman, H. C., 260(54), 321(54), *333*
Brooks, W. V. F., 211(28), *215*
Brown, G. J. N., 227(9), *256*
Brown, W. F., 264(7), 266(7), 277(6), 279(6), 284(7), 286–287(7), 294(8), 397(8), 304(7), 306(8), 318(7), 319(8), 321(8), 323(8), 328(8), *332*
Bunker, P. R., 189(7), 190(12), 210(12), *215*
Burnett, K., 251(27), 254(27), *257*

Carruthers, P., 54(20), *135*
Chamma, D., 125(21), *135*
Chan, H. S., 362(1), *389*
Chandrasekhar, S., 294(9), 307(10), *332*
Chang, C. R., 326(48), *333*
Child, M. S., 218(10), *256*
Coffey, W., 135 (22)
Coffey, W. T., 260(18,20), 261(18,20,28), 262(11–12,17,20), 263(20), 264(12,18,20), 265(20), 266(12,20), 268(20), 269(12,20), 272(16), 273(12,16), 274(12), 275(12,16–20,26), 276(19–20,26,29), 277(12–13,16,18–20), 278(20), 284(20), 285(17,20), 287(18), 288(13,18–20,26), 291(18), 293(16,19), 294(13,26), 295(13,15,20), 297(20), 298(18,28), 301(20), 302(18), 304(12,15–19),

391

SUBJECT INDEX